Analog

Analog Circuit Design

....

Art, Science, and Personalities

Edited by
Jim Williams

Butterworth–Heinemann
An Imprint of Elsevier
Boston London Oxford Singapore Sydney Toronto Wellington

Butterworth Heinemann is an imprint of Elsevier

♾ This book is printed on acid-free paper.

Library of Congress Cataloging-in-Publication Data

Analog circuit design: art, science, and personalities / edited by
  Jim Williams.
        p.   cm. – (The EDN series for design engineers)
  Includes bibliographical references and index.
   ISBN-13: 978-0-7506-9640-1 ISBN-10: 0-7506-9640-0
   1.   Linear integrated circuits-Maintenance and repair.
2. Electronic circuit design. I. Williams, Jim, 1948-.
II. Series
TK7874.A548   1991   91-9930
621.381'5-dc20   CIP

British Library Cataloguing-in-Publication Data

  1.   Analogue circuits, Design
  I.   Williams, Jim 1948- II. Series
  621.3815
ISBN-13: 978-0-7506-9640-1
ISBN-10: 0-7506-9640-0
The publisher offers special discounts on bulk orders of this book.
For information, please contact:
Manager of Special Sales
Elsevier
200 Wheeler Road
Burlington, MA 01803
Tel: 781-313-4700
Fax: 781-313-4802

Transferred to Digital Printing, 2010

Printed and bound in the United Kingdom

Editorial, design and production services provided by HighText Publications, Inc.
Solana Beach, California.

For Celia, Michael, and Bonillas.
These are my friends, and this is what we do.

Bob Widlar's contributions, albeit not received
for this book, are acknowledged by all.

Contents

··

Preface   xi
Contributors   xiii

One   Introduction

··

1. Barometers and Analog Design   3
   Alexander Calandra
2. Analogs Yesterday, Today, and Tomorrow, or
   Metaphors of the Continuum   5
   George A. Philbrick
3. It's An Analog World—Or Is it?   15
   Dan Sheingold
4. Is Analog Circuit Design Dead?   17
   Jim Williams

Two   What Is Analog Design?

··

5. On Being the Machine   23
   Bernard Gordon
6. Reflections of a Dinosaur   31
   Samuel Wilensky
7. Max Wien, Mr. Hewlett, and a Rainy Sunday Afternoon   43
   Jim Williams

Three   The Making of an Analog Engineer

··

8. True Analog Circuit Design   59
   Tom Hornak
9. The Story of the P2 (The First Successful Solid-State Operational
   Amplifier With Picoampere Input Currents)   67
   Bob Pease

Contents

10. Propagation of the Race (of Analog Circuit Designers)   79
    Jim Roberge

11. The Process of Analog Design   89
    Rod Russell

12. Analog Design Discipline: A Tale of Three Diodes   93
    Milton Wilcox

13. Should Ohm's Law Be Repealed?   99
    Jim Williams

Four   Intuitions and Insights

14. Good Engineering and Fast Vertical Amplifiers   107
    John Addis

15. Understanding Why Things Don't Work   123
    Bob Blauschild

16. Building Blocks for the Linear IC Designer:
    Linear Synthesis for Monolithic Circuits   127
    A. Paul Brokaw

17. How to Design Analog Circuits Without a Computer
    or a Lot of Paper   149
    Richard S. Burwen

18. Starting to Like Electronics in Your Twenties   169
    George Erdi

19. Where Do Little Circuits Come From?   177
    Barrie Gilbert

20. The Process of Analog Design   187
    Garry Gillette

21. The Art of Good Analog Circuit Design—Some Basic Problems
    and Possible Solutions   193
    Barry Hilton

22. My Approach to Feedback Loop Design   199
    Phil Perkins

23. The Zoo Circuit: History, Mistakes, and
    Some Monkeys Design a Circuit   215
    Jim Williams

Five   Techniques, Tips, and Applications

24. Reality-Driven Analog Integrated Circuit Design   233
    Derek F. Bowers

25. Current-Feedback Amplifiers   261
    Sergio Franco

26. Analog Extensions of Digital Time and
    Frequency Generation   277
    Garry Gillette

27. Some Practical Aspects of SPICE Modeling for Analog
    Circuits   299
    E. J. Kennedy

28. Design of Crystal Oscillator Circuits   333
    Robert J. Matthys

29. A Tale of Voltage-to-Frequency Converters
    (Ancient History)   349
    Bob Pease

30. Op Amps and Their Characteristics   361
    Dan Sheingold

Index   387

Preface

This is a weird book. When I was asked to write it I refused, because I didn't believe anybody could, or should, try to explain how to do analog design. Later, I decided the book might be possible, but only if it was written by many authors, each with their own style, topic, and opinions. There should be an absolute minimum of editing, no subject or style requirements, no planned page count, no outline, no nothing! I wanted the book's construction to reflect its subject. What I asked for was essentially a mandate for chaos. To my utter astonishment the publisher agreed and we lurched hopefully forward.

A meeting at my home in February 1989 was well-attended by potential participants. What we concluded went something like this: everyone would go off and write about anything that could remotely be construed as relevant to analog design. Additionally, no author would tell any other author what they were writing about. The hope was that the reader would see many different styles and approaches to analog design, along with some commonalities. Hopefully, this would lend courage to someone seeking to do analog work. There are many very different ways to proceed, and every designer has to find a way that feels right.

This evolution of a style, of getting to know oneself, is critical to doing good design. The single greatest asset a designer has is self-knowledge. Knowing when your thinking feels right, and when you're trying to fool yourself. Recognizing when the design is where you want it to be, and when you're pretending it is because you're only human. Knowing your strengths and weaknesses, prowesses and prejudices. Learning to recognize when to ask questions and when to believe your answers.

Formal training can augment all this, but cannot replace it or obviate its necessity. I think that factor is responsible for some of the mystique associated with analog design. Further, I think that someone approaching the field needs to see that there are lots of ways to do this stuff. They should be made to feel comfortable experimenting and evolving their own methods.

The risk in this book, that it will come across as an exercise in discord, is also its promise. As it went together, I began to feel less nervous. People wrote about all kinds of things in all kinds of ways. They had some very different views of the world. But also detectable were commonalities many found essential. It is our hope that readers will see this somewhat discordant book as a reflection of the analog design process. Take what you like, cook it any way you want to, and leave the rest.

Things wouldn't be complete without a special thanks to Carol Lewis and Harry Helms at HighText Publications, and John Martindale at Butterworth-Heinemann Publishers. They took on a book with an amorphous charter and no rudder and made it work. A midstream change of publishers didn't bother Carol and Harry, and John didn't seem to get nervous over a pretty risky approach to book writing.

I hope this book is as interesting and fun to read as it was to put together. Have a good time.

Contributors

∙∙

JIM WILLIAMS is the editor-in-chief of this book. In this role, Jim developed the basic concept of the book, identified, contacted, and cajoled potential contributors, and served as the "guiding light" of the entire project. Jim was at the Massachusetts Institute of Technology from 1969 to 1979, concentrating exclusively on analog circuit design. His teaching and research interests involved application of analog circuit techniques to biochemical and biomedical problems. Concurrently, he consulted U.S. and foreign concerns and governments, specializing in analog circuits. In 1979, he moved to National Semiconductor Corp., continuing work in the analog area with the Linear Integrated Circuits Group. In 1983, he joined Linear Technology Corp. as staff scientist, where he is presently employed. Interests include product definition, development, and support. Jim has authored over 250 publications relating to analog circuit design. His spare time interests include sports cars, collecting antique scientific instruments, art, and restoring and using old Tektronix oscilloscopes. He lives in Belmont, California, with his wife Celia, son Michael, a dog named Bonillas and 14 Tektronix oscilloscopes.

OHN ADDIS received his B.S.E.E. from the Massachusetts Institute of Technology n 1963 and joined Tektronix that same year. His career at Tektronix has been spent ın the design of various vertical amplifiers and pulse sources. The products John has engineered include the 1A7, 10A2A, 7A11, 485 vertical preamplifier, 7A29, and the analog paths of the 11A32, 11A34, and 11A52. He holds 14 U.S. patents, and was formerly responsible for analog integrated circuit design for high-speed oscilloscopes at Tektronix. He is now a consultant on analog design. John has traveled widely, including three trips to the Soviet Union and two to South America.

BOB BLAUSCHILD received his B.S.E.E. from Columbia University in 1971 and his M.S.E.E. from the University of California at Berkeley in 1973. He is also proud of his diploma from Ridgefield Memorial High School in New Jersey. Bob is currently manager of advanced analog development for Signetics, and has previously been an independent design consultant. He holds 12 patents in the area of analog circuit design, served ten years on the program committee for the International Solid State Circuits Conference, and is the author of numerous technical papers. His hobbies include running, visiting with old friends, coaching in the Special Olympics, and daydreaming of retirement on a warm beach.

DEREK F. BOWERS was born in Maesteg, Wales in 1954 and received a B.Sc. in physics and mathematics from the University of Sheffield in 1976. His first positions were with the University Space Physics Group and Triad Computing Systems. In 1978, he joined Precision Monolithics, Inc.'s U.K. division. In 1980, he transferred to Santa Clara as a senior design engineer. Since then he has held various positions within the company and is currently staff vice president, design. He has over thirty integrated circuit designs in volume production, including op amps, instrumentation amplifiers, audio products, and data conversion circuits. Derek has authored 35 technical articles and publications and holds ten patents. He is also a senior member of the IEEE and a member of the Audio Engineering Society. In his spare time, he enjoys music and regards himself as a connoisseur of beer and exorbitantly spicy food.

A. PAUL BROKAW spent his early years investigating flashlight workings and disemboweling toasters. After obtaining his B.S. in physics from Oklahoma State University, he developed electronics for scientific satellites with Labko Scientific, Inc. He also worked with Arthur D. Little, Inc., as a circuit design consultant. In 1971, he joined Nova Devices, which eventually became the semiconductor division of Analog Devices. He has held several positions at Analog, generally related to design, and is now an Analog Fellow. He holds over 50 U.S. patents in such areas as analog-to-digital and digital-to-analog converters, references, amplifiers, and application-specific integrated circuits. He has published technical papers in several IEEE journals, and is an IEEE Fellow.

RICHARD S. BURWEN received a S.B. (cum laude) in physics in 1949 and an A.M. in engineering sciences and applied physics in 1950 from Harvard. He was one of the three founders of Analog Devices and worked as a consultant to the company, designing several of the circuits for its initial product lines. Other companies with which he was associated in their beginning phases have included Mark Levinson Audio Systems, Cello Ltd., and Novametrix Medical Systems. He became a founder of Copley Controls in 1984 and has designed many of the company's products. In the case of all companies he has helped start, Richard maintains his independence by working as a consultant from his own laboratory. His home in Lexington, Massachusetts is designed around his 20,000-watt, 169-speaker recording and reproducing studio. He continues independent research in digital audio.

GEORGE ERDI has been designing linear integrated circuits for a quarter-century. In the 1960s, he designed the first precision op amp and and codesigned the first monolithic digital-to-analog converter while at Fairchild Semiconductor. In 1969, he cofounded Precision Monolithics, Inc., and created such industry standards as the OP-07 and REF-01 analog circuits. In 1981, George was a cofounder of Linear Technology where he designed 30 new circuits, including the popular LT1012, LT1013, LT1028, and LT1078. He has also presented six papers at the International Solid-State Circuits Conference. In September 1988, *Electronic Engineering Times* cited George as one of the "thirty who made a difference" in the history of integrated circuits.

SERGIO FRANCO is a professor of electrical engineering at San Francisco State University, where he teaches microelectronics courses and acts as an industry consultant. Prior to assuming his current professorship, Sergio was employed at Zeltron, Zanussi's Electronics Institute (Udine, Italy). He received a B.S. in physics from the University of Rome, a M.S. in physics from Clark University, and a Ph.D. in computer science from the University of Illinois. Sergio is a member of the IEEE, and in his spare time enjoys classical music, gardening, and mountain hiking.

BARRIE GILBERT has spent most of his life designing analog circuits, beginning with four-pin vacuum tubes in the late 1940s. Work on speech encoding and synthesis at the Signals Research and Development Establishment in Britain began a love affair with the bipolar transistor that shows no signs of cooling off. Barrie joined Analog Devices in 1972, where he is now a Division Fellow working on a wide variety of IC products and processes while managing the Northwest Labs in Beaverton, Oregon. He has published over 40 technical papers and been awarded 20 patents. Barrie received the IEEE Outstanding Achievement Award in 1970, was named an IEEE Fellow in 1984, and received the IEEE Solid-State Circuits Council Outstanding Development Award in 1986. For recreation, Barrie used to climb mountains, but nowadays stays home and tries to write music in a classical style for performance on a cluster of eight computer-controlled synthesizers and other toys.

GARRY GILLETTE received a B.S.E.E. from Stanford in 1961 and a M.S.E.E. from the University of California–Irvine in 1968. While a student at Stanford, his summer employment at Electro-Instruments Corp. in San Diego exposed him to a group of pioneer transistor circuit designers, leaving him with an indelible respect for intellectual honesty, highest technical standards, lightning empiricism, and the fun of creating something efficient and elegant. Since 1974, he has been employed by the Semiconductor Test Division of Teradyne, Inc., and is currently their manager of advanced technology. Garry holds several patents.

BERNARD GORDON is president and chairman of the board of Analogic Corporation, a high technology company specializing in the design and development of precision measuring instrumentation and high-speed computing equipment. He is the holder of over 200 patents worldwide in such fields as data converters, array processing computers, industrial controllers, diagnostic imaging, and automated test equipment. An IEEE Fellow, Bernard received the National Medal of Technology in 1986. He is also the founder of The Gordon Institute to enhance the leadership potential of engineers.

BARRY HILTON was born and educated in Britain and received a Higher National Certificate in Applied Physics from Kingston College of Advanced Technology. Early in his career, he was employed by Solartron Ltd. as a designer of digital voltmeters. In 1969, Analog Devices hired him to help design the first converter modules in Boston, and in 1973 Barry became director of engineering for Analog Devices. In 1975, he decided to establish his own design consulting company, A.I.M., Inc. Since that time, Analog Devices has kept him very busy as a consultant designing hybrid converters and numerous integrated circuits. In 1989, Barry established a second company, Acculin Inc., for the design and manufacture of very high speed analog integrated circuits. In his leisure time, Barry enjoys golf, swimming, traveling, and classical music.

TOM HORNAK was born in Bratislava, Czechoslovakia. He received his Dipl.Ing. degree from the Bratislava Technical University and his Ph.D. from the Prague Technical University, both in electrical engineering. From 1947 to 1961 he worked in Prague at the Tesla Corp.'s Radio Research Laboratory and from 1962 to 1968 in the Prague Computer Research Institute. His work in Czechoslovakia involved development of television cameras, ferrite and thin film memories, high-speed pulse generators, and sampling oscilloscopes. In 1968, Tom joined Hewlett-Packard's Corporate Research Laboratory and is presently head of their high-speed electronics department. He is responsible for applied research of high-speed data communication circuits, high-speed analog/digital interfaces, and electronic instrumentation utilizing advanced Si and GaAs IC processes. Tom has published 50 papers and holds 40 patents. He has served as guest and associate editor of the *IEEE Journal of Solid State Circuits* and as chairman of the IEEE Solid State Circuits and Technology Committee. Tom has been named an IEEE Fellow.

E. J. (ED) KENNEDY received his Ph.D. in electrical engineering in 1967 from the University of Tennessee at Knoxville. Before joining the faculty of the University of Tennessee, he held positions at the Arnold Engineering Development Center and the Oak Ridge National Laboratory. Ed's research interests include nuclear instrumentation, strapdown gyroscope rebalance electronics, hybrid thick-film electronics, switching regulators, and analog integrated circuits. He was appointed Fulton Professor of Electrical Engineering in 1983, has been a Ford Foundation Teaching Fellow, and has received the NASA Commendation Award. Ed's books include *Operational Amplifier Circuits* (1988) and the forthcoming *Semiconductor Devices and Circuits*. He is married, has three daughters, and enjoys tennis, gardening, and growing roses.

ROBERT J. MATTHYS retired from Honeywell's Systems & Research Center in Minneapolis, Minnesota as a research engineer, and is still associated with the Center on a part-time basis. He has over 38 years of experience in optical, mechanical, and electronic design. He has written a book on crystal oscillator circuits, holds five patents (with two pending), and has published 17 technical papers. Among his other achievements of which he is proud are seven children and four grandchildren. His interest in crystal oscillators began when he was asked to design one, and found the reference articles disagreed with each other and some were even obviously wrong.

PHIL PERKINS is a Fellow of LTX Corp. in Westwood, Massachusetts. His work includes analog instrumentation and system design for the LTX semiconductor test systems. Most recently, he has developed test heads for the Synchromaster line of mixed-signal semiconductor test systems. Prior to co-founding LTX, Phil worked eight years at Teradyne, Inc. in Boston. He received his degrees in electrical engineering from the Massachusetts Institute of Technology. Phil's interests include local and national activities in the United Methodist Church, home computer hobbying plus consulting for friends, vegetable gardening, and bicycling. He lives in Needham, Massachusetts, with his lovely wife Laurie.

BOB PEASE graduated from the Massachusetts Institute of Technology in 1961 with a B.S.E.E. He was employed at George A. Philbrick Researches from 1961 to 1975, where he designed many operational amplifiers, analog computing modules, and voltage-to-frequency converters. Bob joined National Semiconductor in 1976. Since then, he has designed several ICs, including regulators, references, voltage-to-frequency converters, temperature sensors, and amplifiers. He has written about 60 magazine articles and holds eight patents. Bob has been the self-declared Czar of Bandgaps since 1986, and enjoys hiking, backpacking, and following abandoned railroad roadbeds. He also designs voltage-to-frequency converters in his spare time. Bob wrote an award-winning series of articles on troubleshooting analog circuits which appeared in *EDN* Magazine in 1989, and which will be expanded into a book to be published by Butterworth-Heinemann. Bob currently writes a column about analog circuits which appears in *Electronic Design* Magazine.

JIM K. ROBERGE has been at the Massachusetts Institute of Technology since 1956, initially as a freshman and currently as professor of electrical engineering. In between, he received the S.B., S.M., and Sc.D. degrees in electrical engineering and held various research and academic staff appointments. His teaching and research interests are in the areas of electronic circuits and system design. Much of his research is conducted at M.I.T. Lincoln Laboratory and is involved with communications satellites. He is the author of *Operational Amplifiers: Theory and Practice* and co-author of *Electronic Components and Measurements*. He has made a twenty-lecture video course entitled *Electronic Feedback Systems*. He has served as consultant to more than 90 organizations, and has eight patents awarded or in process. For recreation, he plays with his toys, which include a large collection of Lionel electric trains and a 1973 E-type Jaguar roadster.

ROD RUSSELL is president of Custom Linear Corp. He got turned on to analog electronics while serving in the U.S. Navy, where he repaired and maintained VHF and UHF transceivers. During his last semester at New Mexico State University, a professor fleetingly mentioned that an operational amplifier had just been fabricated in silicon. After obtaining his B.S., he joined Motorola Semiconductor and also obtained his M.S.E.E. from Arizona State University. He says the vast number of possibilities (some are called problems) in analog electronics is what makes it interesting.

DAN SHEINGOLD received his B.S. with distinction from Worcester Polytechnic Institute in 1948 and a M.S.E.E. from Columbia University in 1949. He then joined George A. Philbrick Researches as their second engineer (the other being George A. Philbrick). Dan eventually became vice president for marketing, and was present at the development of the world's first commercial differential plug-in operational amplifier, the vacuum tube K2-W. He also served as editor of *The Lightning Empiricist* while at Philbrick. In 1969, Dan joined Analog Devices as manager of technical marketing. He's currently involved in the writing and editing of their popular *Analog Dialogue* magazine, and has developed an extensive list of tutorial books on Analog's technologies and products, including such classics as *Analog-Digital Conversion Handbook* and the *Transducer Interfacing Handbook*. He was elected an IEEE Fellow in 1990. He and his wife Ann have two children, Mark (an engineer) and Laura (a physician). Dan enjoys music, walking, running, cross-country skiing, and has an airplane pilot's license.

MILTON WILCOX has been interested in electronics since high school. He received his B.S.E.E. in 1968 and his M.S.E.E. in 1971 from Arizona State University. From 1968 to 1975 he was employed by Motorola as an analog design engineer designing consumer linear integrated circuits. In 1975, Milt moved to National Semiconductor where he was section head of an RF and video IC design group for over 14 years. He currently heads a small group designing new power control ICs at Linear Technology Corporation. Milt holds 25 patents, has authored seven technical papers, and continues to actively design.

SAMUEL WILENSKY was first exposed to Ohm's Law at the Massachusetts Institute of Technology, where he received his B.S.E.E. He did graduate work at the M.I.T. department of nuclear engineering, where his thesis project was the measurement of non-elastic neutron cross-sections using a pulsed neutron source (i.e., the Rockefeller Accelerator). Samuel was one of the founders of Hybrid Systems, now Sypex. During the early years of Hybrid Systems, he became—of necessity—an analog designer. His main efforts have been in the design of data conversion devices, with detours into consumer products. He recently used his nuclear training to study the effects of nuclear radiation on data conversion products. He enjoys playing pick-up basketball, sailing, coaching youth soccer, being embarrassed by his son and daughter on ski slopes, and supplying muscle for his wife's gardening.

Part One

Introduction

Most books have a single introduction. This one has four. Why?

Analog circuit design is a very "personalized" discipline. To be sure, everyone's bound by the same physics and mathematics, but there's no single "right way" for those tools to be applied to solve a problem. Practitioners of analog design are noted for their individuality. Three of the four introductions that follow are by acknowledged masters of the analog art and deal with analog's place in a world that seems overwhelmed by digital electronics. Each of those three authors gives a highly personal viewpoint that can't be objectively proven "right" or "wrong," but that's the way it is in many aspects of analog design. The remaining introduction, which appears first, doesn't directly deal with analog electronics at all. However, it does illustrate the "matrix of thought" that so many successful analog designers bring to their efforts.

Analog design is often less a collection of specific techniques and methods than it is a way of looking at things. Dr. Calandra's thoughts originally appeared in the January, 1970 issue of "The Lightning Empiricist," then published by Teledyne Philbrick Nexus, and is reprinted by permission of Teledyne Corporation. We don't know if the student described ever became interested in analog electronics, but he clearly had all the necessary attributes of a good analog design engineer.

The name of George Philbrick will be invoked several times in this book, and in each instance some awe and reverence is noticeable. This is because if contemporary analog design has a founding father, it would have to be George Philbrick. Many of the top names in the field today either worked under or were influenced by him. Although he passed away several years ago, his wisdom is still relevant to many current situations. Here's a sample from the October 1963 issue of "The Lightning Empiricist," published by the company he founded, Teledyne Philbrick. We're grateful for the company's kind permission to reprint the following, since it's difficult to imagine a *real* guide to analog design without George Philbrick!

Let's face it: analog electronics isn't very sexy these days. The announcement of a new microprocessor or high-capacity DRAM is what makes headlines in the industry and business press; no one seems to care about new precision op amps or voltage-to-frequency converters. Sometimes it seems if digital electronics is the only place in electronics where anything's going on. Not so, says Jim Williams, as he tells why analog electronics is more than still important—it's unavoidable.

Dan Sheingold's essay originated as a letter to the editor of *Electronic Engineering Times*. In its original form (with a slightly different message), it appeared on December 4, 1989. Often electronics engineers draw clear distinctions between "analog electronics" and "digital electronics," implying clear barriers between the two disciplines that only the very brave (or very foolish) dare cross. However, as Dan points out, the differences between them might not be quite what we think.

Introductions are normally read before the rest of the book, and so should these. But you might want to return and read them again after you've finished this book. It's likely that you might have a different reaction to them then than the one you'll have now.

1. Barometers and Analog Design

Some time ago I received a call from a colleague, who asked if I would be the referee on the grading of an examination question. He was about to give a student a zero for his answer to a physics question, while the student claimed he should receive a perfect score and would if the system were not set up against the student. The instructor and the student agreed to an impartial arbiter, and I was selected. I went to my colleague's office and read the examination question: "Show how it is possible to determine the height of a tall building with the aid of a barometer."

The student had answered: "Take the barometer to the top of the building, attach a long rope to it, lower the barometer to the street, and then bring it up, measuring the length of the rope. The length of the rope is the height of the building."

I pointed out that the student really had a strong case for full credit since he had really answered the question completely and correctly. On the other hand, if full credit were given, it could well contribute to a high grade in his physics course. A high grade is supposed to certify competence in physics, but the answer did not confirm this. I suggested that the student have another try at answering the question. I was not surprised that my colleague agreed, but I was surprised that the student did.

I gave the student six minutes to answer the question with the warning that the answer should show some knowledge of physics. At the end of five minutes, he had not written anything. I asked if he wished to give up, but he said no. He had many answers to this problem; he was just thinking of the best one. I excused myself for interrupting him and asked him to please go on. In the next minute he dashed off his answer which read:

"Take the barometer to the top of the building and lean over the edge of the roof. Drop the barometer, timing its fall with a stopwatch. Then using the formula $S = 0.5\ at^2$, calculate the height of the building."

At this point, I asked my colleague if he would give up. He conceded, and gave the student almost full credit.

In leaving my colleague's office, I recalled that the student had said he had other answers to the problem, so I asked him what they were. "Oh, yes," said the student. "There are many ways of getting the height of a tall building with the aid of a barometer. For example, you could take the barometer out on a sunny day and measure the height of the barometer, the length of its shadow, and the length of the shadow of the building, and by the use of simple proportion, determine the height of the building."

"Fine," I said, "and the others?"

"Yes," said the student. "There is a very basic measurement method you will like. In this method, you take the barometer and begin to walk up the stairs. As you

Reprinted with permission of Teledyne Components.

climb the stairs, you mark off the length of the barometer along the wall. You then count the number of marks, and this will give you the height of the building in barometer units. A very direct method.

"Of course, if you want a more sophisticated method, you can tie the barometer to the end of a string, swing it as a pendulum, and determine the value of g at the street level and at the top of the building. From the difference between the two values of g, the height of the building, in principle, can be calculated.

"Finally," he concluded, "there are many other ways of solving the problem. Probably the best," he said, "is to take the barometer to the basement and knock on the superintendent's door. When the superintendent answers, you speak to him as follows: 'Mr. Superintendent, here I have a fine barometer. If you will tell me the height of this building, I will give you this barometer.'"

At this point, I asked the student if he really did not know the conventional answer to this question. He admitted that he did, but said that he was fed up with high school and college instructors trying to teach him how to think, to use the "scientific method," and to explore the deep inner logic of the subject in a pedantic way, as is often done in the new mathematics, rather than teaching him the structure of the subject. With this in mind, he decided to revive scholasticism as an academic lark to challenge the Sputnik-panicked classrooms of America.

George A. Philbrick

2. Analogs Yesterday, Today, and Tomorrow, or Metaphors of the Continuum

It was naturally pleasurable for me to have been approached by the Simulation Councillors to write an article, substantially under the above super-title, for their new magazine. This euphoria persists even now, when my performance has in fact begun, and is only moderately tempered by the haunting suspicion of what their real reason might have been for so honoring me. It certainly could not be because my views on analog computing and simulation are somewhat eccentric in relation to much of the contemporary doctrine, although I accept and actually relish this characterization. It could conceivably be in recognition of my relatively early start in the field of electronic analog technology; this again is not denied by me, but here we may have found the clue. The fact that I began a long time ago in this sort of activity doesn't mean at all that I am either oracle or authority in it. The truth of the matter is subtler still: it only means that I am getting old. So we have it out at last. They are showing respect for the aged. Here then, steeped in mellow nostalgia, are the musing of a well-meaning and harmless Old Timer.

Since truth will out, I might as well admit immediately that I do not claim to be the original inventor of the operational amplifier. It is true, however, that I did build some of them more than four years before hearing of anyone else's, and that their purpose was truly simulative. These amplifiers were indeed DC feedback units, used to perform mathematical operations in an analog structure, but the very first such amplifier *itself* began as a model builder, even at that stage, loomed larger than my possible role as an inventor, and I have been dealing continually with models and analogs ever since. Hereafter in this context I shall not speak of what I may have invented or originated, and in fact shall not much longer continue in the first person singular. By the same token I shall make no pretense in this article of assigning credit to other individuals or to other institutions. There are far too many of both, hundreds and thousands, stretching from this point back into history, to give any accurate and fair account of the brainpower and perspiration which have made analog computing what it is today, without leaving out many who have put vital links in the chain.

While electronic analog equipment, using this phrase in the modern sense, certainly existed in the thirties, and in the forties became available on the open market in several forms, its roots really went still further back in time. It is doubted that a completely exhaustive chronology of the contributory precursor technologies could ever be produced, let alone by one amateur historian. Nothing even approximating such a feat will be attempted, but it is hoped that an outline of the tools and techniques which were on hand in the previous era will show that the ingredients were already there, and that the modern analog machine was almost inevitable. As is usual in such surges of progress, several fields of science and engineering over-

Reprinted with permission of Teledyne Components.

lapped to breathe life into this department. Among others were Physics and Scientific Instruments, Communications and Electronics, Controls and Servo-mechanisms, Mathematics, and Aeronautical plus Electrical plus Mechanical Engineering. It is recognized that these fields are not mutually exclusive, and that each realm constitutes a multidimensional cross-section which has interpenetrated the other realms enumerated.

There is one thread, come to think of it, which appears to run through the whole background of the analog doctrine, and which may be said to belong to it more intrinsically that it does to the other major branch of computation; that thread is *feedback*. It will appear again frequently in what follows.

The clearest anticipation of analog machines was in the differential analyzer. This primarily mechanical device could handle total differential equations at least as well as we can now, and in some ways better. One such analyzer afforded auto-matic establishment of its interconnections and parameters, tape storage of these data, and automatic readout: both numerical and graphical. Although slower than newer electronic equivalents, nonetheless for a 19-integrator problem which was run on it in 1945, a thoroughly non-linear problem by the way, the analyzer time scale was only twice as slow as the real scale for the remotely controlled glide vehicle which was being simulated. The disc integrators of this machine were things of beauty, with accuracies approaching, and resolution exceeding, 5 decimals. They could integrate with respect to dependent variables, thus enabling multiplication with only two integrators, logarithms without approximation, and so on. Integrators of this same general type were also applied in astronomical and military computing devices, in which less elaborate but still legitimate differential equations were em-bodied and solved. This sort of equipment inspired many of the electronic analog devices which followed, as well as the digital differential analyzers which have come much later. Although the electronic integrators of analog equipment prefer time as the direct variable of integration, they have shown extreme flexibility of operating speed. One imagines the mechanical discs of the older analyzers running at millions of rpm trying to keep up with their progeny!

The disc integrators of the differential analyzer worked without feedback, as did its other basic parts. Where then did feedback appear in these analyzers? In the differential equations acted out within it. Any equation requiring solution involves at least one causal loop. But for feedback in its more exuberant forms we nominate the next discipline to be considered, namely automatic controls.

Regulatory mechanisms such as those which are found in industrial control sys-tems have been around for a long time. Roughly in historical sequence, they have been mechanical, hydraulic, pneumatic, electric, and electronic. Translating as they do from the unbalance or error in a controlled condition to the manipulation which is intended to reduce that unbalance, they close a feedback loop which includes some sort of plant. In typical cases these mechanisms have embodied mathematical laws with continuous fidelity, and in order to attain fidelity they have resorted to internal feedbacks precisely analogous to those employed in a modern amplifier. It may not be widely known, particularly among the younger computing set, that this sort of local feedback was applied in standard controller mechanisms of the twen-ties and even earlier. These antecedent regulatory devices qualify as DC feedback and even null-seeking at two distinct levels, and with mathematical capabilities, it is not difficult to trace the logical paths of evolution from these devices to analog computing as it is now enjoyed. Furthermore it is not uncommon in the thirties to build simulators embodying convenient models of plants, into which the real regu-latory mechanism could be connected. Both developmental and educational purposes were served by these structures, just as with simulators today. The next stage, in which the real control mechanisms were replaced by models, permitted the whole loop to be electronic and hence vastly more flexible and greatly accelerated. In simulators of this sort, several plants might be interconnected under control, so that the newer stability problems thus encountered could be studied conveniently. Again, plants with multiple inputs and outputs having internally interacting paths were included, and regulatory loops in hierarchies where master controls manipulated the desired conditions of subordinate controls, all could be simulated in an analog. Note the ascending succession of feedback loops, which are most dramatically represented in control systems of this sort: within amplifiers to attain promptness and stability; locally around amplifiers to give the desired mathematical performance for regulatory mechanisms; in control loops to promote the minimum difference between desired and existing conditions; in more comprehensive control loops which include complete but subordinate loops in cascade; in still more comprehensive loops for supervisory or evaluative purposes; and finally in the experimental design and optimizing operations, using models or computational structures to evolve most effective system operation.

Servomechanisms are also part of the lore which preceded and inspired the modern analog machines. Though not as old as the governors, pressure regulators, and controllers of temperature, flow, level, etcetera of the last paragraph, servos as positional followers were functionally similar as regards control philosophy and feedback loops. Further, being more modern, they benefited from the increasingly mathematical technologies of development and design. Perhaps most relevant was the simultaneity and parallelism between servo theory and that of feedback amplifiers in communications. Stability criteria for the latter were seen as applicable to the former, at least in the linear realm. Analysis in the frequency domain, a natural procedure for linear communications equipment, was carried over rather directly to servomechanisms. This debt has since been partially repaid, as servomechanisms have helped to furnish nonlinear analog elements and other items in computing equipment for the study of nonlinear phenomena, generally in the time domain, as they occur in communications and elsewhere. Thus do the various doctrines and practical disciplines feed on each other to mutual benefit, and (if you will forgive the liberty) feedback sideways as well as back and forth.

We pick up servomechanisms again, much further back along the trail, and usually in relatively low-performance embodiments. Though scientific instruments do practically everything today, including computation, synthesis, manipulation, and regulation, on every scale, they were once used principally for measurement, in the laboratory or the observatory. For accurate measurement it was found that feedback methods, when possible, were surpassingly effective. While the underlying philosophical reasons for this circumstance are of vital importance, we shall take them here on faith. Note, however, that the observation of balance in a measurement, and the manipulation which may be made to achieve balance, is still a feedback process even if done by a human agency. The slave can be the experimenter himself. Precise weighing with a beam balance may stand as a clear example of this procedure, but a myriad of others may readily be spread forth. Succinctly, the process is reduced by feedback to dependency on only one or a few reliable elements. Automation of the loop-closing, null-seeking action merely replaces one slave by another. In this light the venerable self-balancing slidewire potentiometer recorder stands with the latest feedback operational amplifier, and so we see yet another plausible path from then to now.

Antedating but partly anticipating the development of active analogs was the use of models which depended much more directly on the analogies between phenomena as they appear in widely differing physical media. Of main concern here are those cases in which the modelling medium has been electric, but quite accurate and articulate models have also been mechanical and hydraulic, and many of these are hoary with age indeed. Ever since accurate and dependable circuit elements have been available, and this has been for many decades, notably for resistors and capacitors, highly successful passive models have been built for the study and solution of such problems as those which occur in heat conduction. Dynamic as well as steady state phenomena may be handled, often in the same model. Again, vibrations have been studied with direct models having all three kinds of circuit element, plus transformers. Furthermore very large and complete simulative structures, called network analyzers and based heavily on passive elements, were used in particular for—though not limited to—AC power distribution and communication lines. Even today one finds such continuous conductive models as electrolytic tanks still in use and under development. Many of these tools have specialized capabilities which are hard to match with the more familiar sort of modern apparatus. The similitude conditions and principles which accompanied and abetted the application of such models have been carried over to, and guided the users of, the newer computing means. It should be added that the very demanding doctrines of "lumping," which must take place when continuous systems are represented by separate but connected analog operations, are substantially unchanged as compared to those in passive models. Here is another branch of knowledge and effort, then, to which we own recognition as contributing to present day simulation and computing.

From a different direction, in terms of need and application, came another practical model-building technique which is woven into the analog fabric which surrounds us today. This one is straight down the simulation highway; we refer to trainers of the sort used for many years to indoctrinate pilots of aircraft. These trainers modelled just about everything except nonangular spatial accelerations. They presented, to a human operator, a simulated environment resembling the real one in many important ways, as regards his manipulations and the responses returned to him as a consequence thereof. Of course the later counterparts of the first training aids have become tremendously more refined, and similar structures have been adapted to other man–machine collaborations, but the inspiration to analog enthusiasts on a broader scale seems rather obvious. Here was an operative model, in real time and undelayed, where to the sensory and motor periphery of the trainee the real environment was presented in a safe and pedagogically corrective atmosphere. Now it is true that training devices for physical skills are even more numerous today, and analog simulative equipment finds important applications in these, but a somewhat extended simile might be in order. For system design in its larger implications we are all trainees; analog simulation to teach us how a proposed system might work when at least part of it is new, to guarantee safety if we try out a poor idea, and to offer peripheral communication at the deliberative level, projects the trainer concept to an advanced modern setting. The task of simulating the trained pilot and even the learning pilot, or other human operators, provided a challenge which has been partly met, and which is still relevant. Simulating the system designer, as a logical extension, leads as far as you might care to travel.

Overlook

Things are looking up all over for the analog profession. Substantially every branch of engineering now applies analog computing equipment: in theory, experiment,

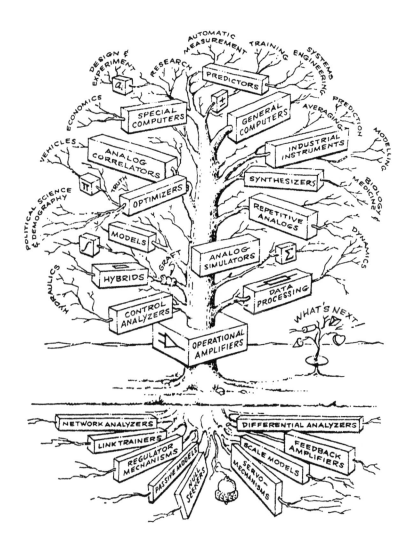

Figure 2-1.
This was George's vision of the mighty analog tree. It remains relevant almost three decades later. Courtesy of Teledyne Components.

design, manufacture, and test. Applications are even on the increase for scientific research, where in a sense such equipment began. We shall not try to list the many embodiments and applications in this text, but have included some of them in a figure to be found nearby, which has been prepared to bear out the morphology of our burgeoning field.

Analog representation in terms of modern apparatus is a far cry from scale models, but the model concepts still seem incomparably fruitful. In direct models, which retain the physical medium of their prototypes, scaling is the biggest part of the game. Similitude conditions must be faithfully adhered to, and an appreciation of these conditions imparts a feeling for models which is never lost. Actually the use of direct scale models has not decreased, and is still a powerful technique in such areas as hydraulics and structures: natural and man-made. Much ingenuity has been lavished on such models; they must by no means be looked down upon by the users and designers of more fashionable modelling items.

In a scale model the transformation of dimensions is typically direct and simple, especially if shape is preserved. Even when the scaling involves distortions of shape, such as relative compression and bending, the transformations generally carry distance into distance, velocity into velocity, and so on, with only numerical scale factors relating them in pairs. Basic parameters, when the scale ratios are properly assigned, turn out to be numerical, and apply equally to model and to prototype. This doctrine, whereby characteristic system parameters are dimensionless, is applicable to all modelling procedures. The transformation concept, so clear and concise for scale models, carries over with little confusion to modelling in which the physical form is changed, and ultimately to electronic analogs where transformation includes transmogrification. The scale ratios in general, however, are no longer numbers, but the basic parameters may be. This sort of introduction is recommended for physicists and applied mathematicians who may be coming suddenly into modern analog contacts, since it utilizes some of the ideas and precepts, however badly expressed here, of the more classical fields.

Another sort who is momentarily taken aback by the liberties permitted in analog models is typified by an engineer who has been too long away from the time domain. Often brought up, pedagogically, on linear systems and frequency analysis, he (or she) may even be suspicious of a mechanism which gives solutions as functions of time, perhaps not realizing that it will provide amplitude and phase spectra as well if one merely applies a different stimulus to the same model structure. It is frequently worthwhile, in these cases, to introduce the analog from the viewpoint of the frequency domain, shifting later from the familiar to the strange and magical. Oddly enough, the most confirmed practical and the most profoundly theoretical of engineers will both be found to favor the time domain, with or without computing equipment. In the former case this is by virtue of convenience in handling real equipment, and in the latter it is since—among other reasons—he finds it better to approach nonlinear problems in the time domain than in the frequency domain.

Analog engines have not always been as respected as they are now becoming. Analogy itself we have been warned against, in proverb and in folklore, as being dangerous and requiring proof. Parenthetically, this is good advice. Simulation has had connotations of deceit, empiricism of quackery. It was stylish, even recently, to say that the only good electronics is that which says Yes or No. There is nothing to be gained in disputing these allegations, least of all by excited rejoinder. The continuous active analog is in its infancy, and time is (literally) running in its favor.

Time as an independent variable, given at low cost by Nature, has the advantage of nearly, if not actually, infinite resolution. This continuity, coupled with the continuity of voltage and charge, leads to the ability to close loops at very high frequency, or with short time intervals. As a consequence one may approach the ideals of differentiability which are inherent in the infinitesimal calculus, which postulates the existence of a continuum. While most contemporary analog apparatus does not press these limits, it is comforting to know that there is room left to maneuver in.

In modest applications to on-line measurement and data-processing, it is quite generally conceded that the advantages of continuous analog apparatus make it irresistible. This is partly owing to the simplicity and speed which its continuity makes possible, and partly to the fact that almost every input transducer is also "analog" in character, that is to say continuous in excursion and time. Storage and sampling, for example are frequently unnecessary in such applications, as in many others. When we turn from simpler to more involved data processing, to ambitious simulation, or when in general we pass from modest to more pretentious computations, there has been some feeling that digital means should automatically be substituted, especially if funds are available. In this connection we should like to quote, on the other side of the argument, no less a figure than Dr. Simon Ramo, writing on Systems Engineering in a collected volume called *Parts and Wholes* (edited by Daniel Lerner; Macmillan, New York, 1963). The following is admittedly taken out of context:

Digital computers, however, cannot be used conveniently or efficiently to obtain answers to all of the problems. In some cases, even they cannot solve the equations in any reasonable time, and in other cases the problems are not understood well enough for satisfactory mathematical formulation. Under these circumstances we can often turn to analog, real-time, simulation devices to predict the behaviour of the system. No engineering computing center is well equipped without such devices.

One should certainly be happy to settle for this, even though the text continues in a discussion of other kinds of equipment than analog with which the latter may be associated. Only the most hard-shelled of analog champions would suggest that all simulative and computational equipment be undiluted by numerical or logical adjuncts. Certainly many of the best known individuals and organizations in the analog field are now willing and able to talk about hybrids. This term, by the way, is too broad to have much meaning at this stage of the game. Is an analog apparatus hybridized by adding a digital voltmeter? The possibilities are far too numerous. The present treatment does not even contemplate giving a complete account of analog computing machines themselves, let alone the combination they may form with other machines. A large and growing library of good books cover these areas quite completely. Many of these are written by officials of the Simulation Councils, who typically have the sort of university connections which should give them appropriately unbiased viewpoints: viewpoints which a mere company man can only envy. Perhaps, however, an example or two might be appended here which will amuse and even edify.

At a large Eastern university, under the guidance of a well-known and gifted computationalist, a successful project has been reported on whereby the scaling for an analog installation is done entirely by rote on a digital machine. No guessing or trial runs at all are involved. Straight from the equations, the digital solution dictates the analog settings which will bring the maximum excursion of every variable analog voltage to within 20% of the limiting value. Local wags thus proclaim the discovery at last of a practical contribution by the digital apparatus. Seriously, they enjoy the ability to "get at" the solutions of the analog during operation.

Some analog men, perhaps over-fond and defensive as regards continuous functions, really believe that analog operations are generalizations of digital ones, or that conversely digital operations are special cases of analog ones. What can be done with such people? They depreciate the importance of the fact that discrete measure-scales approach continuity in the limit, alleging that infinite processes are already tacit and available, without passing to the limit, in an analog variable. Pointing for example to analog selector circuits which can pick out and transmit whichever of a set of variables is algebraically the greatest of the least, they cite this capability as broader than the logical sum or the logical product, amounting in fact to infinitely-many-valued logic. Selectors followed, for example, by bounding operations serve directly in the rudimentary case of two-valued logic. On the basis of such reasoning it is surprising, the argument runs, that analog apparatus is not permitted to make decisions for itself. It is hard to answer these arguments, especially when dealing with confirmed analog partisans. When cornered on some point of superior digital accomplishment, they simply claim the whole digital province as part of their analogs.

Predictions are scheduled for the Tomorrow part of this article, but one such properly belongs here. While it is agreed that analog and digital techniques will increasingly cross-fertilize and inter-relate, it is predicted that the controversy between their camps will rage on, good natured but unabated, for years to come in spite of hybrid attachments. The serious issue of reliability has recently arisen as between the two ideologies referring for example to instruments for interplanetary exploration. It is preferred here to avoid an opinion of judgment on this very important issue, but it is suggested that others similarly withhold judgment. At all costs we must not go down the wrong road. There are quite powerful and rational and experienced brains in which the reliability vote would be cast for analog, or at least against the exclusion of continuous variability. We must cooperate in a dispassionate but devoted study to determine the likeliest facts and fancies in this affair. If one believes that Nature is ahead in reliability, and there would appear to be justification for this belief in recognition of the redundancy, repairability, and adaptability of animal organisms, then conclusions may follow which are based on how one views such organisms. It has been standard practice to view the details of animal nervous systems as evidence that they are digital, but there are major reasons to question this.[1] The central nervous system itself seems digital to digital men, and analog to analog men. If it is both, then it is more intimately and profoundly intermingled hybrid than any of the artificial structures which have come to light. One thing is pretty sure, and that is that the brain builds models. We are in good company.

Back on reliability, at least in the sense of predictability, there is a duality to be noted in the relation between analog and digital techniques. If one must predictably manipulate an imperfectly accessible entity, he may proceed by arranging a discrete set of states for that entity, then transmit a prearranged number of command signals to it. Alternatively, with a nonquantitized feedback indicating the state of the entity, one commands changes outwardly by whatever means until the desired state is shown to have been attained. What one achieves by quantitizing, the other does by feedback. This is oversimplified, and does not immediately enable an evaluation of reliability. For the moment, it is only a point in (practical) philosophy, but as with many other continuous/discrete instrumental relations it is reminiscent of the wave–particle dualism.

Auguries

It has been predicted above that the analog–digital struggle will persist, and this will mean some wear and tear as the proponents contend, but on balance such contention will probably be beneficial since it will assure that the maximum potential of each technique will be realized. As to some mixtures, all the obvious ones will soon be seen somewhere. More intimate mixtures, which might offer something approaching universal applicability, will depend on the appearance of new instrumental tools. But also note that urgent needs provide as potent a force for development as does the availability of new and startling techniques. Hasty prediction from either angle would be hazardous; certainly anything specific on our part would be irresponsible as well as foolhardy. There do seem to be possibilities, however, in recognition of the ability of continuous analog instruments to operate quickly and smoothly in closing feedback loops, plus the abitrary accuracy and permanency of discrete processes. Graphical computation may give a clue of sorts here, since anyone who deals with geometrical plots is prone to appeal alternately to continuous criteria and to numerical coincidences in calibration. Coordinates in general may have both of these meanings simultaneously. Are they any better than we are?

As to analogs themselves, it is evident that some forms of instrument, though not all, will become progressively smaller and handier in solid state incarnations. It is

---

1. R.W. Jones, *Science* **140**, 3566 (1963). See also the companion article by J. S. Gray.

also evident that optimizing and search operations will be made increasingly automatic, as the deliberative functions of the user are encroached on more and more by deliberately imposed autonomous controls. But one of the principal lessons from the past is that substantially all the earlier techniques will continue to be used, and will grow and improve horizontally. Possibly you have a slide rule in your pocket, though admittedly you may have turned in your abacus for a desk calculator. All the older apparatus of the above section on origins are in current usage, and will continue so. As an example may we consider passive models?

It would be a big surprise if passive electric models do not expand in application and in technical excellence. More adept peripheral instruments, to drive and to measure them, are either in the cards or on the table. Passive circuit elements, adjustable as well as fixed, are gradually but surely improving as to accuracy, bandwidth, and stability. In this category are included not only resistors and capacitors, and less insistently inductors and transformers, but also certain nonlinear elements. A combination of compensation and regulation can cut the parametric effects of temperature down to size, especially with the advent of flexible devices for thermoelectric heat pumping. Relatively little work has been done on passive networks for model building, even for linear systems, compared to that expended for communications. The challenges introduced in the nonlinear cases are considerable, but with newer analytical techniques and instrumental tools it would be unwise to put limits on what might be accomplished. Part of the lure is that many biological structures appear to have been designed along these lines, though not of course without active adjuncts.

Another trend which is evident, and which will probably gain in momentum, is that of the unification of assorted instrumental techniques based on analog feedback operations. When it is considered how fundamental is the function of the operational amplifier, and how its benefits are continually being rediscovered in new fields of technology, it seems likely that multipurpose modular structures will perform the tasks of a number of specialized measuring and manipulative instruments. Beyond its classical and celebrated mathematical operations, comprising addition, algebraic and functional inversion, linear combination, differentiation, integration, etcetera, are the abilities to store and to isolate, among a number of others which are less well known. Since it *is* well known, on the other hand, where information of this kind is available, there is no need or propriety to elaborate here on the application of this basic tool. However, the philosophy of this sort of amplifier as an electrical null-seeking or balancing agent carries its own impact once it is understood. When basically similar methods and equipment are found to be effective in each, such fields as computing, data processing, testing, regulation, and model building will not be kept separate, but will diffuse and perhaps ultimately fuse with one another. One key to the future appears to lie in the quasi-paradox of special-purpose instrumental assemblages based on general-purpose analog modules.

Systems engineers are coming along now in greater numbers and of higher average caliber, and they are not now so brutally divided into disparate camps of practical and theoretical people. More mutual respect, at least seems to obtain between these two sides of the track. Analog models will be increasingly resorted to by both groups in studying the formidable problems of system engineering they must attack. It is getting around generally that the modelling approach may best be taken in stages. Not only should subsystems be separately modelled and carefully confirmed, but a given model need not represent all the aspects of a given subsystem or system at once. Linear approximations usually represent only a crude beginning, but may be confirmed by relatively simple analysis. Nonlinear models are harder to build but much harder to analyze, so that frequently the approach to nonlinear structures should begin with drastic approximations to the nonlinear features, which are refined in stages as the project develops. Each step should be simple and well defined, with continual checking of the assumptions, and of those portions which are assumed to be complete, before forging ahead. Of course the parallel development of rudimentary overall models is in order if it is understood that they should be taken with a grain of salt: they may impart some idea of the flavor of the final concoction. Aspects of a system suitable for separate analog study will depend on the nature of the system; this is the age of broadness of system definition, extending even to all of Society. Taking such a case, one might study population density, political stability, wealth and commerce, considering these somewhat independently before they are all joined in one model. Again, the study in each case might be from the viewpoints of transients, or cycles, or statistics (possibly introducing random perturbations from independent sources). Still further, the item of interest might be tolerance to parametric changes, transitions from one regime to another, extrapolations backward and forward in time, and so on. But my prognostications have turned into a ramble.

As an offshoot of specialized training applications, analogs should find growing applications to pedagogy of a more general kind. This is partly owing to the personal experience which the subject may be afforded, but also to the interest which is induced by living analogies. The speed at which dynamic models may be operated is another factor in maintaining interest, and in saving time as well. If fast repetitive operations are employed, an introductory step may involve slower demonstrations, better to enable the mental transformation of time scale. Block diagrams or signal flow graphs become immediately more meaningful if tangible analog apparatus is made available to fulfill them. The innate property of causality, for example, is given memorable and dramatic emphasis. Feedback is of course the big thrill to the innocent in its general framework, along with its embodiment in differential equations, automatic controls including servomechanisms, and vibrations.

Models and analogs, even as concepts, are powerful teaching means in any case. Symbols themselves are rudimentary analogs, striving close to reality in mathematical operators. Words and languages are analogs right down to the ground. Physicists think and talk in models, the very best of them saying that models are their most powerful tools. Similitude conditions apply equally to all physical phenomena, along with dimensional analysis, so called. The unification of a set of properties in one structure, suggestive of an underlying organization and beauty, gives power and appeal to the model concept in the education of students: and students we all should remain, every one. So we close with a student's recollection.

Emerging many years ago from the old Jefferson Physical Laboratory at Harvard, one could read on the Music Building opposite, cut into the stone under the eaves, an inscription which should still be there:

> To charm, to strengthen, and to teach,
> These are the three great chords of truth.

3. It's an Analog World—Or Is It?

Back in the 1950s, I once heard George Philbrick say, "Digital is a special case of analog." He was a passionate believer in the analog nature of the world. (He was also skeptical about frequency, though he understood transform theory—Laplace, Fourier, and especially Heaviside—better than most. But that's a matter for another essay.)

Now that we've had a few more decades to reflect on nature, to observe convergences between organisms and computer programs, and to see ways of simulating electrical behavior of organisms with computers (e.g., neural nets), it's possible to make some definite statements about what's analog and what's digital.

First of all, though, we have to dispose of *nonlinearity* and *discontinuity* in nature as arguments for digital.

Linearity of real-world phenomena has nothing to do with the analog versus digital question. The real (analog) world is full of nonlinearities. My employer and others manufacture a number of purposely, predictably, and accurately nonlinear devices—for example, devices with natural logarithmic or trigonometric (instead of linear) responses. They are all *analog* devices.

Second, *discreteness* and *discontinuity* really have little to do with the analog versus digital question. You don't have to go to microscopic phenomena to find discrete analog devices. My employer also manufactures analog switches and comparators. They are discontinuous (hence discrete) devices. The switches are fundamental digital to analog converters: the comparators are fundamental analog to digital converters. But voltage or current, *representing digital quantities,* operates the switches; and the outputs of the comparators are voltages, *representing the choice of a digital 1 or 0.* Thus, these basic data converters are *analog to analog* devices.

Perhaps nature is *discrete* at the limits; current could, in a sense, be counted as a flow of discrete charge carriers; time could be counted as ticks of a clock. And noise limits the resolution of continuous measurements, which some might use to argue against the continuous case. But these arguments also work against the discrete case. The uncertainty principle says we can't locate a charge carrier and at the same time say accurately how fast it's going. So we *measure* current as the *average* number of charge carriers that flow in a circuit and call the individual carriers *noise.* Similarly, a clock that ticked with every event would be useless because it would tick irregularly, so again we choose a clock that *averages* the basic ticks, and call the basic ticks *jitter.*

Perhaps it's useful to accept the *duality* of discrete and continuous in the analog real world, even as most people accept that natural phenomena are both particles (discrete) and waves (continuous).

The important point is that "digital" is irrelevant to all that. Digital in the quantitative sense applied to physical phenomena is a human concept; it didn't exist before people, while voltages did (e.g., lightning, which fixed nitrogen, thus fertilizing plants without human intervention). Digital as a quantitative idea first occurred when people learned how to count—using their God-given *digits*. Digital as a computational idea is the human-invented number system. Digital is the numbers marked on an *analog* meter. Except for the natural phenomena shaped to embody it, digital is everything having to do with logic, microprocessors, computers, and so on. But such natural phenomena, and the quantitative equations governing them, are *analog* in nature, because they are analogs for one another.

As a clincher, note that Voyager II's information was digitally encoded; but to find the "digital" signal you had to resort to analog processes, such as amplification, demodulation, and filtering, to recover some sort of pulses representing the noisy information before sophisticated digital signal-processing could be employed to actually pry the information out of the noise. The pulses carrying the digital information were analog quantities. The hardware to do all that (the DSP, too) used real-world analog quantities like voltage and current. The *software* was truly digital.

Have you now been convinced that everything in the world, except for human creations, is analog? Well, I'm not! Apart from logic and number systems, there's another feature of digital that we have to consider: the ability to encode and decode, to program, to store in memory, and to execute.

That ability existed in nature long before humankind. It exists in the genes of all living beings, the strings and interconnections of DNA elements A, G, C, and T that encode, remember, and carry the program for the nature and development of life. They permit biochemical processes to differentiate between flora and fauna and, within these, all the many phyla, species, and individuals.

So perhaps, if we are to generalize, we might say that the vibrant world of life is based on digital phenomena; the physical world is analog and basically noncreative, except as its random, chaotic, and analog-programmed behaviors act on—and are acted upon by—living creatures.

4. Is Analog Circuit Design Dead?

Rumor has it that analog circuit design is dead. Indeed, it is widely reported and accepted that *rigor mortis* has set in. Precious filters, integrators, and the like seem to have been buried beneath an avalanche of microprocessors, ROMs, RAMs, and bits and bytes. As some analog people see it (peering out from behind their barricades), a digital monster has been turned loose, destroying the elegance of continuous functions with a blitzing array of flipping and flopping waveforms. The introduction of a "computerized" oscilloscope—the most analog of all instruments—with *no knobs* would seem to be the *coup de grâce*.

These events have produced some bizarre behavior. It has been kindly suggested, for instance, that the few remaining analog types be rounded up and protected as an endangered species. Colleges and universities offer few analog design courses. And some localities have defined copies of Korn and Korn publications, the *Philbrick Applications Manual*, and the *Linear Applications Handbook* as pornographic material, to be kept away from engineering students' innocent and impressionable minds. Sadly, a few well-known practitioners of the art are slipping across the border (James E. Solomon has stated, for example, that "all classical analog techniques are dead"), while more principled ones are simply leaving town.

Can all this be happening? Is it really so? Is analog dead? Or has the hysteria of the moment given rise to exaggeration and distorted judgment?

To answer these questions with any degree of intelligence and sensitivity, it is necessary to consult history. And to start this process, we must examine the patient's body.

Analog circuit design is described using such terms as subtractor, integrator, differentiator, and summing junction. These mathematical operations are performed by that pillar of analoggery, the operational amplifier. The use of an amplifier as a computing tool is not entirely obvious and was first investigated before World War II. Practical "computing amplifiers" found their first real niche inside electronic analog computers (as opposed to mechanical analog computers such as the Norden bombsight or Bush's Differential Analyzer), which were developed in the late 1940s and 1950s. These machines were, by current standards, monstrous assemblages made up of large numbers of amplifiers that could be programmed to integrate, sum, differentiate, and perform a host of mathematical operations. Individual amplifiers performed singular functions, but complex operations were performed when all the amplifiers were interconnected in any desired configuration.

The analog computer's forte was its ability to model or simulate events. Analog computers did not die out because analog simulations are no longer useful or do not approximate truth; rather, the rise of digital machines made it enticingly easy to use digital fakery to *simulate the simulations.*

Adapted from the July 22, 1991, issue of *EDN Magazine*.

Figure 4-1.
Some analog
types are merely
leaving town.

As digital systems came on line in the late 1950s and early 1960s, a protracted and brutally partisan dispute (some recall it as more of a war) arose between the analog and digital camps. Digital methods offered high precision at the cost of circuit complexity. The analog way achieved sophisticated results at lower accuracy and with comparatively simple circuit configurations. One good op amp (eight transistors) could do the work of 100 digitally configured 2N404s. It seemed that digital circuitry was an accurate but inelegant and overcomplex albatross. Digital types insisted that analog techniques could never achieve any significant accuracy, regardless of how adept they were at modeling and simulating real systems.

This battle was not without its editorializing. One eloquent speaker was George A. Philbrick, a decided analog man, who wrote in 1963 (in *The Lightning Empiricist,* Volume II, No. 4, October, "Analogs Yesterday, Today, and Tomorrow," pp. 3–8), "In modest applications to on-line measurement and data processing, it is quite generally conceded that the advantage of continuous analog apparatus make it irresistible. This is partly owing to the simplicity and speed which its continuity makes possible, and partly to the fact that almost every input transducer is also 'analog' in character, that is to say, continuous in excursion and time."

Philbrick, however, a brilliant man, was aware enough to see that digital had at least some place in the lab: "Only the most hard-shelled of analog champions would suggest that all simulative and computational equipment be undiluted by numerical or logical adjuncts."

He continued by noting that "some analog men, perhaps overfond and defensive as regards continuous functions, really believe that analog operations are generalizations of digital ones, or that conversely digital operations are special cases of analog ones. What can be done with such people?

"While it is agreed that analog and digital techniques will increasingly cross-fertilize and interrelate," Philbrick concluded, "it is predicted that the controversy between their camps will rage on, good natured but unabated, for years to come in spite of hybrid attachments."

Although Philbrick and others were intelligent enough to prevent their analog passions from obscuring their reasoning powers, they could not possibly see what was coming in a very few years.

Figure 4-2.
Is this the fate of oscilloscopes whose innards are controlled by knobs instead of microchips?

Jack Kilby built his IC in 1958. By the middle 1960s, RTL and DTL were in common use.

While almost everyone agreed that digital approximations weren't as elegant as "the real thing," they were becoming eminently workable, increasingly inexpensive, and physically more compactable. With their computing business slipping away, the analog people pulled their amplifiers out of computers, threw the racks away, and scurried into the measurement and control business. (For a nostalgic, if not tearful, look at analog computers at the zenith of their glory, read *A Palimpsest on the Electronic Analog Art,* edited by Henry M. Paynter.)

If you have read thoughtfully to this point, it should be obvious that analog is not dead, rather just badly shaken and overshadowed in the aftermath of the war. Although measurement and control are certainly still around, the really glamorous and publicized territory has been staked out by the digital troops for some time. Hard-core guerrilla resistance to this state of affairs, while heroic, is guaranteed suicide. To stay alive, and even prosper, calls for skillful bargaining based on thorough analysis of the competition's need.

The understanding that analog is *not* dead lies in two key observations. First, to do any useful work, the digital world requires information to perform its operations upon. The information must come from something loosely referred to as "the real world." Deleting quantum mechanics, the "real world" is analog. Supermarket scales, automobile engines, blast furnaces, and the human body are all examples of systems that furnish the analog information that the silicon abacus requires to justify its existence. So long as transduction remains analog in nature, the conversion process will be required.

A further observation is that many microprocessors are being used not to replace but to enhance a fundamentally analog measurement or process. The current spate of microprocessor-controlled digital voltmeters furnishes one good example; others include digital storage oscilloscopes and smart thermometers.

If one insists on bringing ego into the arena, the digital devotee will argue that the analog content of these things is an unfortunate nuisance that must be tolerated. The analog aficionado, if permitted to speak, will counter that digital techniques exist only to aid in getting a better grip on a fundamentally analog existence. The question of who is most correct is subject to endless debate and is not really germane.

The point is that although analog is not dead, its remaining practitioners must be more systems creatures and less circuit addicts. To be sure, circuits are required to build systems, but analog technicians can only make themselves indispensable in a digital world by their recognized ability to supply what it needs to accomplish its mission.

That this is the case can be easily proven. Consider the effect on the major digital powers of a complete embargo of data converters and signal-conditioning components by the small analog nations. How can a supermarket scale compute the cost of goods it can't get weight information on? Of what use is a process controller without inputs or outputs? Think of the long lines of microprocessors waiting at the distributors for what few DIPs of analog I/O might be available! Imagine rationing of instrumentation amplifiers and V/F converters and alternate D/A and A/D days.

So it seems that analog is not so dead after all but really playing possum. By occupying this position, analoggers will stay healthy, very much alive, and need not leave town.

An uneasy but workable harmony has thus been negotiated with the dominating numerical nemesis. This compromise is not optimal, but it's certainly a more desirable and useful existence than being dead and is worthy of praise and respect by everyone.

Do all you bit pushers out there get the message?

Figure 4-3.
Analoggers can stay very much alive and need not leave town.

What Is Analog Design?

Everyone knows analog design is different from other branches of electronics. But just what is analog design? There's no definitive answer in this section, but three authors do offer insights that point the way toward an answer.

Bernard Gordon, president of Analogic Corporation, discusses a key part of analog design—the requirement that designers be able to visualize and manipulate, both on a conscious and unconscious level, the multiple factors and interrelationships between those factors present in every analog design. As he notes, this is more an art than a science.

Digital electronics can be thought of as dealing with a world that's either black or white (or 0/1 or true/false), with no fuzzy gray areas between those levels. Samuel Wilensky tells how analog design is the art of working in those gray areas, with designers required to optimize a circuit by sacrificing one parameter so another can be enhanced. He uses the evolution of the digital to analog converter to show how advances in analog design come through intuition and "feel" as much as through rigid application of fixed rules.

Maybe the best way to understand what analog design is all about would be to "walk through" an analog design task. Jim Williams retraces William R. Hewlett's footsteps a half-century later and discovers that, while the components may have changed, the basic principles and philosophy are still intact.

5. On Being the Machine

The art of analog design per se is not generally very different from that of other engineering endeavors. It is the purpose of this chapter to convey a visceral sense of the art of engineering, particularly as related to creative or innovative conceptions.

Assume the engineer possesses, as a necessary minimum requisite for being an analog designer, a broad and general knowledge of electronic circuit physics and mathematics, the characteristics of available componentry, and the capabilities of modern manufacturing processes. Furthermore, to produce competent designs capable of being manufactured in quantity and of retaining their desired performance specifications, the engineer must have developed a thorough understanding, sensitivity to, and appreciation of tolerances and error budgeting.

There remains, however, an additional criterion for being truly creative and not merely competent . . . the development of sufficient art and skills to synthesize innovative, inventive new devices (machines). What is needed is the ability to envision the purpose and totality of the device as a whole, in order to be able to synergistically relate the parts of the design, minimize the number of elements, and produce what must be described as an elegantly simple solution.

The creative designer must be able to develop the mindset of "being the machine," in order to become the "mental and living embodiment" of the circuit or system. The ability to do so is less dependent on textbook learning and analysis than on developing the capacity, by experiencing a succession of increasingly complex problems, to simultaneously conceive, pursue, examine, and compare multiple possible solutions. The designer must then be able to envision the interrelationships, tolerances, and weaknesses of components and processes and then *consciously and subconsciously* recognize what suddenly appears as a realization and visualization of an elegantly simple coherent solution of interacting, self-supporting componentry.

As a first simple example, consider the design of the acoustic memory that was incorporated into the first commercial digital computer, Univac I, circa 1949. While it was part of a digital computer and was employed for the storage of serial digital words, the design requirements were basically analog in nature. The recirculating loop, shown in Figure 5-1, was to consist of an input gate structure whose output was applied to an RF modulator circuit, which in turn drove a piezoelectric transducer, which converted the electrical signal into mechanical vibrations. These vibrations propagated acoustically through the mercury channel and impinged upon an identical piezoelectric transducer, reciprocally producing the RF signal at highly attenuated levels. This attenuated signal was to be amplified and demodulated and returned to the gating structure for reclocking to pass through the loop again.

Univac I operated at a clock frequency of 2.25 MHz, or approximately 0.444 msec per pulse. The system design called for a rise time of approximately 0.2 msec, cor-

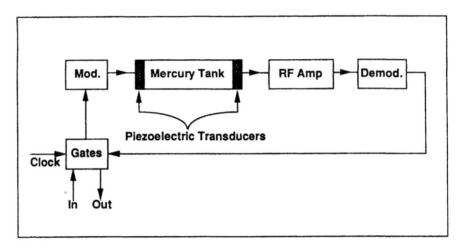

Figure 5-1.
Block diagram of
"recirculating
loop" mercury
acoustic memory
used in the
Univac I circa
1949.

responding to a video bandwidth of about 2.5 MHz or an RF bandwidth of about 5 MHz.

On the surface, this might seem like a straightforward problem. Indeed, referring to Figure 5-2, the initial brute force design, in part based on competent considerations of practical transducers and ultrasonic attenuation characteristics, called for a synchronously tuned system with each component of the system tuned to 11.25 MHz. It might have seemed quite obvious to cut the transducers to the frequencies that they would be expected to vibrate at and to tune the RF amplifiers to the same frequency. However, the designers of the system found that they could not obtain even a close approximation to the transient rise times needed, for the mechanical physics of the transducers established their bandwidth, and therefore, regardless of the width of the individual stages of the RF amplifier or the number of stages employed, the desired performance could not be obtained.

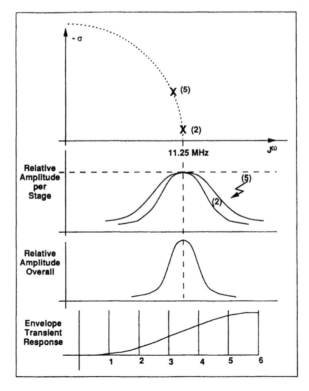

Figure 5-2.
Results of "brute
force" approach
to design of
the circuit in
Figure 5-1.

Now, consider another approach in which the characteristics of each of the individual elements of the systems and their limitations are internalized by the designer. With an understanding of the physics limiting the bandwidths of the crystal transducers, allow for the possibility that the crystals should not be cut to the transmission frequency but to frequencies both well below and above that frequency.

Assume that the designer has not only a knowledge of the physics and therefore the equivalent circuit of the mechanical transducer but also has a broad background in the mathematics related to functions of a complex variable and an ability to compute the transient response of a system characterized by a complex set of conjugate poles. Further assume that the designer is intimately familiar with the responses of conventional textbook filters, such as Butterworth maximally flat or Bessel maximally linear-phase filters, and recognizes that with the Butterworth the transient response will ring too much and with the Bessel the economy of the system will be inadequate due to the requirement for too many gain stages.

Now further suppose that the designer "fills" his head with many, many possible other relevant factors and design possibilities, orders them in his head, thinks of little else . . . and *goes to sleep.*

The designer then subconsciously conceives of making not a good flat-response, amplitude-wise or linearity-wise, amplifier, but rather of making an amplifier which on its own cannot do the job. In concert with the displaced poles of the crystal transducers, however, the modulator, amplifier, and transducers make up a system whose transfer function, characterized by the concerted totality of the pole positions, provides a performance significantly better than any single part of the system could have done individually. That is, the *whole* is better than the sum of its parts—see Figure 5-3.

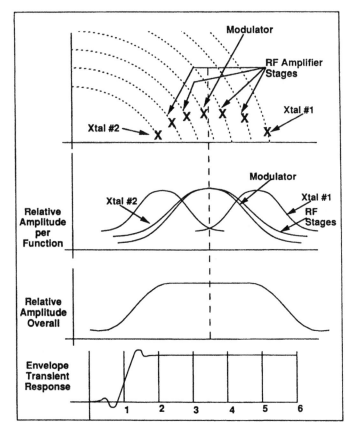

Figure 5-3. Results of an integrated system design.

Having considered a solution to a problem that existed over four decades ago and recognizing what may now appear to be an obvious solution, the reader may be inclined to shrug and say, "So what?" But recall that major engineering efforts, based on what was, at the time, the obvious brute-force approach were expended by highly competent engineers before the more apparently sophisticated, but actually simpler, approach based on broader considerations and visceral recognition was brought to bear.

Consider now another example, perhaps somewhat more complex, related to the development, in the early 1970s, of what were then called "Computer Assisted Tomography" systems, which today are known as CAT scanners. In one generation of these systems (later the most popular called "Fan Beam Rotate-Rotate" machines), a multiplicity of detectors, indeed many hundreds, simultaneously convert impinging X-ray photons into low-level currents on the order of a few hundred nanoamperes full scale. In order to be able to compute a high-quality image with great detail and minimum artifacts, it is necessary to integrate and measure these currents over a dynamic range of one million to one and with every channel tracking every one of the hundreds of other channels to within a few parts in a million over that entire range. Experienced analog engineers will recognize this requirement to be a formidable task.

Early designs made by competent specialists resembled Figure 5-4. In a conventional way, the designers placed a preamplifier at the output of the detector, converted the output of the preamp into a current source, whose output in turn was applied to an integrator, which was periodically reset between pulses of X-ray. In an attempt to achieve performance approaching that needed, engineers searched catalogs for the lowest leakage current, lowest input offset voltage, and most stable amplifiers available. They obtained the most stable resistors and best capacitors. But no designs were made that could achieve, within perhaps two orders of magnitude, the necessary accuracy, stability, and linearity. Objective calculations, based on components available two decades ago, indeed even now at the time of the writing of this essay, would indicate that no error budget could be drawn that would imply that there was a practical solution.

However, the practical circuit of Figure 5-5 resulted from the concepts of engineers who had studied the entire tomography process, understood the physics of X-ray attenuation and statistical noise, who knew control loop theory, and particularly understood and accepted the limitations of components. They conceived that it should be possible to make a circuit which simultaneously autozeroed out both voltage and current drift errors. If this could be achieved, they could not only compensate for errors within the amplifier and integrator amplifiers but also for leakage currents in the detector, dielectric absorption in connecting cables, and also for other dielectric absorption in the integrator capacitor, following reset.

On the surface it would appear as if the block diagram of Figure 5-5 is more complicated than that of Figure 5-4 and that the costs for such a circuit might be greater. But if such a circuit could be practically designed, then the total cost of its

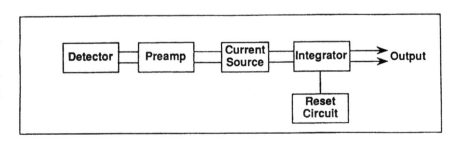

Figure 5-4.
Block diagram of
a precision
"pulsed" X-ray
detector current
integrator.

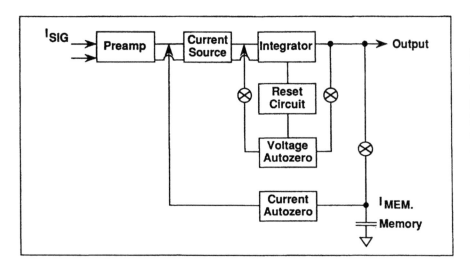

Figure 5-5.
Current and
voltage auto-
zeroing pulsed
X-ray detector
current integrator
circuit.

components could be substantially less than those "very best" components that had
been selected for the block diagram of Figure 5-4 . . . and yet produce significantly
superior performance.

Now examine the circuit of Figure 5-6. Its operation is not obvious. This circuit,
conceived and invented by Hans J. Weedon and awarded U.S. Patent No. 4,163,947,
is entitled "Current and Voltage Autozeroing Integrator." Its parts cost at the time
of design was nearly five times less than the parts cost of a high-quality implementation of Figure 5-4. In this circuit there are two low-cost operational amplifiers
having moderately common input current and offset voltage specifications. Study
the configuration of the four switches. Notice the following combination of actions.
Assume that the switch labeled V_{az} is connected to ground and the switch labeled
Int is open, so that the right side of the integrating capacitor of C1 must be at ground
potential. If, at the same time, the switch labeled Reset/V_{az} is closed, C1 will be
discharged. If also at the same time the switch labeled I_{az} is closed, the error voltage
that would otherwise appear at the output of the second amplifier is servoed to yield
a net zero current sum into the integrating capacitor. Thus, there is established at C2
a potential which compensates for detector leakage currents, cable leakages, offset
current in the first amplifier, offset voltage in the first amplifier, offset voltage in
the second amplifier, and input current in the second amplifier. When switch Int is
reconnected and all other switches opened, the right-hand side of C1 must be at

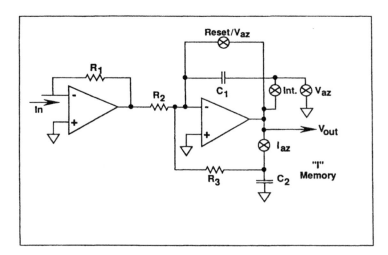

Figure 5-6.
Current and
voltage auto-
zeroing integrator
developed by
Hans J. Weedon
of Analogic
Corporation
(U.S. patent
#4,163,947).

"zero" voltage, and the net current to be integrated in C1 other than from the detector source is "zero."

With the realization of this circuit nearly two decades ago, it was quite possible to produce, in large-scale manufacture and at quite low cost, a low-level pulsed integrator whose effective input current was about 0.01 picoamp! At the same time, the effective output voltage drift was less than 0.1 microvolt per degree Celsius! But, perhaps more important, this elegantly simple circuit enabled the system designer to realize significant savings in detector costs. Due to the ability of the circuit to cancel the effects of detector leakage, the construction of the detector could be based on substantially lower-cost materials and testing.

Now consider a modern digital signal processing system requirement. Figure 5-7 shows the sequence of functions in what might be a high-speed, high-accuracy mathematical waveform generator. The sequence of events is that a primary computed function undergoes secondary digital filtering, the output of which is applied to a high-speed digital-to-analog converter, whose output is transferred to a holding circuit, the output of which is applied to a recovery filter, and thence to an output amplifier. The problem is to determine the "optimum characteristics" of each of the building blocks. Presume that a project engineer were to write a specification for each of these building blocks. Is there, in fact, a set of individual specifications which, independent of each other, can provide a reasonably optimum technical economic result?

Assume that the computed function is to cover a frequency range from 0 to 100 MHz, that the rate of computed words is to be about four hundred million per second, and that analog integrity is to be preserved to the 12-bit level. A multiplicity of interacting, conflicting requirements arises if the architect project engineer attempts to assign to a group of designers the task of preparing a specification for each individual piece. Characteristics of the recovery filter depend on the characteristics of, at least, the holding circuit, the digital prefilter, and the computing function. The characteristics of the holding circuit in turn certainly depend on the D/A converter and the output filter. The nature of the computed function depends on the algorithms and the computing capability, and this in turn will be directly related to the capabilities and characteristics of the other building blocks.

How, then, will the project engineer realize a superior solution unless an integrated sequence of operations can be visualized and internalized? One part of the engineer must be the computing function, linked to another part which is the digital filter, in turn linked to yet another part acting as the D/A converter, and so on. The designer must, in a certain sense, interactively play these internal-self parts, balancing off in various combinations the possible capabilities and limitations of the individual functions. The designer must fully understand the error-producing effects of these interrelationships. The design decisions for each function cannot be explicitly com-

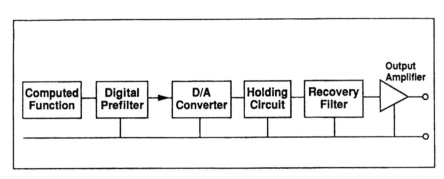

Figure 5-7. Functional sequence of a mathematical wave form generator.

puted in a linear manner but rather must be arrived at via a series of interactive decisions. The designer must be as an artist—he must cogitate, smell, and feel the system. The problem must dominate the engineer's mind and being, such as when a concert pianist or violinist performs a work.

This is a trying and complex task. A breadth of knowledge is required, an intensity of concentration or attack is required, the designer must live and sleep the problem, and eventually, a set of compromises, clever relationships, and compensating effects will yield, in their totality, a result significantly superior to that which might be obtained by the mere summation of a subset of independent specifications.

There has always been, and there probably always will be, the need for analog designers to find competitive and clever solutions, superior to those which might merely be just competently arrived at. In the examples given, the common denominator is the need for the designer to mentally and physically, consciously and unconsciously, relate to the problem. In the case of the Univac memory, it was the need to understand the behavior and limitations of the piezoelectric transducers by attempting to think, "I am a transducer; what can I do, what can I not do, how can I best be used?" And then, to become a modulator or an amplifier and understand, as such, one's limitations and possibilities in gain bandwidth relations. In the second example, it is the need to become an X-ray detector that converts photons to electrons or to understand the limitations of analog components and then to try to arrange one's being into a new form and to mentally play until a new solution is recognized.

As postulated at the beginning of this chapter, one cannot write an equation for the mental state or methodology by which any individual designer might go about "Being the Machine," but the ability to do so, however it is derived, generally produces superior results. It transcends simple competent knowledge and its application and becomes an act of creativity, or an art form.

6. Reflections of a Dinosaur

Sixty five million years ago, at the end of the Cretaceous period, the dinosaur vanished from the earth. Some scientist believe that the disappearance was due to the cataclysmic collision of a large body with Earth.

The explosive growth of digital technology is the cataclysmic event that has threatened the analog designer with extinction. The linear circuit engineer has been added to the list of endangered species. For the past twenty years the focus of the engineering curriculum has shifted priority from analog to digital technology. The result of this shift is that only a small fraction of recently trained engineers have the analog design skills necessary to attack "real world" problems. The microprocessor has revolutionized the area of measurement and control, but the transducers used to measure and control temperature, pressure, and displacement are analog instruments. Until sensors and actuators are available that can convert a physical parameter such as temperature directly to digital information, the analog designer will still be in demand.

Analog design is a challenging field because most projects require the designer to optimize a circuit by surrendering one performance parameter to enhance another. As an old analog guru once said when comparing the analog and digital disciplines, "Any idiot can count to one, but analog design requires the engineer to make intelligent trade-offs to optimize a circuit." Analog design is not black or white as in "ones" and "zeros"; analog design is shades of gray.

This essay contains the reflections, thoughts, and design philosophies of a nearly extinct species of electrical engineer, the analog circuit designer. Digital technology has reduced our population to a small fraction of those that existed twenty or thirty years ago. This is unfortunate since the need for, and the challenge of, analog design is still with us. This chapter relates experiences I have had as an electrical engineer since I received my degree in June 1959. I hope these reflections will in some way encourage and help the recently initiated and entertain those of you who remember filament transformers and B^+ power supplies.

My undergraduate electrical engineering education covered mainly vacuum tube technology, but there were two "new" areas that the department felt were of significant enough importance to include in the curriculum. As a result, we received a one-hour lecture on transistors and a one-hour lecture on crysistors. For those of you who are unfamiliar with the crysistor, it is a superconducting magnetic memory element that showed promise of revolutionizing the computer world.

It would have been difficult to predict in 1960 that the vacuum tube would become a relic of the past, transistor technology would rule, and the crysistor would completely disappear from the scene. Although the crysistors never made it, the discovery of new low-temperature superconductors may give it a second chance.

It amazes me that most of the technology I work with today did not even exist in

Figure 6-1.
Basic digital to
analog converter
(DAC).

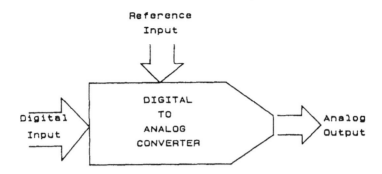

the late '50s and early '60s. I mention this to emphasize that a firm understanding of fundamental principles is much more important to one's long-term success in engineering, or any field for that matter, than the learning of some specific skill. For example, without a thorough understanding of Maxwell's equations and Ohm's law and how they are applied, it would be difficult, if not impossible, to progress with new technologies. My approach to troubleshooting circuits is, "The circuit will not violate Ohm's law." If I make measurements that suggest the opposite, I look for oscillations. But I digress—back to the "early years."

The late 1950s were the times of vacuum tube digital computers with 16 K of memory. Computing power that today fits on a desktop occupied hundreds of square feet of space. The mechanical desktop calculators we used required several seconds to multiply two 10-digit numbers. They were not portable, so everyone carried slide rules that were quicker to use, easier to carry around, and didn't need 110 V electrical power. The slide rule only produced an answer to three or four significant digits, but this was not a real limitation since electrical engineering was only a 1% or at best a 0.1% science. Measuring instruments were all analog and even a General Radio meter with the black crinkle finish and a mirrored scale (now that shows my age) would only yield a voltage measurement of three significant digits at best.

During the mid 1950s a 12-ounce container of Coke (which at that time referred to a soft drink) cost a dime. The top-of-the-line Chevrolet and a year at a private university cost about the same—$2,000. As an economist friend of mine once pointed out, inflation is a relative thing, since the price of the Chevrolet and a year's tuition for a private university have remained constant over the years.

The thirty or so years between the late 1950s and the present have brought many changes. The vacuum tube digital computer which once occupied a room is now fabricated on a silicon chip the size of your thumbnail. The mechanical calculator and slide rule have disappeared and been replaced by the solar powered scientific calculator. Electrical measurements are made with digital instruments that are accurate to six or seven significant digits, and Coke is no longer just a soft drink. To those of us in the analog world, digital technology is a two-edged sword. Digital technology has created powerful tools for the analog designer to use, but it has also depleted our ranks by attracting some of the most promising students. This is unfortunate since some of the most challenging problems are analog in nature, and fewer and fewer graduating engineers are equipped to solve them.

I classify analog designers into one of two categories. There are those who do truly original work, and these I consider the artists of our profession. These individuals, as in most fields, are very rare. Then there are the rest of us, who are indeed

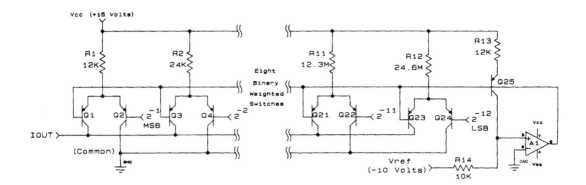

Figure 6-2.
Transistor–
transistor
switched binary
weighted 12-bit
DAC.

creative, but do it by building on the present base of knowledge. A quote from Sir Isaac Newton beautifully describes how this design process works:

> If I have seen farther than others
> it is by standing on the shoulders of giants.
> —Sir Isaac Newton to Robert Hooke, February 5, 1675

A less tasteful, but some would say more honest, illustration of how electronic circuits are designed is contained in a humorous 1950s song by Tom Lehrer:

> Plagiarize, Plagiarize,
> Let no one else's work evade your eyes.
> Remember why the good Lord made your eyes.
> So don't shade your eyes but,
> Plagiarize, Plagiarize.
> Only be sure always to call it please,
> Research.
> —Song by Tom Lehrer

I quoted the lyrics of the Lehrer song tongue-in-cheek, but circuit design is an evolutionary process where one must draw on past developments. The digital-to-analog converter (DAC) of the early 1960s is a classic example of how a circuit develops, changes, and improves as it moves through the hands of different designers.

For those of you not familiar with DACs a quick explanation is in order. The DAC is a device whose input is a digital number, usually in a binary format, and whose output is an analog signal. The analog output is usually a voltage or a current whose value is a function of the digital input and a reference voltage (see Figure 6-1). The DAC was one of the first circuits developed for linking the analog and digital domains, and even today the DAC plays a large role in computer graphic terminals, music synthesizers, and the many other applications in which a digital processor must communicate with the analog world.

During the early 1960s transistors were replacing vacuum tubes, and digital integrated circuits were just becoming available. Analog integrated circuits were not widely available, and those that were available were expensive. Almost all analog circuit design was carried out with discrete components and an occasional integrated amplifier. The transistor was becoming available, and since it closely approximates an ideal current source, it was an excellent candidate for the switch in a current output DAC. The first DACs built with transistors used emitter coupled current

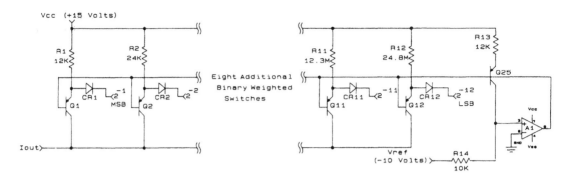

Figure 6-3.
Transistor-diode
switched binary
12-bit weighted
DAC.

sources. The emitter coupled transistors (see Figure 6-2) steered the current to the output bus (I_{out}) or common (GND), depending on the level of the digital input.

The most significant bit (MSB)[1] current source consist of a resistor (R1) and two pnp transistors (Q1, Q2). The servo amplifier (A1) biases the base of Q1 to approximately 1.4 V. When the base of Q2 is above 2.0 V (a digital logic "1"), all current through R1 is steered to I_{out} through Q1, since Q2 is cutoff. Conversely, when the base of Q2 is lower than 0.8 V (a digital logic "0"), all the current is steered to GND through Q2, since Q1 is cutoff.

The reference loop Q25, A1, R13, and R14 biases the bases of the transistors (Q1, Q2, ..., Q21, Q23) connected to I_{out}, maintaining a constant voltage across the current setting resistors R1 through R12. The values of the components are selected for a nominal base bias voltage of 1.4 V. It will be left as an exercise for the student to show that when the digital input bit is a logic "1" the servo amplifier (A1) will maintain the same voltage across resistors R1 through R12 by adjusting the base voltages of all the transistors connected to I_{out}. The magnitude of the constant voltage across the resistors will be $V_{ref} \times$ (R13/R14). Since each current setting resistor is twice the value of the resistor to its left, the currents from each switch will be binary weighted. That is, the current of each switch will be ½ the current of the switch to its left.

If the operation of the reference loop is not clear, don't spend serious time trying to understand it, as it is not necessary for the discussion that follows. A detailed discussion of DAC reference loops can be found in one of the data conversion handbooks that are available from converter manufacturers.

The analog output of this DAC is a current that can be converted to a voltage by connecting a resistor from the I_{out} terminal to ground. To ensure that the transistors remain biased in the correct operating range, the I_{out} terminal should not exceed +1 V. For a DAC that produced a 2 mA full scale output current, a 500 Ω resistor connected from I_{out} to ground would produce a 0 to +1 V output swing. A –1 V to +1 V output swing could be obtained by terminating the I_{out} terminal with a 1000 Ω resistor to –1 V source instead of ground.

As stated before, the current setting resistors of each switch pair increases in a binary sequence. The current from each transistor pair is twice that of the transistor pair on its right and half of the current of the transistor pair on its left. If the MSB

---

1. Bit is an acronym for a digit of a binary number. It is derived from *Binary InTeger*. The highest order digit of the binary number is usually called the MSB or *Most Significant Bit*. The Bit's are also labeled to indicate their relative weight in the binary number. For example the MSB is also called the 2^{-1} bit because it contributes ½ of the full scale output current the next lower order bit is labeled 2^{-2} since it contributes ¼ of the full scale output current. The lowest order bit of a binary number is called the LSB or Least Significant Bit.

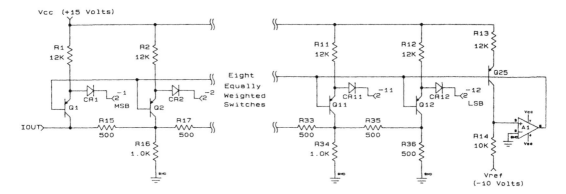

Figure 6-4.
Transistor–diode
switched R-2R
current division
12-bit DAC.

of the digital input is a logic "1" and all the other digital inputs are "0," the output current would be ½ its full scale value. If the MSB-1 (2^{-2}) is a logic "1" and all the other digital inputs are "0," the output current would be ¼ of full scale. If both the MSB and the MSB-1 are logic "1"s and all the other digital inputs are "0," the output current would be ¾ (½ + ¼) of full scale. In this manner any combination of digital "1"s and "0"s can be converted to a current.

This circuit topology functioned fine but used two transistors per switch. In the late 1960s transistors were expensive and occupied significant space on the printed circuit board. In an effort to reduce cost and size, an imaginative engineer realized that the transistors that steered the current to ground could be replaced with simple diodes (see Figure 6-3). The substitution was possible because converters were usually driven with digital logic capable of sinking several milliamps of current to ground.

The diode is smaller and less expensive than a transistor, reducing the cost and size of the converter with no degradation in performance. The trade-off that the designer made to obtain a decrease in cost and size was the requirement that the converter's digital drive sink several milliamps of current to ground. At the time this did not represent a serious compromise, because digital CMOS logic was not widely used. The most popular logic used bipolar transistors and could easily sink the necessary several milliamps of current.

The circuits of Figures 6-2 and 6-3, although very simple, possessed one major drawback, that of speed. The currents of the LSBs are so much less that the currents of the MSBs, that the switching times of the LSBs are significantly slower than the MSBs. This difference in switching time results in large switching transients or "glitches." In the case of a 12-bit converter the ratio of the MSB current to the LSB current is 2048 to 1. For a 12-bit converter with a 1 mA MSB, the LSB would only switch 500 nA and the LSB switching time would be at least an order of magnitude slower than the MSB. In many slow speed applications the switching transients are not important, but for high speed applications, such as drivers for graphic terminals, glitch-free operation is essential.

I don't know who first had the idea, but someone formulated the concept of operating all the switches at the same current and performing the binary division of each bit at the output of the appropriate current source (see Figure 6-4).

The binary current division is accomplished with the R-2R[2] ladder connected to

2. The current divider of Figure 6-4 is called an R-2R ladder because of the similarity of the resistor configuration to a ladder laid on its side. the rungs of the ladder are the even numbered resistors R16 through R32. The top side of the ladder is formed by the odd numbered resistors R15 throught R35. The bottom side of the ladder is Common (GND). The ratio of the values of the even numbered resistors to the odd numbered resistors is 2:1. Thus the current divider is called and R-2R ladder.

The termination resistor is a special case and has a value of R since the ladder is finite in length.

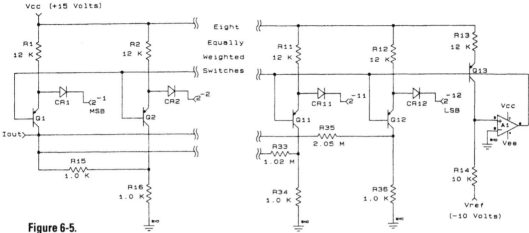

Figure 6-5.
Transistor–diode
switched
individual current
division 12-bit
DAC.

the outputs of the current steering switches. For those unfamiliar with the R-2R ladder, it is an interesting exercise to calculate what fraction of current introduced into the nth R-2R-R node will reach a load resistor R_L connected from the I_{out} node to ground. A little mathematical manipulation will show that the current introduced at any node of the R-2R ladder is attenuated by

$$2^{-n} \times (2R/(R_L+2R))$$

when it reaches the load resistor R_L; where n = the number of stages between the node where the current is introduced and the I_{out} node.

$n = 0$ for the MSB
$n = 1$ for the MSB-1
$n = 2$ for the MSB-2

.

.

.

$n = n - 2$ for the LSB+1
$n = n - 1$ for the LSB

An interesting property of the R-2R ladder is that the resistance of the ladder at any R-2R node looking to the right is always 2R. Using Figure 6-4 as an example, the resistance looking into R35 is 1000 Ω, R35 + R36. The resistance looking into R33 is also 1000 Ω, (R33 added to the parallel combination of R34 with the sum of R35 and R36). This calculation can be repeated at each node, and you will find that the resistance looking into I_{out} is also 2R.

When all the current sources are made equal and the current division is done with the R-2R ladder, the switching times of each bit are matched. The physical length of the R-2R ladder will introduce a differential delay from each bit to the I_{out} node, but this is a very small effect and turns out not to be important if the resistance of the R-2R ladder is low. Even the small delay due to the propagation time through the R-2R ladder can be reduced by providing a separate divider for each bit (see Figure 6-5). This scheme has been tried, and the results are almost identical to the dynamic performance of the R-2R divider.

The use of equal current sources and a resistive divider, either the R-2R or the individual, improves the dynamic performance. The improved performance is gained at the expense of complexity and power consumption. The R-2R and the individual divider circuits use three resistors per bit instead of the one resistor per bit of the

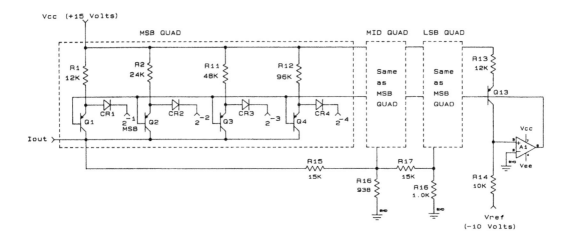

binary weighted converters. The total current switched in the binary converters is the full scale output current of the converter. The total current switched by the resistive divider converters is ½ the full scale output current of the converter multiplied by the number of bits, since each bit switches ½ of the full scale current. A 12-bit binary weighted converter with a 2 mA full scale output current would switch 2 mA. A 12-bit resistive divider converter with a 2 mA full scale output current would switch 12 mA. The dynamic performance of the slower binary weighted circuit is improved by increasing its complexity and power consumption.

The binary weighted configuration and the current division configuration can be combined to form a converter that is faster and slightly more complex than the binary weighted scheme but less complex and only slightly slower than current division. The two combined topologies, binary weighting and current division, are shown in Figure 6-6.

The first four bits of this hybrid converter are binary weighted. The four-bit configuration is repeated two more times to obtain 12 bits. The four-bit sections are coupled with a 16 to 1 divider so that the proper fraction of current from each bit will appear at the I_{out} node. Using this scheme, the ratio of the highest to lowest switched current is now 8 to 1 instead of the 2048 to 1 ratio of the binary weighted converter. The 8 to 1 ratio is not as ideal as the 1 to 1 ratio of current division scheme, but the total switched current is halved from 12 mA, for the current division to 6 mA for the hybrid configuration. The 8 to 1 current ratio yields switching times that are matched closely enough for all but the most demanding applications.

The hybrid configuration averages 4/3 resistors per switch, which is slightly more

Figure 6-6.
Transistor–diode switched binary weighted quad current division 12-bit DAC.

Figure 6-7.
Diode–diode switched binary weighted 8-bit DAC.

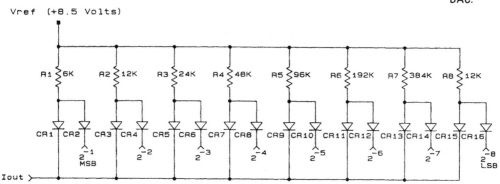

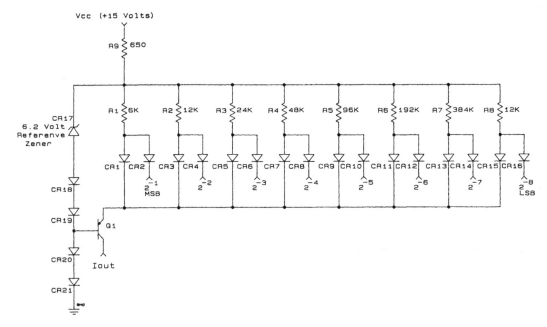

Figure 6-8.
Hybrid Systems'
DAC 371-8
8-bit DAC.

than the binary weighted converter (one resistor per switch) and significantly less than the current division configuration (three resistors per switch).

By combining and modifying existing circuits, new circuits can be created that are better suited for a particular application than the circuits from which they were derived.

Performance is not the only parameter that can be optimized by modifying existing circuits. Performance can sometimes be traded off for other variables such as cost and size.

In the early 1970s Hybrid Systems (now Sipex) was looking for a technique to build a "cheap and dirty" digital-to-analog converter that could be given away as a promotional item. At the time, the circuit of Figure 6-6, or some slight variation, was the configuration used by most converter manufacturers. This circuit was too expensive to give away, so a modification was in order. We modified the circuit of Figure 6-2 by replacing all the switching transistors with diodes (see Figure 6-7).

Figure 6-9.
Internals of
Hybrid Systems's
DAC 371-8.

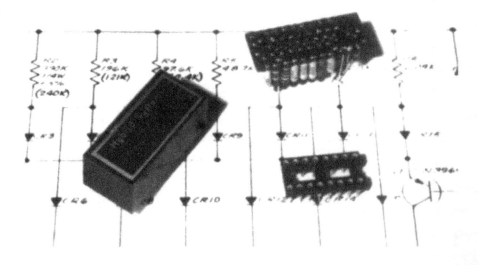

This design works as long the I_{out} is maintained at 2 V. The current from each bit would flow to either the right- or left-hand diode, depending on the state of the digital input. To maintain the proper digital switching level and to keep the current through each bit constant, it is necessary to hold the I_{out} node at 2 V. This is accomplished by using a transistor (Q1) as a constant voltage node to sum the currents from each bit. The reference loop of Figure 6-2 is replaced by four diodes (CR18,CR19,CR20,CR21), a zener reference (CR17), and a resistor (R9) (see Figure 6-8). The reference circuit depends on the forward voltage across a diode (CR19) tracking the V_{be} of the transistor (Q1). This circuit compensates for V_{be} changes of the transistor with temperature, but it does not compensate for changes in transistor beta. The reference circuit does not adjust as well as the servo loop, but it is good enough. The reference circuit maintains a constant voltage across the resistors (R1,R2, . . ., R8), and the transistor sums the bit currents to the I_{out} node. Since the emitter-to-base voltage of the transistor varies with emitter current, the linearity of the circuit was limited to slightly better than 8 bits (0.2%).

A schematic of the design of what became Hybrid Systems' DAC 371-8 is shown in Figure 6-8. The mechanical construction of the DAC 371-8 was also distinctive. The diodes and resistors were mounted on end, resulting in a DAC footprint only slightly larger than a 16 pin dual in-line integrated circuit package. The pins for the unit were configured to plug into a 16 pin DIP socket (see Figure 6-9).

The HS 371-8, an 8-bit current output converter, was used as a promotional give-away, but the demand was so great we added it to our catalog as a standard product. It ultimately became our best-selling product of the early 70s, averaging about 40,000 units a year for 10 years. The product was developed as a gimmick and turned out to be a real winner. Even today, 20 years later, units are still being sold.

This trip through DAC history is an example of how a circuit evolves by modifying and improving an old design. One does not have to reinvent the wheel with each new project. You should keep up to date on recent developments and not be afraid to research how a particular function was implemented in the past. You can benefit from the accomplishments and the mistakes of others. Fight the NIH (*Not Invented Here*) attitude and improve on the work of others with your own original ideas.

Manufacturing technology is also an area that gives the designer an opportunity to exercise innovation and creativity. The early DACs (vintage 1960s) were all built on printed circuit boards with discrete components. To keep the DAC as small as possible, designers used the fewest number of the smallest components. This meant, as we have seen, that diodes were substituted for transistors whenever possible. The two-terminal diode occupies less space than a three-terminal transistor. The modifi-

Figure 6-10.
Chip and wire hybrid construction.

cation of the transistor–transistor switch (Figure 6-2) to the transistor–diode switch (Figure 6-3) is an illustration of replacing a transistor with a diode to reduce cost and save space. If switching time is not a consideration, one would choose a binary weighted DAC (Figures 6-2 and 6-3) over the current divider configuration (Figures 6-4 and 6-5) since fewer resistors are required. The value of the resistors is not important when working with discrete components since all resistors of the same power rating are the same physical size. A 1/4 W 10 Ω resistor is the same size as a 1/4 W 100 MΩ resistor. Since the least number of components minimizes size, the circuit with the least number of resistors is preferred. The "minimum component count" strategy is the one to use when the assembly is discrete components on a printed circuit board, but when chip and wire hybrid construction is used, a different approach is necessary.

Chip and wire hybrid assemblies are constructed by attaching individual semiconductor dice to a ceramic substrate. The surface of the ceramic substrate contains a gold conductor pattern that interconnects the semiconductor dice attached to the substrate. Electrical connections are made from the pads of the semiconductor dice to the gold conductors on the substrate with gold wire (see Figure 6-10).

Precision resistors for the hybrid are made by depositing a thin film of resistive material such as nickel-chromium on a silicon wafer. Using standard semiconductor technology, the unwanted resistive film is etched away, leaving the desired resistor geometries on the silicon wafer. The wafer, with many identical circuits on it, is cut into individual dice. Each resistor die contains from one to several dozen resistors, depending on the design. The resistance value of the thin film resistor is determined by its geometry. The larger the value of the resistor the more area it occupies on the silicon die. Therefore, the area occupied by resistors in a chip and wire hybrid is determined by the total resistance and not by the number of resistors. In a chip and wire hybrid, the total resistance should be minimized to keep the unit as small as possible.

The size advantage gained with discrete components by using a diode instead of a transistor is lost in a chip and wire hybrid assembly. A transistor die is approximately the same size and cost as a diode die. In fact, when the circuit requires a diode, it can be obtained by using a transistor connected as a diode. The base and collector of the transistor are connected together to form one terminal of the diode, and the emitter of the transistor is the other terminal of the diode. Using a transistor to replace the diode can also help your purchasing department by reducing the number of different components it has to buy.

It can be concluded from the last two paragraphs that the circuit topology used for printed circuit construction is not optimum for a chip and wire hybrid. Printed circuit construction places a high priority on minimizing the number of resistors to reduce the size of the unit. The total value of the resistance is the parameter that determines the size of a thin film resistor die used in a hybrid. For example, five 10 KΩ resistors on a thin film die occupy less than 1/10th the area that one 500 KΩ thin film resistor die will occupy. But five 10 KΩ discrete resistors on a printed circuit board will occupy five times the space of one 500 KΩ discrete resistor. To minimize the size of a chip and wire hybrid, one must minimize the total resistance, even though a greater number of resistors is used.

The optimum topology for a 12-bit chip and wire hybrid DAC is different from the optimum topology for a discrete component version of a 12-bit DAC. Table 6-1 shows the number of resistors required in the switching section for both the binary weighted and the current division DACs constructed using discrete components on a printed circuit board and the total resistance for both versions constructed using chip and wire hybrid technology.

Table 6-1

Construction Technology	Binary Weighted	Current Division
Printed circuit	12 resistors	36 resistors
Chip and wire	20 megohms	0.15 megohms

If size were the only consideration, the binary weighting would be selected for the printed circuit assembly and the current division would be the selected for the hybrid. Since current division is faster than the binary weighted design, one might think that the hybrid possesses the best of both worlds—small size and good performance. But alas, the First Law of Engineering[3] has not been repealed. The hybrid is smaller and has better performance than the discrete component model, but to obtain these improvements the designer must compromise on cost. A precision chip and wire hybrid is always more expensive than an equivalent printed circuit design. If size is important, the user must be willing to pay for the decrease in size with an increase in price.

A designer will usually have several circuit configurations from which to choose to perform a desired function. The designer should evaluate all circuit possibilities and select the configuration best suited for the job. To make the proper selection, a designer must evaluate every component of the circuit and be able to integrate these components into an optimum system.

The paper design of the circuit is only one aspect of product development. Packaging, assembly, documentation, repair, trimming, testing, and last but not least, helping the end user with application problems are all important parts of producing a usable product. A good designer becomes involved in every aspect of product development. The designers name is on the product, and a good designer should do everything possible to assure its success. The designer should feel personally responsible when the product develops a problem.

At some point in the product development process, hardware, in the form of a breadboard, will appear. This is a decisive moment. One now has a circuit to which power can be applied. Before the breadboard was available, the design only existed on paper. You now find out if your theoretical design performs as you predicted. A word of advice: if the breadboard is completed on Friday afternoon, don't power it up until Monday morning. Enjoy the weekend.

The breadboard evaluation is a time to compare the actual performance to the predicted performance. If the predicted and actual results do not agree, *beware*. Don't casually dismiss the difference. Investigate thoroughly until you find the discrepancy between the paper design and the breadboard. I cannot emphasize enough the importance of attaining agreement between the paper design and actual circuit operation.

Occasionally during a breadboard evaluation, even though everything seems to be operating properly, I will get a second sense that something is not right. It's hard to describe, but the feeling is there. It might be a wave form that has an insignificant wiggle or a voltage that is close to the proper value but not exact. When I get this feeling, I investigate the circuit more thoroughly than I normally would. More times than not I find a hidden problem. If the problem is not solved then, it will appear at a later time and really bite me in the rear end. If you sense a circuit is not operating properly, take heed; it probably isn't. Place your trust in the "Force" and investigate.

Working with customers on application problems is challenging and can be

---

3. The First Law of Engineering, "You don't get something for nothing," is a result of the First Law of Thermodynamics. The First Law of Engineering also has applications in economics, business, and politics.

rewarding. Your interface with the customer is usually over the phone, so you have to develop a technique for trouble shooting at a distance. The person on the other end of the phone line usually performs the measurements you request and verbally communicates the results. In these situations take nothing for granted. What is obvious to you is probably not obvious to the person on the other end of the line, or you would not be on the phone in the first place. All the questions should be asked, no matter how mundane they may be. "Did you by-pass the power supplies with both ceramic and tantalum capacitors to ground?" Answers to such questions as this will give you a better feel for the level of expertise at the other end of the line. Customer interface can be rewarding, as you can sometimes solve a problem that the customer has struggled with for some time. Occasionally a situation will arise that can make you a legend in your own time.

Several years ago I was testing a 12-bit DAC in the lab and obtaining some very strange results. After performing the usual checks I found the -15 V supply had become disconnected. The loss of the negative supply voltage resulted in the strange behavior of the DAC. I reconnected the supply and the unit worked fine. The next day I was sitting in our application engineer's office when he received a call from a customer who was having a problem. The customer was testing the same model DAC that had given me the strange problem the previous day. As luck would have it, the problem he described was exactly the strange behavior I had witnessed the day before. I played a hunch and told our application engineer to have the customer check for a cold solder joint on the -15 V supply. The application engineer, looking a little skeptical, conveyed the information. About 15 minutes later the customer called back, verifying that his technician did find a bad connection on the -15 V supply. He fixed the cold solder joint and the unit worked fine. I never told our application engineer the whole story. Situations like that happen very seldom, so when they do, milk them for all they are worth. That is how legends are born.

Even though digital technology has become the glamor segment of the electronics industry, analog design still provides excitement and challenge for those of us who enjoy the color gray. Integrated circuit technology has allowed the development of complex analog circuits on a single silicon die. It is ironic that digital technology has played a major role in making the new innovations in analog design possible. Without simulators for design, CAD systems for layout, and digital measurement systems for testing, analog technology could not have advanced to its present state. The design process has been highly automated, but a creative and innovative mind is still a requirement for good circuit design. It was once said that, "Anyone who can be replaced by a computer should be." The number of analog designers is fewer, but until the world is quantized into "ones" and "zeros," the analog circuit designer still has a place in the electronic industry.

I will close with an old story I first heard from Don Bruck, one of the founders of Hybrid Systems. The story defines the difference between an analog and a digital engineer. In keeping with contemporary demands, the story can be made less gender specific by switching the male and female roles.

Two male engineers, one specializing in digital design and the other in analog, are working together in the laboratory. A nude female appears at the door, attracting the attention of both men. The vision of beauty announces that every 10 seconds she will reduce the distance between herself and the engineers by one half. The digital engineer looks disappointed and states, "That's terrible, she will never get here." The analog engineer smiles and then replies, "That's okay, she will get close enough."

That is the essence of analog design—all else is explanation.

7. Max Wien, Mr. Hewlett, and a Rainy Sunday Afternoon

One rainy Sunday afternoon, I found myself with nothing much to do. I've always treasured rainy Sundays that come supplied with spare time. With my first child on the way, I've taken a particular devotion to them lately. So I wandered off to my lab (no true home is complete without a lab).

I surveyed several breadboards in various states of inexplicable nonfunction and some newly acquired power transistors that needed putting away. Neither option offered irresistibly alluring possibilities. My attention drifted, softly coming to rest on the instrument storage area. On the left side of the third shelf sat a Hewlett-Packard series 200 oscillator. (No lab is complete without an HP series 200 oscillator, see Figure 7-1.)

The HP 200, directly descended from HP cofounder William R. Hewlett's master's degree thesis, is not simply a great instrument. Nor was it simply mighty HP's first product.[1] This machine is history. It provided a direction, methods, and standards that have been reflected in HP products to this day. There is a fundamental honesty about the thing, a sense of trustworthiness and integrity. The little box is a remarkable amalgam of elegant theoretical ideas, inspired design, careful engineering, dedicated execution, and capitalism. It answered a market need with a superior solution. The contribution was genuine, with the rewards evenly divided between Hewlett-Packard and its customers. The HP 200 is the way mother said things are supposed to be—the good guys won and nobody lost.

Digging in the lab library (no lab is complete without a library), I found my copy of William Redington Hewlett's 1939 Stanford thesis, "A New Type Resistance-Capacity Oscillator" (no lab library is complete without a copy).

Hewlett concisely stated the thesis objective (aside from graduating):

```
    The author has felt that there is a real need of
a new type oscillator that would combine the stability
of the coil-condenser type, the flexibility of operation
of the beat-frequency type, and still be light and portable
as well as simple in construction and adjustment.
    The object of this research has been develop-
ment, construction, and testing of such an oscillator
```

Hewlett's oscillator used a resonant RC network originated by Max Wien in 1891 (see the references at the end of this chapter). Wien had no source of electronic gain

---

1. Also, incidentally, easily their longest-lived product. The HP 200 series was sold by Hewlett-Packard until the mid-1980s, a production lifetime of almost 50 years.

Figure 7-1.
One of the original Hewlett-Packard Model 200A oscillators —the good guys won and nobody lost. (Photo courtesy of Hewlett-Packard Company.)

(DeForest hadn't even dreamed of adding a third element to Edison's Effect in 1891), so he couldn't readily get anything to oscillate. Anyway, Wien was preoccupied with other problems and developed the network for AC bridge measurements.

Hewlett saw that Wien's network, combined with suitably controlled electronic gain, offered significant potential improvements over approaches then used to make oscillators. These included dynamic tuning range, amplitude and frequency stability, low distortion, and simplicity.

Hewlett had something else besides electronic gain available; he also had the new tools of feedback theory. Harold S. Black's pioneering work, "Stabilized Feedback Amplifier," appears as the fourth reference in the thesis bibliography. Similarly, Nyquist's "Regeneration Theory," a classic describing necessary conditions for oscillation, is reference number three.

Hewlett synthesized all this nicely to show that Wien's network could be made to oscillate. Then he added a single (quite literally) crucial element. The oscillator's gain must be carefully controlled to support stable sinusoidal oscillation. If gain is too low, oscillation will not occur. Conversely, excessive gain forces limit cycles, creating a square wave oscillator. The problem is to introduce an amplitude regulation mechanism that does not generate output waveform distortion. Hewlett describes the elegant solution:

```
The last requirement, an amplitude-limiting
device that will not introduce distortion, is more
difficult to achieve. It is well known that the gain
of an amplifier with negative feedback is 1/ß, providing
Aß is large compared to 1. Thus if a resistance whose
value increases with the current through it is used as
past of the negative feedback network, the gain of the
amplifier may be made to decrease with an increase in
the input voltage. If an amplifier of this type is
used as part of the oscillator, it can be adjusted so
that oscillations will just start. As oscillations
```

build up, the gain of the amplifier will be reduced, thus reducing the tendency to oscillate and causing the amplitude of oscillations to reach a stable value. If this value is low enough, the tubes will operate class A, and no serious distortion will be introduced. Furthermore, any distortion that is produced, due to the nonlinear characteristics of the tubes, will be reduced by a factor of Aß by the action of the negative feedback.

For the variable resistance, a small tungsten lamp may be used. It is a well known property of such lamps that as the current through them increases, the filament warms up, thereby increasing the lamp resistance. Figure 2 shows how the resistance of a 110 volt, 6 watt, tungsten lamp changes with the current through it. It may seem that the maximum rate of change of resistance is when the load current is less than 20 milliamperes, and so to get maximum effect, the lamp should be operated in this region. In Fig. 3 is shown a complete diagram of the oscillator. The negative feedback is applied from the plate of the output tube to the cathode of the input tube. The lamp is placed from cathode to ground, so as to increase the feedback and reduce the gain of the amplifier as the oscillation builds up.

The only requirement placed on the lamp is that it be operated at such a temperature that the time rate of change of cooling be small compared to half the period of the lowest frequency. As the radiation

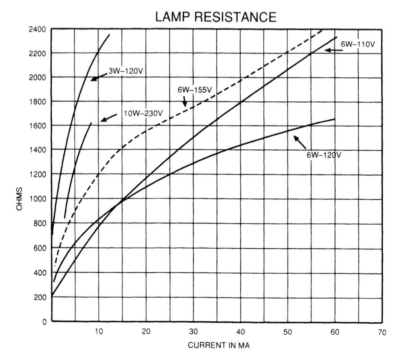

LAMP RESISTANCE

Figure 7-2. Hewlett's Figure 2 plotted lamp I-V characteristics. (Courtesy Stanford University Archives.)

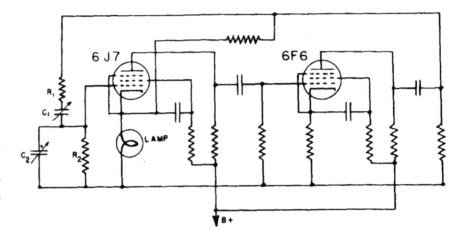

Figure 7-3.
Hewlett's Figure
3 detailed the
oscillator circuit.
Note Wien net-
work and lamp
(Courtesy
Stanford
University
Archives.)

```
is proportional to the fourth power of the absolute
temperature, and as most of the energy is lost through
radiation, this requirement may be easily met by not
operating the lamp at too high a current. Under these
conditions, the life of the lamp should be almost infinite.
```

Hewlett's use of the lamp is elegant because of its hardware simplicity.[2] More importantly, it is elegant because it is a beautiful example of lateral thinking. The *whole* problem was considered in an interdisciplinary spirit, not just as an electronic one. This is the signature of superior problem solving and good engineering.

The lamp solved a tricky problem, completing the requirements for a practical instrument. The design worked very well. It covered a frequency range of 20 to 20,000 cycles (it was cycles then, not Hertz) in three decade ranges with dial cali-

2. Hewlett may have adapted this technique from Meacham, who published it in 1938 as a way to stabilize a quartz crystal oscillator. Meacham's paper, ''The Bridge Stabilized Oscillator,'' is in reference number five in Hewlett's thesis.

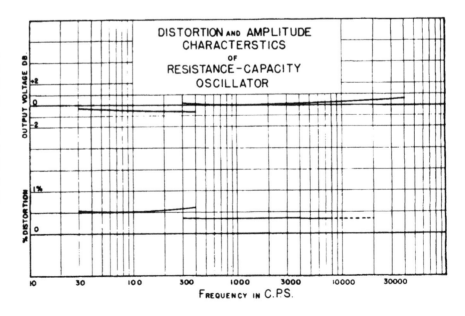

Figure 7-4.
Hewlett's Figure
4 showed good
distortion
performance.
What limited it?
(Courtesy
Stanford
University
Archives.)

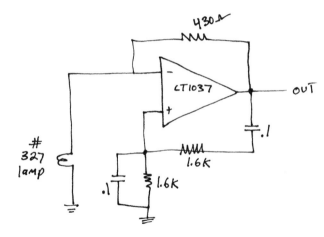

Figure 7-5.
My version of
Hewlett's circuit.
Distortion was
much better, but I
was fifty years
too late.

bration accuracy of 1%. The lamp maintained output amplitude stability within 0.2% at 100 cycles, varying only 1 dB from 20 to 20,000 cycles. Peering into my HP 201, I can see the light bulb, just where Hewlett, or one of his assistants, left it.

Hewlett's Figure 4 showed distortion well within 0.5% over the output range. This distortion figure caught my attention. By contemporary standards, Hewlett's 6J7/6F6-based "op amp" had major performance limitations.[3] How good, I wondered, would Hewlett's basic circuit be with a modern op amp?

And so, some fifty years after Hewlett finished, I sat down and breadboarded the oscillator to the meter of that Sunday afternoon rain. My circuit is shown in Figure 7-5.

This circuit is identical to Hewlett's, except that I have managed to replace two vacuum tubes with 94 monolithic transistors, resistors, and capacitors.[4] (I suppose this constitutes progress.) After establishing the 430 Ω value, the circuit produced a very nice sine wave. Connecting my (HP) distortion analyzer, I was pleased to measure only 0.0025% distortion (Figure 7-6). Then, I went ahead and endowed the basic circuit with multiple output ranges as shown in Figure 7-7.

This also worked out well. As Hewlett warned, distortion increases as oscillator

---

3. For those tender in years, the 6J7 and 6F6 are thermionically activated FETs, descended from Lee DeForest.

4. To be precise, there are 50 transistors, 40 resistors, and 4 capacitors in the device.

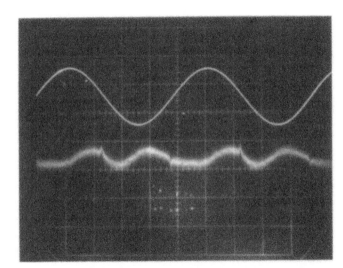

Output 10V/DIV

Distortion .003%

Horiz. =
100μsec/DIV

Figure 7-6.
Output waveform
and distortion for
my first oscillator.
Distortion was
0.0025%.

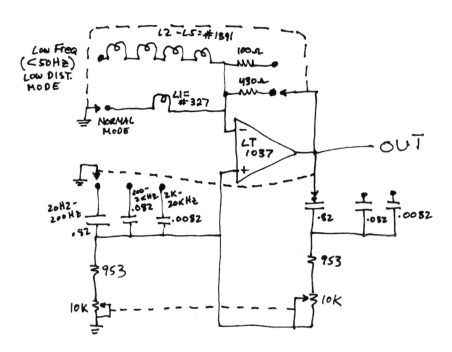

frequency descends towards the lamp's thermal time constant. This effect can be attenuated by increasing the lamp's thermal time constant. The easiest way to do this is to add more and bigger lamps. This causes longer amplitude settling times, but low frequency distortion is reduced. Plotting distortion versus frequency clearly shows this (see Figure 7-8).

Looking at the plot, I wondered just how far distortion performance could be pushed using Hewlett's suppositions and conclusions as a guide. The multi-lamp experiment indicates that distortion rise at low frequencies is almost certainly due to the lamp's thermal time constant. But what causes the slight upward tilt around 15 to 20 kc? And just what limits distortion performance? Chasing all this down

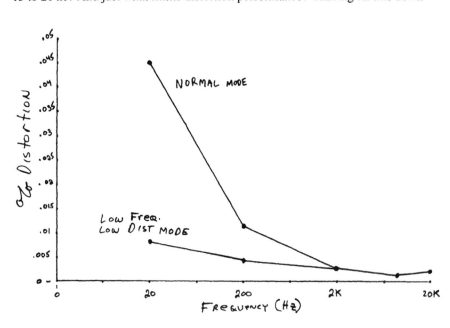

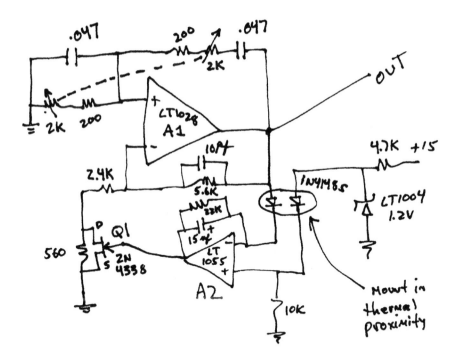

Figure 7-9.
The first attempt
at improving
distortion. A2 and
Q1 replace the
lamp.

seemed an entertaining way to stay out of the rain. Of course, I couldn't ignore that
I was already perilously near my analyzer's 0.0018% specification limit when inter-
preting results. Not to worry.

The next circuit is shown in Figure 7-9.

A1, a low noise wideband amplifier, is the oscillator. The variable resistor's
decreased value maintains low noise performance by minimizing bias current induced noise. The 10 pF capacitor suppresses parasitic high frequency oscillation.
A2 and associated components replace the lamp(s). A2 compares the oscillator's
positive peaks with a DC reference and servo-controls Q1 to establish proper loop
gain. The diode in series with the DC reference temperature compensates the rectifier
diode. The large feedback capacitor sets a long time constant for A2, minimizing
output ripple.

When I turned this circuit on, it oscillated, but distortion increased to a whopping
0.15%! The analyzer output showed a fierce second harmonic (twice the oscillator
frequency), although A2's output seemed relatively clean (see Figure 7-10).

So, I might have gotten away with dumping the two tubes for 94 transistors,
capacitors, and resistors, but replacing the lamp with a bunch of stuff was another
matter! I looked apologetically at the forsaken light bulbs.

What happened? The Wien network is the same, and it's hard to believe A1 is so
bad. A2's output shows some residual rectification peaking, but nothing that would
unleash such a monster.

The culprit turns out to be Q1. In a FET, the channel resistance is ideally fixed by
the gate–channel bias. In fact, slight modulation of channel resistance occurs as the
voltage across the channel varies. Unfortunately, Q1's drain sees significant swing
at the oscillator fundamental. The gate is nominally at DC, and the source
grounded. This causes unwanted modulation of the amplitude stabilization loop by
the oscillator's fundamental, creating distortion. The humble light bulb was beginning to look pretty good.

If you stare at this state of affairs long enough, the needed Band-Aid presents

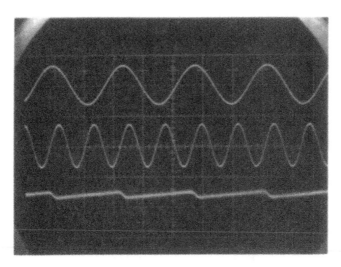

Figure 7-10.
Performance for the "lampless" oscillator. Modern technology is almost 100 times worse!

Output 2V/DIV

Distortion .15%

A2 Output (AC coupled) .1V/DIV

Horiz. = 200μsec/DIV

itself and is (thank the gods) refreshingly simple. The JFET is a fairly symmetrical structure, although this circuit drives it asymmetrically from gate to source. If you arrange things so the gate is driven with a signal halfway between the drain and source, symmetry is reestablished. This symmetrical drive eliminates all even-order harmonics. Q1's new companions make things look like Figure 7-11.

With the trimmer set to provide the optimum amount of feedback, distortion dropped to just 0.0018%—the analyzer's specified limit (see Figure 7-12).

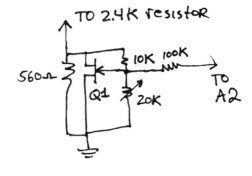

Figure 7-11.
The local feedback network around Q1, intended to cure channel resistance modulation effect.

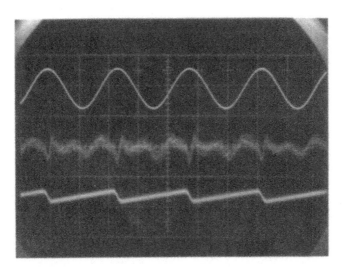

Figure 7-12.
Results of Q1's local feedback fix. Distortion improves to 0.0018%....about as good as the light bulb.

Output 2V/DIV

Distortion .0018%

A2 Output (AC coupled) .1V/DIV

Horiz. = 200μsec/DIV

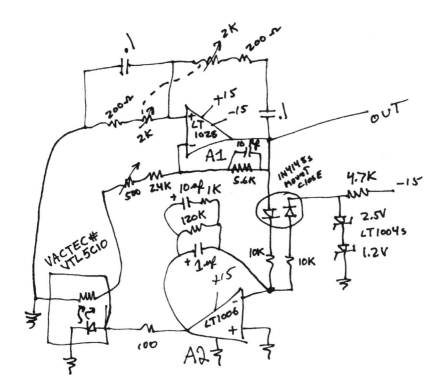

Figure 7-13.
Replacing Q1 with an optically driven photocell eliminates the resistance modulation trim. A2 is now a ground-referenced integrator.

While praying that the analyzer was better than it had to be, I looked at what it was saying. Some of the first harmonic was visible, along with artifacts of the amplitude control loop's rectification peaking. No amount of fiddling with the distortion trimmer could reduce the first harmonic, although increasing A2's feedback time constant reduced rectification related content.

I didn't like the trimmer, and A2's feedback capacitor was a big dog. Also, A2 is not a true integrator and has noise gain from its positive input. This seemed more irritating than obviously relevant. Similarly annoying was the notion that if A2 ever swings positive (start-up, whatever), the electrolytic reverse biases. This ain't pertinent either but still is bad manners!

The next iteration attempted to deal with some of these issues (see Figure 7-13).

The most noticeable change is that Q1 has been replaced with an optically driven CdS photocell. These devices don't suffer from parasitic resistivity modulation, offering a way to eliminate the trim. A2, running single supply, is now a ground-sensing type configured as a true integrator. The feedback components are arranged in a weak attempt to get a long time constant with improved settling time. Lastly, the DC reference has been increased, forcing greater oscillator swing. This is a brute force play for a more favorable signal/noise ratio.

This experiment provided useful information. A2's modifications eliminated rectifier peaking artifacts from the distortion analyzer's output. The LED-driven photocell really did work, and I tossed the trimmer down to the end of the bench. The analyzer indicated 0.0015%, but I wasn't sure if I could take this "improvement" seriously. Interestingly, the second harmonic distortion product looked the same, although perhaps less noisy. It increased a bit with higher frequencies and more or less ratioed with shifts in output amplitude (facilitated by clip-leading across one of the LT1004 references). The analyzer seemed to give readings a few parts-per-million (ppm) lower for higher oscillator amplitude, suggesting signal/noise issues with the circuit, the analyzer, or both. But understanding the source of the second harmonic distortion product was clearly the key to squeezing

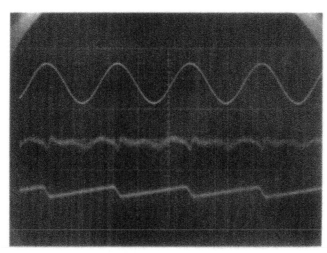

Figure 7-14.
A2's increased time constant reduces rectification related distortion content.

Output 10V/DIV

Distortion .0015%

A2 Output (AC coupled) .1V/DIV

Horiz. = 200μsec/DIV

more performance. The circuit was talking, and I was trying to listen, but I wasn't hearing (see Figure 7-14).

All this seemed to exonerate the gain control loop. That left the Wien network, the op amp, or some parasitic that wasn't on the schematic as the villain.

I considered the possible effects of voltage coefficient in the Wien network resistors and ESR or dielectric absorption in the capacitors. Sometimes when you don't know how to make things better you can learn by trying to make them worse. So I added tiny, controlled parasitic RC terms to the Wien R's and C's to test their sensitivity to component imperfections. What I found indicated that the reasonably good grades of R and C I was using were not the problem. I bolstered this conclusion by trying different R's and C's in the Wien network. Various decent grades of components all produced about the same result. That kinda left A1. Open loop gain, which degrades with frequency, could be a problem, so I decided to add a buffer to unload the amplifier. Beyond this, I couldn't do much else to increase available gain.

Now that I had license to accuse the op amp, the answer quickly seemed apparent. This circuit was in violation of a little known tenet of precision op amp circuits: Williams's Rule. Williams's Rule is simple: *always invert* (except when you can't). This rule, promulgated after countless wars with bizarre, mysterious, and stubborn effects in a variety of circuits, is designed to avoid the mercurial results of imperfect op amp common mode rejection. Common mode–induced effects are often difficult to predict and diagnose, let alone cure. A zero volt summing point is a very friendly, very reassuring place. It is (nominally) predictable, mathematically docile, and immune from the sneaky common mode dragons.

All present amplifiers have decreasing common mode rejection with frequency, and A1 is no exception. Its common mode rejection ratio (CMRR) versus frequency plot is shown in Figure 7-15.

The oscillator forces large common mode swings at A1. Since CMRR degrades with frequency, it's not surprising that I saw somewhat increased distortion at higher frequencies. This seemed at least a plausible explanation. Now I had to test the notion. Doing so required bringing the circuit into alignment with Williams's Rule. Committing A1's positive input to ground seems an enormous sacrifice in this circuit. I considered various hideous schemes to accomplish this goal. One abomination coupled the Wien network to A1's remaining input via a transformer. This approach wasn't confined to technical ugliness; in all likelihood, it would be considered obscene in some locales. I won't even sketch it, lest the publisher be hauled

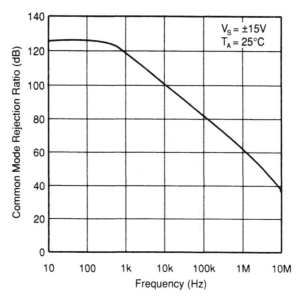

Figure 7-15. Common mode rejection ratio versus frequency for A1.

into court by some fundamentalist op amp group. Even if I could have gotten the whole perverse hulking thing to work, it just didn't feel right. I could hear Hewlett's simple, elegant little light bulb, which worked so well, laughing at me.

Somewhere in the venerable *Philbrick Applications Manual*, the writer counsels that "there is always a Way Out." The last circuit (Figure 7-16) shows what it was.

This configuration is identical to the previous one, except A3 appears along with buffer A4. A3 maintains A2's positive input at virtual ground by servocontrolling the formerly grounded nodes of the Wien network and the gain control loop. This adds a third control loop to Hewlett's basic design (this is getting to be a very busy

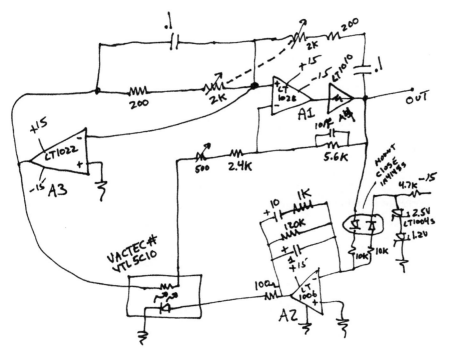

Figure 7-16. The final circuit. A3 eliminates common mode swing, allowing 0.0003% (3 ppm) distortion performance.

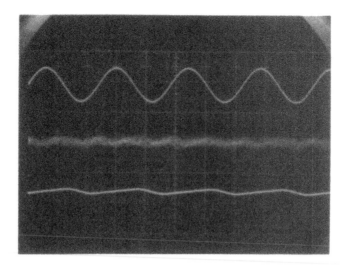

Figure 7-17.
Common mode
suppression runs
distortion
analyzer into its
noise floor.

Output 10V/DIV

Distortion
(analyzer limited)
.0003%

A2 Output (AC
coupled) .1V/DIV

Horiz. =
200μsec/DIV

oscillator—pity poor A1, besieged by three masters) but does not adversely affect its operation. With its positive input forced to virtual ground, A1 sees no common mode swing. Williams's Rule is satisfied, and ostensibly, good things should happen.

To my utter amazement, this whole thing did not explode when I finally summoned the nerve to turn it on. Even more astonishing was the distortion analyzer's 0.0008% reading (Figure 7-17).

Its output showed only faint traces of the first harmonic outlined in noise. The analyzer was indicating more than a factor of two beyond specification, which was really asking a lot. While it's unlikely that the oscillator and analyzer have compensatory errors, it's dangerous to conclude anything. As such, I turned to some very specialized equipment to get at the truth.

The Audio Precision System One will read distortion down to 0.0003% (3 ppm). I was quite pleased to see that it couldn't find anything above this level.

After Hewlett finished *his* oscillator, he and David Packard went into their garage and built a few into boxes and then made some more kinds of instruments

Figure 7-18.
Bill Hewlett and
David Packard
building
oscillators at the
Hewlett-Packard
Company,
located in their
garage.
(Photo courtesy
Hewlett-Packard
Company)

(Figure 7-18). After I finished *my* oscillator, I went into the kitchen and made a few hot dogs for dinner (mustard, chili sauce, no beans) and then made some other stuff. So, not only was Hewlett a lot cleverer than me, he also had somewhat different priorities. However, he did eventually get around to dinner, and I understand he ate pretty well. My hot dogs tasted pretty good.

Acknowledgment

The author gratefully acknowledges William R. Hewlett's review of the manuscript and his commentary. Additionally, the cooperation of the Hewlett-Packard Company and Stanford University was beyond value.

References

Wein, Max (1891). "Measung der induction constanten mit dern 'Optischen Telephon', " *Ann. der. Phs.* **44**, 704–707.

Hewlett, William R. (1939). "A New Type Resistance-Capacity Oscillator," M.S. Thesis, Stanford University, Palo Alto, California.

Bauer, Brunton (1949). "Design Notes on the Resistance-Capacity Oscillator Circuit," Parts I and II, *Hewlett-Packard Journal,* Volume 1, Nos. 3 and 4 (November–December 1949).

Hewlett, William R. (1983). "Inventions of Opportunity: Matching Technology with Market Needs," Hewlett-Packard Company, Palo Alto, California.

Williams, Jim (1984). "Thermal Techniques in Measurement and Control Circuitry," Linear Technology Corporation Application Note 5, Linear Technology Corporation, Milpitas, California.

Williams, Jim (1990). "Bridge Circuits: Marrying Gain and Balance," Linear Technology Corporation Application Note 43, Linear Technology Corporation, Milpitas, California.

Part Three

The Making of an Analog Engineer

If we accept the premise that analog engineers are made rather than born, then how do we go about making a good one? The contributors to this book are certainly "good ones," and here they explore some of the influences that shaped themselves and others.

Tom Hornak started down the analog path as a boy when he tried to figure out the difference between voltage and amperage. As part of this effort, he learned how to "visualize" the operation of circuits. In his contribution, Tom shows the utility of visualization and how others can learn to do it.

Bob Pease was fortunate to spend his early years as an engineer under the wing of George Philbrick. Perhaps the best way to learn analog design is to do it. The next best way is to watch and mentor under some master analog engineers. In Chapter 9, Bob tells what he learned watching and participating in the development of the P7 and P2 operational amplifier modules.

James K. Roberge is a professor at the Massachusetts Institute of Technology, the alma mater of several of this book's contributors. Here, James describes how M.I.T. attempts to train the next generation of analog wizards through a hefty diet of problem solving and design tasks to supplement the theoretical coursework.

There's a certain philosophy of analog design that analog designers need to learn. Rod Russell describes that philosophy and the elements composing it, showing that success in analog design often depends as much on how you approach a task as what you know.

Experience, even of a negative sort, is a big factor in the making (or breaking!) of an analog designer. Milton Wilcox relates what he learned about the importance of adhering to detail while designing analog integrated circuits. The "three out of three" rule Milton develops in his contribution may be not the sort of thing that's easily expressed mathematically or as an elegant theory, but it does manifest itself in such eminently objective forms as "the bottom line."

Are there any shortcuts to mastery of the analog art? Is it possible to buy a computer-aided design software package, load it on a workstation, input the desired parameters, "point and click" with a mouse, and come up with a working analog design a few minutes later? Some people say so. Jim Williams disagrees, and in the final chapter of this section he makes an eloquent case why breadboards and fingertips will still be part of the analog designer's arsenal for the foreseeable future.

8. True Analog Circuit Design

There is no analog vs. digital circuit design. All circuits are designed paying attention to their speed, power, accuracy, size, reliability, cost, and so forth. It is only the relative importance of these individual circuit parameters (and, of course, the mode of their application) that is different from case to case. However, there is something that can (with a slight twist of tongue) truly be called "analog circuit design," i.e., circuit design by using analogs. This is what this chapter is all about. But first, the story of how it all started.

In Chapter 10, you can read how the eight-year-old Jim Williams got hooked forever on electronics by being close to a hobbyist who owned an oscilloscope. I would like to share a quite different experience, which nevertheless, had a long-lasting influence too. It took place much earlier, around the year 10 BT (Before Transistors).

A long time before taking physics in high school, a good friend of mine and I tried desperately to understand what volts and amps really meant. No one in our families or among our family friends was technically inclined enough to help. One day we noticed that on one floor of the apartment house where I lived, the label on a kilowatt-hour-meter listed 15 A, while on the floor above it said 5 A. We deduced that the amps were something like water pressure, decreasing with elevation above ground. This theory survived only until we climbed up one more floor and found again a 15-A label there. Many weeks later it began to dawn on us that volts are like pressure and amps like strength of flow. Meanwhile, our apartment house got a new janitor in whom we naively saw a technical expert. We asked him to confirm our analogy. He said: "Yes, you are close, but you have the volts and amps mixed up." This was a setback which took weeks to overcome.

Our first hands-on experiments took place in my friend's home and dealt with electric arcs. The ability to generate intense light and heat was fascinating. We used a 1 kW smoothing iron as a series resistor and large iron nails as electrodes. When first joining the two nails and then pulling them apart, we were able to pull arcs of up to 1 cm in length. The heat of the arc was so intense that the nail tips melted into iron droplets. We loved it.

Our experiments were always interrupted during the summer when my friend and I were taken out of town to separate places for vacations. That year, when school started again and we, happily rejoined, wanted to pull arcs again, it simply did not work anymore. We were mystified: the same wall outlet, the same smoothing iron, the same nails, but no arc. We found out after some time that, during that summer, the local power company had converted my friend's home from DC to AC. A new chapter in our "education" began.

Our getting acquainted with AC started by learning that electrical power was delivered to my friend's home by four wires, three "hot" and one "safe." We were told that the voltage between any one of the "hots" and the "safe" was 220 V, while the voltage between any two of the "hots" was 380 V. Being used only to "positive" and "negative" from our DC experience, this was again a mystery. If two "hots" are each 220 V away from the "safe," then the two "hots" must be either 0 V or 440 V away from each other. Wasn't that crystal clear? This time we found somebody who knew the right answer, but he did not help us too much. Instead of using a simple analog such as a phasor diagram, he started to talk sine and cosine. We accused him of not knowing the answer either and covering up his ignorance by muttering some mumbo-jumbo. It again took considerable time and effort before we got the picture.

Why was our progress so difficult and slow? Was it lack of intelligence? That friend of mine is one of the world's leading mathematicians today. At least in his case, lack of intelligence is not a feasible explanation. I think we were slow because our young minds needed to see the invisible electrical processes translated into easy-to-envision analogs. And we had to develop these analogs ourselves, step-by-step.

I know that trying to "understand electricity" early in life had a lasting benefit to me. I got used to "seeing electricity" in analogs and I am still seeing it that way. I believe every electronic circuit designer could benefit from thinking in analogs, and it is never too late to start. This belief made me write this chapter.

It is mainly during the initial, qualitative phase of designing an electronic circuit that it is most helpful to translate the circuit's operation into a more transparent process. The same applies when one has to quickly comprehend the operation of a circuit designed by somebody else during presentations, design reviews, or the like. I am, of course, not against exact mathematical analysis or computer simulation, but those have their justification in the next phase, if and when exact quantitative verification of a design is required. I find that mainly circuit operation described in the time domain is easy to synthesize or analyze this way.

My process of visualization is quite simple. Circuit diagrams are commonly drawn with current flowing from the top to the bottom and with the individual circuit elements in the schematic positioned approximately according to their voltage. When imagining a circuit's operation, I am, in my mind, moving and shifting parts of the circuit schematic up and down following the time-varying voltage they carry. This helps me to "see" in time sequence which diodes or transistors are turning on, which are turning off, how op-amp circuits behave, and so forth. I never draw these distorted circuit diagrams on paper; rather, I see the moves and bends only in my mind. (An excellent way to avoid boredom in many situations in life!) Of course, these imagined moves and shifts are only approximate and cannot be applied consistently because that would often distort the schematic beyond recognition.

To illustrate what I mean, I will describe the process as best I can in the following few examples. Unfortunately, the static nature of printed figures in general, combined with the black-only print in this book, will make it somewhat difficult to get the message across. I wish I could attach a videotape on which I could perhaps convey my mental images much better. Because of this difficulty, I had to reverse the process. As examples I picked conventional circuits of well-known operation to describe the method of visualization. Normally it would be the other way around.

Example 1. The Astable Multivibrator

The first example is an astable multivibrator shown by its conventional circuit diagram in Figure 8-1. The idealized wave forms on the collectors and bases of transistors Q1 and Q2 are shown in Figure 8-2. At time t_0 transistor Q1 is in saturation,

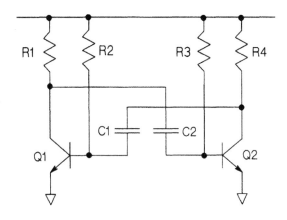

Figure 8-1.
Conventional diagram of an astable multivibrator.

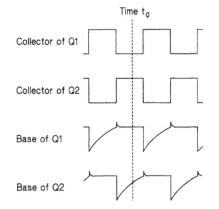

Figure 8-2.
The waveforms of an astable multivibrator.

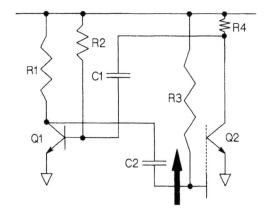

Figure 8-3.
The astable multivibrator at time t_0.

and its collector voltage is close to ground. Transistor Q2 is cut off by a negative voltage on its base, so its collector voltage is high above ground. The voltage on Q2's base is changing in a positive direction due to resistor R3 charging capacitor C2. In Figure 8-3, my mental image of the multivibrator at time t_0, this is represented by the different heights of Q1's and Q2's collector nodes, by the base of Q2 being shown "below" its emitter, and by the arrow indicating the base moving in a positive direction. Note how resistors R1 and R3 are more "stretched" than resistors R2 and R4.

Example 2. The Precision Rectifier

The next example, a precision half-wave rectifier, is shown by its conventional circuit diagram in Figure 8-4. Node X between resistors R1 and R2 is held at ground level by the feedback action from the operational amplifier's output. When a positive input voltage V_{in} is applied at R1 (see Figure 8-5), the output of the operational amplifier goes negative and pulls via diode D1 the output end of R2 "down," so that the current flowing in R1 continues via R2 into the operational amplifier's output. Diode D2 is off, with its anode being more negative ("lower") than its cathode. As long as the input voltage V_{in} is positive, resistors R1 and R2 behave like a seesaw with the fulcrum at node X. When the input voltage V_{in} applied to resistor R1 is negative (see Figure 8-6), the operational amplifier's output goes "up" until diode D2 conducts and delivers via D2 to fulcrum X the current required by R1. Diode D1 is off, because its cathode is "above" its anode. For negative input voltages, R1 and R2 do not behave as a seesaw; R2 and the circuit's output remain at ground level.

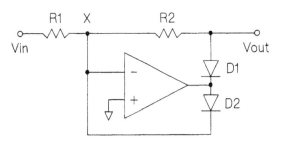

Figure 8-4.
Conventional diagram of a precision rectifier.

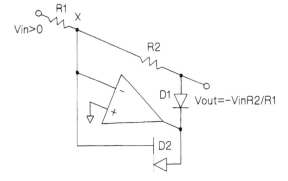

Figure 8-5.
The precision rectifier with positive input voltage.

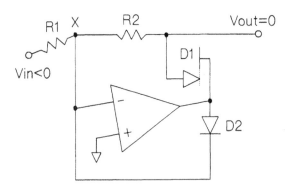

Figure 8-6.
The precision rectifier with a negative input voltage.

Example 3. The Transition Pulse Generator

The last example is a transition pulse generator used in digital communication links in their clock recovery portion. The input of the circuit is a data stream in "non-return-to-zero" (NRZ) format, in which logic ones and logic zeros are represented by "high" and "low" levels, respectively, each lasting over the whole bit period. The purpose of the transition pulse generator is to generate a pulse of uniform polarity whenever a transition from level to level occurs in the input data. The conventional diagram of a commonly used circuit for this purpose is shown in Figure 8-7. Transistors Q1 and Q2 with capacitor C constitute a differentiator, transistors Q3 and Q4 act as a full-wave rectifier. The transition pulses delivered by this circuit have a uniform, positive polarity. The NRZ data input and transition pulse output of the circuit are shown in Figure 8-8. Time instants t_1–t_4 represent four distinct states in the circuit: t_1 and t_3 when the circuit is ready for the next transition, and t_2 and t_4 the state immediately after a data transition has occurred.

Figure 8-9 represents my vision of the circuit at t_1, with logic zero level $V(0) < V_{bias}$ at its input, waiting for a positive transition. Resistors R1 and R2 carry equal currents set by the two matched current sinks, CS1 and CS2. The voltage on the collectors of Q1 and Q2 and on the bases of Q3 and Q4 are the same. Voltage V_{out} is at the LOW level of the transition pulse. The voltage across capacitor C is essentially V_{bias}–$V(0)$ with its positive terminal facing Q2.

Figure 8-10 shows the state of the circuit at t_2, shortly after a positive transition at the data input. During the positive data transition, the voltage across capacitor C changes only very little. This means that Q1 "lifts" the emitter of Q2 via capacitor C essentially by $V(1)$–$V(0)$ and Q2 is shut off. The voltage at collector Q2 and base Q4 goes "up," and the emitter of Q4 takes V_{out} to the HIGH level. The current through resistor R1 is now the sum of the currents of CS1 and CS2, and the collector of Q1 "drops down." The current of CS2 is now discharging capacitor C and "pulling" emitter Q2 "down," as indicated by the vertical arrow.

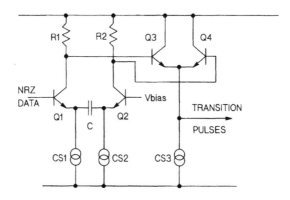

Figure 8-7.
Conventional diagram of a transition pulse generator.

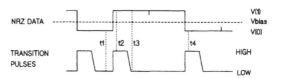

Figure 8-8.
Input and output signals of the transition pulse generator.

Figure 8-9.
The transition
pulse generator
at time t_1.

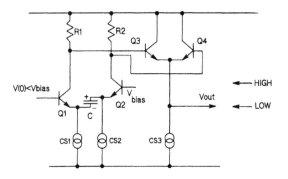

Figure 8-10.
The transition
pulse generator
at time t_2.

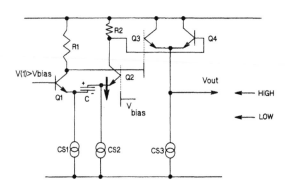

When, at time t_3, the current of CS2 reverses the polarity of capacitor C and "pulls" the emitter of Q2 sufficiently "down" to make Q2 conduct again, resistors R1 and R2 carry again the same currents set by CS1 and CS2, respectively, and V_{out} returns to the LOW level. This is depicted in Figure 8-11. Capacitor C is charged now essentially to $V(1)-V_{bias}$ with the positive terminal facing Q1.

Finally, Figure 8-12 shows the state of the circuit at time t_4, shortly after a negative transition at the data input. The negative data transition has "pulled" the base of Q1 "down," its emitter is being held "high" by capacitor C, and Q1 is cut off. The voltage of collector Q1 is "up" and V_{out} is held HIGH by Q3. Resistor R2 is now carrying the currents of both CS1 and CS2. Capacitor C is being discharged by the current of CS1 and the emitter of Q1 is moving "down" as indicated by the arrow.

I hope that these three simple examples were sufficient to illustrate my message and that from now on many readers of this chapter will twist and bend circuit schematics in their minds.

In conclusion, I will list one more reason for writing this chapter. I'm convinced that my childhood experience is not unique. I'm sure there are tens of thousands of

Figure 8-11.
The transition
pulse generator
at time t_3.

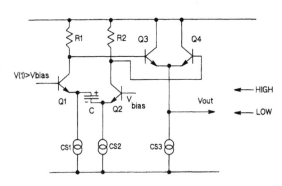

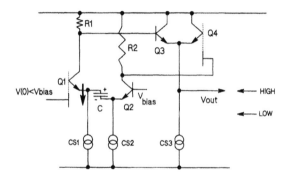

Figure 8-12.
The transition
pulse generator
at time t_4.

young children in our country who are intelligent and attracted to electronics but
who have nobody around to supply the basics in a form that is easy to digest on
their level of comprehension. Many laudable efforts are taking place that attempt to
help children to visualize natural phenomena: the San Francisco and San Diego
Exploratoria are two. But there, when attempting to cover all sciences, electronics
is necessarily a small part of the whole. There are plans to build a similar permanent
exposition devoted mostly to electronics in Silicon Valley. I pleaded for the installation of simple visualizations of, for example, how a transistor works, as opposed
to trying to impress the young visitors with giant models of million-transistor
chips. I believe that an indifferent child visiting out of superficial curiosity or com-
pulsion will not get hooked by either, while the information-starved gifted kid
could be helped very much by pushing him or her one rung higher on the ladder of
understanding.

We are constantly reminded that, due to the expected demographic development
in our country, we will be short of electronic talent in the near future if we don't
succeed in exciting interest in electronics in more children. By supplying easy-to-
visualize basic information through properly written books, proper expositions,
and last but not least, personal interaction, we could perhaps increase the number of
talented children hooked by electronics in their early age, and, it is hoped, turn them
into devoted executors of this art, as we are ourselves.

9. The Story of the P2—The First Successful Solid-State Operational Amplifier with Picoampere Input Currents

First, let us start with—

A Fable

Once upon a time there were two wizards who decided they wanted to play golf. The first wizard stepped up to the tee, addressed the ball, and drove the ball right down the middle of the fairway; the ball then bounced twice, and rolled, and rolled, and rolled, and rolled right into the cup. The second wizard looked at the first wizard. Then he stepped up to the ball and drove a wicked screaming slice off to the right, which hit a tree, bounced back toward the green, ricocheted off a rock, and plopped into the cup. The first wizard turned to the second wizard and said, "Okay, now let's play golf."

End of Fable

Once upon a time, back in the ancient days of the electronics art, about 1958, there were two wizards, George A. Philbrick and Robert H. Malter, and they enjoyed designing operational amplifiers. In those days, that's what they called them—operational amplifiers, not "op-amps." George had the idea to use some of those new "transistors" to amplify the error signal from a balanced bridge, up to a good level where it could then be demodulated and amplified some more and used to form an operational amplifier. Ah, but what kind of balanced bridge would this be? A ring of conducting diodes? Heavens, no—George proposed a bridge made of 100-pF varactor diodes, so that when the bridge was driven with perhaps 100 mV of RF drive, the diodes would not really conduct very much and would still look like a high impedance—perhaps 10,000 MΩ. Then just a few millivolts of DC signal could imbalance the bridge and permit many microvolts of radio frequency signals to be fed to the AC amplifier. Now, back in 1958, just about the only available transistors at any reasonable price were leaky germaniums, and you certainly could not build a decent operational amplifier out of those. But George got some of the new 2N344 "drift" transistors that still had some decent current gain at 5 Mcps. He ran his oscillator at 5 Mcps, and after running his signal through the whole path of the modulator and then four stages of 2N344 RF amplifier, and a demodulator, he fed it into a DC amplifier stage with push–pull drive to a class AB output. And it was all built as a quasi-cordwood assembly, with seven or eight little PC (Printed Circuit) boards strung between two long PC boards. Since each little PC board had about six wires connecting into the long PC boards, this was a kluge of a *very high order,* and *not* fun to assemble or test, or to evaluate, or to experiment with, or to troubleshoot. George called this amplifier the P7. Please refer to the schematic in Figure 9-1. I know this is the right schematic because I still have a P7. Also, see the photograph of a P7's inner workings in Figure 9-2.

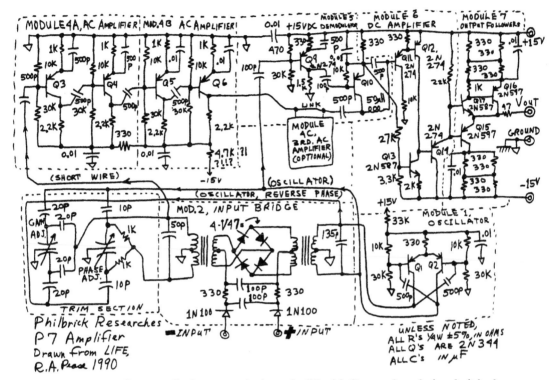

Figure 9-1.
Bob Pease's
rendition of the
schematic for the
P7 amplifier.

I never talked very much about the P7 with George, but obviously it had many problems. I don't think it was ever tested successfully as an operational amplifier, not as a working amplifier, nor as a product. Still, it had some promise. After all, if you could get it to work, this little circuit that used about 15 inexpensive germanium transistors could (it was hoped) have a better input current than even the better vacuum-tube amplifiers of that era—less than 10^{-9} A—better than a nA. (Heck, even a 12AX7 had 10 nA of grid current, and that was sort of the standard for ordinary operational amplifiers.) So George Philbrick concentrated on the P7 principle and the P7 design. Some people would use the word *obsessed*. He spent most of his

Figure 9-2.
The guts of the P7
amplifier. It was
photographed in
front of a mirror
so both sides
could be seen.

time for a couple years, and a lot of the company's resources, trying to get the P7 working.

All this experimentation was going on at George A. Philbrick Researches, first at 230 Congress Street and then at 127/129 Clarendon Street, and then at 221 Columbus Avenue and 285 Columbus Avenue, in Boston, Massachusetts, back about 32 years ago. George had started a business to sell analog computers, but even in the 1960 era, the business in operational amplifiers (such as the K2-W) was starting to grow and overshadow the analog computer business. Imagine that—people actually buying op amps so they could build their own instruments!

When Bob Malter arrived at Philbrick in 1957, he was already a smart and accomplished engineer. He was a native of Chicago, and he had served in the army at Dugway Proving Ground. After designing several analog computer modules (which were the flagships of the Philbrick catalog), he became intrigued with the concept of the varactor amplifier, about the time that George was getting frustrated. Now, Bob Malter was a very pragmatic, hard-headed engineer. You would *not* want to bet him that he could not do something, because he would determinedly go out and do it, and prove that he was right—that you were wrong. Bob had his own ideas on how to simplify the P7, down to a level that would be practical. I do not know how many false starts and wild experiments Bob made on what he called the P2, but when I arrived at Philbrick as a green kid engineer in 1960, Bob was just getting the P2 into production.

Instead of George's 10 PC boards, Bob had put his circuits all on 2 PC boards that lay back-to-back. Instead of 14 transistors, he had a basic circuit of 7 transistors—just one more device than the little 6-transistor AM radios of the day. He actually had two little transformers—one to do the coupling from the oscillator down into the bridge and one to couple out of the balanced bridge into the first RF amplifier. A third inductance was connected in the emitter of the output transistor, to help tailor the frequency response. Please refer to the schematic diagram of the P2 in Figure 9-3. I mean, just because *everybody else* used only capacitors to roll off the frequency response of *their* operational amplifiers—well, that did not scare

Figure 9-3.
The schematic for the P2, as drawn by Bob Pease.

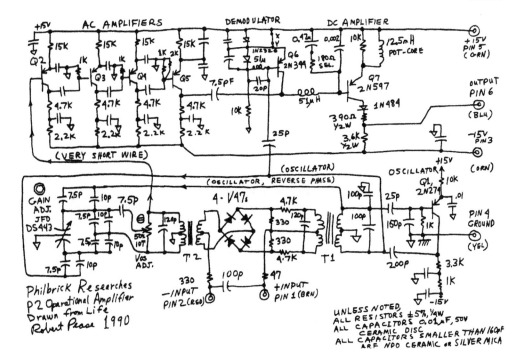

or impress Bob. He had to tailor the response of this operational amplifier with about 75 kcps of unity-gain bandwidth, and he had to roll it off at about 11 dB per octave or else have no output swing past 10 cps. As it was, he got the full 20 V p-p out to 500 cps, and even that was a struggle to accomplish. So Bob used the inductors and anything else he could think of that would help, in addition to various capacitive damping circuits. And he got it all to work. He got it to work quite well.

So, what's the big deal? Here's a pretty crude operational amplifier with a voltage gain of 10,000, and an output of ±1 mA at ±10 V, with a *vicious* slew rate of 0.03 V/μsec. Who would buy an amplifier like that? It turned out that thousands and *thousands* and *thousands* of people bought this amplifier, because the input bias current at either input was just a few *picoamperes*. What the heck is a *picoampere*? Most electrical engineers in 1960 didn't even know what a picofarad was, not to mention a picoampere, but they figured out it was a heck of a small fraction of a microampere—at 10^{-12} A, a picoampere is only 1 millionth of a microampere—and for many high-impedance instrumentation applications, the P2 was by far the only amplifier you could buy that would do the job. And it had this low bias current, only a few picoamperes, because all those germanium transistors were running at 5 Mcps, and their 5 or 10 μA of DC base current had no effect on the precision of the input current. The input current was low, thanks to a well-matched bridge of four V47 varicaps. These were sold by Pacific Semiconductor, Inc. (PSI) for use as varactors in parametric amplifiers, up in the hundreds of "megacycles," in low-noise communications receivers, mixers, and front-end amplifiers—parametric amplifiers. The "V47" designation meant that they had a nominal capacitance of 47 pF at 4 V reverse bias, which is where most RF engineers would bias them. But Bob Malter biased them right around 0 V DC, with a minuscule ±60 mV of AC drive.

At this level of drive, each diode would leak only 20 or 40 pA. But Bob had a gang of technicians working day and night to match up the forward conduction characteristics and the reverse capacitance voltage coefficients, and he was able to make sets of four varactors that would cancel out their offset drift versus temperature, and also their reverse leakage. Of course, there was plenty of experimenting and hacking around, plenty of experiments that didn't work, but eventually a lot of things that worked okay. After all, when you buy 10,000 V47s, *some* of them have to match pretty well.

So, here's a little do-hickey, a little circuit made up of just about as much parts as a cheap $12 transistor radio, but there was quite a lot of demand for this kind of precision. How much demand? Would you believe $227 of demand? Yes! The P2 originally started out selling for $185, but when the supply/demand situation heated up, it was obvious that at $185, the P2 was underpriced, so the price was pushed up to $227 to ensure that the people who got them were people who really *wanted* and *needed* them. So, the people who really wanted a P2 had to pay a price that was more than ⅛ the price of a Volkswagen Beetle—that was back when $227 was a real chunk of money!

Meanwhile, what other kinds of "transistorized" op amps could you buy? Well, by 1963, for $70 to $100, you could buy a 6- or 8-transistor amplifier, with I_{bias} in the ball-park of 60,000 to 150,000 pA, and a common-mode range of 11 V. The P2 had a quiet stable input current guaranteed less than 100 pA (5 or 10 pA, typical), and a common-mode range of ±200 V. (After all, with transformer coupling, the actual DC level at the balanced bridge could be at any DC level, so there was no reason the common mode rejection ratio could not be infinite.)

Wow. A $227 *gouge*. (You couldn't call it a "rip-off," because the phrase hadn't been invented, but perhaps that is the only reason.) Obviously, this must be a very profitable circuit. Every competitor—and many customers—realized that the P2 must cost a rather small amount to build, even allowing a few hours of work for some special grading and matching and testing. So, people would invest their $227 and buy a P2 and take it home and pull it apart and try to figure out how it worked. The story I heard (it might be partly apocryphal, but most of it probably has a lot of truth) was that Burr Brown hired a bright engineer, handed him a P2 and told him, "Figure out how they do this—figure out how we can do it, too." In a few days the engineer had dismantled the circuit, traced it out, and had drawn up the schematic. Then he analyzed it and began experiments to be able to meet or exceed the P2's performance. But he couldn't get it to work well. He tried every approach, but he never could. After a full year, Burr Brown gave up and put the engineer to work on some other project. Burr Brown never did get into the varactor-input amplifier business, and I believe there is truth behind that story. Let me tell you why.

The P2 had an offset-adjust trim that was a little 20-turn trim-pot—that's not a surprise. But it also had a "gain adjust." This was not any ordinary gain adjust. This was a 20-turn variable trim capacitor—a differential piston capacitor—which the user could trim, per the instruction sheet, with a tiny little Allen wrench or hexagonal key. But it did not just have a linear control over the gain. If you trimmed the pot over to one end, the gain might be at 300 or 500, and then as you trimmed it closer to the center, the gain might rise to 700 or 900—and then, suddenly, the gain would pop up to 7,000 and then to 10,000 before the nonlinearity made the gain fall off again, when you turned the adjustment too far. The test techs called this, "going into mode." I used to wonder what they meant by that.

Several years later, George Philbrick brought me in to help him on an up-dated, up-graded version of the P2—the "P2A". We had to redesign the P2 because Philco was stopping the manufacture of those old 2N344s, and we couldn't buy any more, so we had to redesign it to use the more modern (cheaper) Silicon Mesa transistors, such as 2N706 or 2N760 or whatever. When we bought them from Texas Instruments, they were labelled "SM0387." I had a new circuit working pretty well, with the help and advice of George Philbrick, because Bob Malter had passed away, sadly, after a long bout with multiple sclerosis, about 1966. Anyhow, I was getting some results with the silicon-transistor version, but the improvements weren't coming along as well or as fast as I expected, so I went back to fool around with several real P2s, and to study them.

The original P2 had an apparent imbalance at the output of its demodulator. Well, that looked kind of dumb, that the first DC transistor would be turned off unless there was a pretty big signal coming out of the demodulator. To turn on the DC transistor, you had to have a considerable imbalance of the RF. So I took one unit and modified it to bias the demodulator about 1 V_{be} down from the positive supply, so it would not have to handle a great amount of signal just to drive the DC transistor Q7. Refer to the schematic of the P2 (Figure 9-3). Normally, Q6's emitter was connected directly to the +15 V bus. I disconnected it by removing link X–Y and connected it to a bias diode. Yes, the RF amplifier ran with less RF signal at balance—but the gain refused to "come into mode." So that "improvement" scheme was unusable. Now, what was *that* trying to tell me?

After some more study, I planned a few more experiments, and then I tried pulling apart the two PC boards so I could access some of the signals down in between the boards. As I eased the two boards apart (with power ON), the gain "jumped out of mode." I gradually realized the P2 amplifier was running, all these years, as a reflex amplifier. The "gain adjust" consisted of changing the phase between the oscillator and the bridge, so when the amplified signal came down to the end of the RF amplifier (four stages, remember) and was patched back to the other PC board, it would be able to regeneratively amplify even more than the honest gain of the RF amplifier. *That* was why the demodulator wouldn't work right unless a certain constant minimum amount of 5 Mcps signal was always flowing through the amplifier. *That* was why the gain would "pop into mode" (and when it wouldn't "pop into mode," that explained why not). *That* was why the engineer down at Burr Brown couldn't figure out how to get it working right—the gain depended on the two PC boards being spaced just the right distance apart! *That* was the trick that Bob Malter had accidentally built into the P2, and that he had figured out how to take advantage of. To this day, I am not sure if Bob Malter knew exactly what a tiger he had by the tail. But I would never dare to underestimate Malter's tough and pragmatic brilliance, so I guess he probably did know and understand it. (I never did have the brass to ask him exactly how he thought it worked. I bet if I had had the brass to ask him, he would have told me.) I must say, if any engineer was bright enough to grasp and take advantage of a strange interaction like this, well, Bob Malter was that sharp guy.

Now, since my P2A was designed on a single board, with the demodulator far away from the inputs and oscillator, we wouldn't have any "mode" to help us. But that was okay—now that I understood the "mode" business, I could engineer the rest of it okay without any "mode," and I did. But that explained why none of our competitors ever second-sourced the P2. And the P2A and SP2A remained profitable and popular even when the new FET-input amplifiers came along at much lower prices. It was years before these costly and complex parametric amplifiers were truly and finally made obsolete by the inexpensive monolithic BIFET™ (a trademark of National Semiconductor Corporation) amplifiers from National Semiconductor and other IC makers. Even then, the FET amplifiers could not compete when your instrument called for an op amp with a common-mode range of 50 or 200 V.

A friend pointed out that in 1966, Analog Devices came out with a "Model 301," which had a varactor input stage. It did work over a wider temperature range, but it did not use the same package or the same pin-out as the P2.

Still, it is an amazing piece of history, that the old P2 amplifier did so many things right—it manufactured its gain out of thin air, when just throwing more transistors at it would probably have done more harm than good. And it had low noise and extremely good input current errors—traits that made it a lot of friends. The profits from that P2 were big enough to buy us a whole new building down in Dedham, Massachusetts, where Teledyne Philbrick is located to this day. The popularity of the P2 made a lot of friends, who (after they had paid the steep price) were amazed and delighted with the performance of the P2. And the men of Philbrick continued to sell those high-priced operational amplifiers and popularize the whole concept of the op amp as a versatile building block. Then, when good low-cost amplifiers like the μA741 and LM301A came along, they were accepted by most engineers. *Their* popularity swept right along the path that had been paved by those expensive amplifiers from Philbrick. If George Philbrick and Bob Malter and Dan Sheingold and Henry Paynter and Bruce Seddon hadn't written all those applications notes and all those books and stories, heck, Bob Widlar might not have been able to *give* away his μA709s and LM301s! And the P2—the little junk-box made up virtually of parts left over from making cheap transistor radios—that was the profit-engine that enabled and drove and powered the whole operational-amplifier industry.

Since George Philbrick passed away about 1974, and Bob Malter had died earlier, I figured I had an obligation to tell this story as there was nobody else left to tell it. Even though I was not in on the design of the P7 or the P2, I understood their designs better than just about anybody else. So, I just have to express my appreciation to Jim Williams for leading and editing this book. I know he will want to read about the P2, because I know he has one in his lab. (Meanwhile, I agreed to write a chapter for Jim's book, and to support and encourage the book, because I want to read all the *other* stories that will be in here.)

Vignettes: Additional Little Stories about the P2

One time Bob Malter came back from the big WESCON show in Los Angeles. He said, "I made a good bargain for a new spec on the piston capacitors. I got the price down from $2.15 to $1.65. That savings will pay for my trip and then some." It sure did.

One time, there were some P2s that had a lousy tempco. Most of the units had a drift much better than 6 mV from 20 to 45 °C. But this time a couple batches had a lousy yield for drift. So Bob figured out where to install some little thermistors—across one of the legs of the 50 k pot—and his wizardly technicians delved away like mad, and trimmed and tweaked and tested, and sure enough they got the drift to improve enough to meet specs. I said "delved," because they had to dig through the room-temperature-vulcanizing (RTV) potting material to access the places they needed.

One time, just a couple days after Bob went on vacation, the frequency response began to go to pot, and none of the usual tricks would fix it. So the senior technician, Tom Milligan, got on the phone to Bob (who had given him his vacation phone number), and Bob figured out a tweak, and by the time Bob got back from his vacation, the problem was completely cured.

One time, I was standing around in front of the Philbrick booth at the big IEEE show in New York City. A couple engineers were hiking past the booth, and one said to the other, nodding his head toward the booth, ". . . and there's the company that makes a *big* f------ profit." Well, at that time George A. Philbrick Researches was indeed making big profits from the P2. Can't deny it.

On various occasions, customers would ask about how to get the best long-term stability of the offset voltage. It turns out that most parts, if held at a constant temperature, could hold an offset voltage better than 100 µV/hour, and some were as good as 20 µV/hour. We had our little Rustrak meter to prove it. Heavens, we used

Figure 9-4.
It was what was inside that was important!

miles of that Rustrak paper. When the P2A came along, it was able to do as good as 1 or 2 μV p-p for an hour. But Bruce Seddon, one of the senior engineers, was always skeptical about the possibility of a P2 having V_{OS} stability that good. He computed that a single microvolt was worth about 600 electrons on each varactor. Since a varactor diode had really a rather shallow slope, you could compute that a 1 mV DC input would cause a 0.03 pF imbalance in a 200 pF bridge. And a 1 μV DC imbalance would cause a 0.00003 pF imbalance. Needless to say, that was a preposterous situation. You could compute that even a couple of *atoms* of shift on the components nearest to the varactors would cause worse imbalance than that. But we measured a lot of P2s, and a lot of P2As, and some of them would hold better than 1 μV p-p for an hour or two. Bruce always was incredulous about that.

Now, if you trimmed the offset voltage to zero, the input current was pretty small, about 5 or 10 pA typical, and 100 pA guaranteed max. Some people would pay a surcharge for selected units with extra-low input current. But many people would just crank the input offset pot over to the side—perhaps a few millivolts—and get the input current down to less than 1 pA. It wasn't perfectly stable if somebody suddenly turned on the airconditioner, but under constant ambient conditions, it was better than all but the best electrometer-tube amplifiers.

In addition to having low DC errors, the P2 had fairly decent low noise. The P2A was guaranteed to be better than 1 μV rms in the bandwidth 1–100 Hz, and many P2s were almost that good. Now, how can an operational amplifier have noise as good as that, right where most solid-state amplifiers have many microvolts of noise p-p? The fact is that the varactors transform, or down-convert, the noise of the first RF transistor at (5 Mcps to 5.0001 Mcps), down to input noise at the inputs of the P2, in the frequency range (0–100 cps). Those varactors really did provide the advantages of a parametric amplifier. And those germaniums weren't bad at 5 Mcps, so the P2 did a respectably good job for low noise. It took many years before its performance was matched by FET amplifiers.

The P2 was assembled with its two little PC boards rivetted securely together and then installed in a cast aluminum case. Then the whole cavity was filled with room-temperature-vulcanizing (RTV) silastic material. It did seem to keep things at a constant temperature, and if there was much moisture, the RTV did seem to help keep it off the boards. Still, on some moist days, they could not get the P2s to pass a 100 pA final test, so they would just set them aside and wait for drier weather. When we built the P2A, we did not use RTV, because at +85 °C, the RTV would expand and *pop* the P2 right out of its case. We just used several heavy coats of Humiseal, and that gave very good results. I don't think moisture gave us much problem on the P2A.

According to some of Bob's friends, Bob said that he could tell when the women assembling the P2s were menstruating. He thought it was the amount of sweat that would cause corrosion or leaky printed-circuit boards. He could check out failure rates versus serial numbers versus the initials of the assembler, and the yield would go up and down every 28 days. I know I was impressed that there were always two inspectors, inspecting the PC boards after they were hand soldered. They could spot badly soldered joints and cold-soldered joints, and mark them with a red pen, to go back and get resoldered and touched up, because a P2 would sometimes run really badly, noisy and flaky, if there were cold-soldered joints on the board.

To this day, I still have the dismantled carcasses of a few P2s. That is because Bob Malter decreed that if you could not get a P2 to meet spec, after you had tried everything, the technicians would pull off the valuable components—the trim-pot and the piston capacitor—for re-use. Then the transistors and transformers would be removed, so that even a competitor who wanted to raid our trash cans would not learn anything much. And in retrospect, well, Bob had a lot of good hunches, and he probably had a good hunch in this respect also.

Of course, if you wait long enough, any good thing can become obsolete. As of 1989, you could buy low-leakage amplifiers such as the NSC LMC660, with input currents normally less than 0.004 pA, for about 50 cents per amplifier ($2.00 for a quad). But what do you expect after a 30-year wait?

Notes on George Philbrick's P7 Circuit

1. The AC amplifiers are all supplied through a single 4.7 kΩ (?!) resistor! George wanted to run all 4 AC amplifier stages on barely 1 mA total! In the P2, Bob Malter was willing to spend 4 mA. I could never understand why George was so unwilling to spend just 15 mW for the entire four-stage AC amplifier when he spent 30 mW to bias up the output stage (27 k on Module 6). Maybe if he fed any more current through those AC amplifiers, they would break into song and oscillate hopelessly?! Because in the layout, the output of the fourth stage is right next to the input of the first AC stage?! Those of you who are not chess players might like to know that in the notation of chess, "?!" signifies a *blunder*.
2. Likewise, the P7 oscillator was intended to run on less than 0.3 mA (?!), whereas the P2 spent 1.5 mA. I checked the actual P7 circuit to see if these values represented a typo error—but they didn't.
3. The P7 I have uses two AC amplifier modules—four stages of transistor—but the arrangement of the upper and lower (mother and father?) boards left room for three amplifier modules—six stages of AC gain. You can see the gap in the middle of the assembled unit, where another two stages could have fitted in. But if you had six stages, then the possibility of oscillation would become hopelessly bad. No wonder George backed up to go with just four stages.
4. The germanium 1N100 diodes in series with the inputs are intended to act as low-impedances near null but to act as current-limiters (just a few microamperes) when overdriven. That is what George intended—a neat concept. But in actuality, I bet they made bad errors when they rectified any ambient noise, and I bet they had *awful* leakages when operated in ambient light. Furthermore, if you ever got around to running this amplifier with feedback, you would find they add a lot of phase shift. At room temperature, they would cause a lag of perhaps 25 kΩ and 600 pF, or a 10 kHz roll-off. If you get it cold, the break at +5°C would be at 2.5 kHz. This confirms my suspicions that George never really got the DC operation working okay, so he never even *began* to think seriously about AC response. The circuit shows no evidence of a big filter around the output stage to give 6 dB-per-octave rolloff.

Comments on Bob Malter's P2 Circuit

1. Obviously, 30 years ago this was a high-level industrial secret. But as I mentioned, even if I gave you the schematic, that would not help you make a P2 that works. Since the P2 has been out of production for more than 20 years, this is more like industrial archaeology than espionage.
2. The doubled capacitors (several places where you see [7.5 pF in parallel with 10 pF]) were arranged so that the test technicians could do some coarse trims by snipping out one or another of the caps. Much judgment and experience were needed. There were many other places where discretionary trim resistors and

capacitors could be added. To improve the temperature coefficent of V_{OS}, for example, you could install a thermistor from the wiper of the 50 k pot over to one side *or* the other.

3. Unlike the P7, the P2 had a lot of AC roll-off, provided by the 15 millihenry inductor and the 0.47 μF capacitor. It rolled the DC gain off at a steep 10 dB-per-octave rate down to about 15 kcps and then there was a lead (selected resistor in series with the 0.47 μF) so it could cross over at a unity gain frequency of about 75 kcps at about 6 or 7 dB per octave. The frequency response was trimmed and fitted on each individual unit.

However, it is fair to note that the roll-off did not use any Miller integrator around the output transistor. Consequently, the high-frequency open-loop output impedance of the P2 was not a whole lot lower than 3 kΩ. If you combine that statement with the fact that the P2's input capacitance is just about 600 pF, you can see that the output impedance, just trying to drive the input capacitance, gives you an extra phase shift of about 40 degrees. No wonder each unit was hand-fitted for response!

4. The demodulator (Q6) would put out a voltage right near the +15-V bus if you did not feed in any amplitude from the AC amplifiers, and then the DC transistor (Q7) would not turn on. To get the output transistor on, you had to have a minimum amount (perhaps 400 mV p-p) of 5 Mcps signal coming through. And it was the interaction of that signal that talked back from one board to the other and let the gain come "into mode." Look at the coupling capacitor from the fourth AC amplifier into the demodulator! The P7 had a reasonable value—500 pF. But Bob Malter found something magic about the 7.5 pF, probably because it was the right way to get the amplifier into "mode." Surely, Bob Malter was the embodiment of "The Lightning Empiricist."

Comments on Rustrak data

I set up a P2—Jim Williams loaned me his old P2—at a gain of 20. I followed this with a gain of 200 (or 100 or 400) to get the offset voltage's drift up to a decent level, and fed it through 10 kΩ into an old 1–mA Rustrak meter—the kind that goes

Figure 9-5.
A Rustrak strip recorder tracking the offset drift of a P2.

B. Pease

✱ 600 μμf, differential, <u>NOT</u> Z-M.

Figure 9-6.
The original data sheet from Philbrick describing the P2.

77

"—tick—tick—tick—." After some warming up and some clearing away the cobwebs, this P2 began giving pretty good stability. In some hours when the temperature wasn't changing much, it would hold 20 or 60 μV p-p—not bad for a unit with perhaps 200 μV/°C. Also not too bad, considering that the trim pot had an end-to-end range of 100 mV, so that asking it to hold 100 μV—the equivalent of 0.1% of span—was about as optimistic a task as anybody ever demanded of a carbon pot. But sometimes it did a lot better than that.

However, the offset kept drifting to the left. Could it be some kind of chemical interaction, where the RTV is changing slightly after all these years of inactivity? After all, the P2 really is *not* a well-balanced circuit. Maybe the drift rate will slow down if I do some warm-temperature burn-in?!

You never can tell. . . .

10. Propagation of the Race (of Analog Circuit Designers)

This book presents the wisdom, tricks, and philosophies of an impressive collection of analog circuit designers. While I consider myself an engineer, I spend about half my professional time teaching. ("He who can, does; he who cannot, teaches"— George Bernard Shaw.) M.I.T. has given me the opportunity to think about teaching design and to try various approaches on generations of bright, receptive, and motivated students.

Obvious questions surface. How does the race of circuit designers propagate? What characteristics separate the good from the average designers? Can the necessary characteristics be taught in any environment? Can the teaching be effectively accomplished at a university?

While these questions are hard to answer in general, certain patterns emerge. Many designers mention one or two mentors with whom they interned intensively and who had a major impact on their careers. Designers often are more receptive than their analytically inclined colleagues to accept physically plausible arguments without proof. Pragmatism, combined with at least occasional unstructured thinking, facilitates, and possibly enables, the design process.

The abilities required for effective design, while hard to quantify, are common to all disciplines. I believe that a good analog circuit designer could also become a good designer of airplane wings or steam turbines after a relatively short internship in the new field. (It may be fortunate for frequent flyers that this hypothesis is infrequently tested.)

These observations suggest some of the difficulties that are encountered teaching design in an academic setting. The usual mode of teaching is via relatively large classes that preclude much one-to-one interaction. Even in the case of research or thesis supervision, interaction is usually limited to a few hours a week at most, thus precluding the type of mentor relationship that can evolve in other environments.

Classroom education often involves presentations more structured and analytic than those required for design. Many faculty prefer to write a fundamental relationship on the upper-left-hand corner of the blackboard at the start of the hour, and conclude a precise and mathematically detailed development as the end of the hour and blackboard are reached simultaneously.

Also, the art of design, regardless of context, never seems to appear on the ever changing list of academically "hot" topics. Consequently, junior faculty members who practice and teach good design are frequently bypassed when promotion and tenure decisions are made. This reality certainly influences the choice of research area for many potential faculty members.

In spite of these difficulties, a significant fraction of the graduates of many engineering programs become good design engineers. The remainder of this chapter focuses on a few of the ways this educational process is aided at M.I.T.

Subjects that provide the background essential for design are offered by all of our departments. An excellent example is the widely acclaimed Introduction to Design architected by Professor Woodie Flowers of the department of Mechanical Engineering. The culmination of this subject is a spirited contest that finds which student-designed machine best accomplishes a specific task. Although Introduction to Design and many other M.I.T. courses would provide interesting examples of approaches to teaching design skills, I will limit this discussion to those subjects with which I am involved and which I have taught several times. The subjects described have evolved to their current form and content via the contributions, suggestions, teaching, and inspiration of many colleagues, including Professors Hae-Seung Lee, Leonard A. Gould, Winston R. Markey, and Campbell L. Searle, and Drs. Chathan M. Cooke, Thomas H. Lee, and F. Williams Sarles, Jr.

Engineering education at M.I.T. and elsewhere started a fundamental change in the 1950s because of the pioneering effort led by Professor Gordon S. Brown, then head of M.I.T.'s department of Electrical Engineering. Prior to that time, engineering education was generally quite specific, with options that channeled the student into a narrow area early in his or her educational process. However, the technological explosion that followed the second world war made it impossible to predict areas likely to be of interest even at graduation, much less a few years later. The approach that evolved from this dilemma was to provide an education broadly based in mathematics and physics. Regardless of the opportunities available to the graduate, the truths discovered by Fourier, Maxwell, and Laplace would be essential. The new engineer would have a background that permitted easy assimilation of the specifics of any particular area, including design.

A further justification for this approach is that analytic skills are the ones most difficult to acquire through self-study, with the discipline and structure provided by the classroom almost required for success. Few high school students study vector calculus on their own because of the joy it provides. Conversely, hobbies or acquired interests often lead students to "thing"-oriented pursuits such as circuit or computer hacking long before they get to college.

It is impossible to argue with the general success that this approach to education has enjoyed. However, the potential negative impact on the propagation of the race of designers comes when a student spends 4 or 6 or 8 years (depending on the final degree obtained) in an academic program devoid of hardware and design experience. While this student *could* become an innovative and productive design engineer with a very short internship in a specific area, he or she may not *want* to. This bias is particularly likely when none of the academic role models practice design.

The three subjects to be described provide a degree of balance by exploring design-oriented specifics and philosophies. It really doesn't matter if an occasional specific is obsolete when the student leaves M.I.T. The generalized background acquired from other subjects allows easy adaptation for the student whose career has been motivated in this direction. The subjects are electives and thus acquire their enrollment only because of residual interest from earlier, often pre-M.I.T., experiences or because of a favorable student grapevine.

These subjects share a number of features. All have an associated laboratory, with students averaging approximately 2 hours per week in this endeavor. While the details of the laboratory vary depending on the subject, all reflect our belief that it is essential to attempt actual design in order to learn how to do it. None demands much literary effort in the final write up (not because we don't think this aspect is important—but we prefer to exercise other skills in the limited time available). All laboratory exercises require close interaction between the student and a teaching assistant and include an interview as an important component of performance evaluation. This approach ensures that the student's time in the laboratory is spent productively. It also discourages an occasional student from "borrowing" results from a similar topic assigned in a previous year.

The subjects all depend heavily on classroom demonstrations, performed real time, to illustrate concepts as they are introduced. These demonstrations make it very clear that the material is applicable to practical systems. There is also real educational benefit in the (fortunately rare) event that a real-time demonstration fails. The students realize that they are in good company when one of their experimental attempts fails.

Teaching assistants are an important part of the instructional team in all of these courses. Fortunately, assignment to one of these courses is viewed highly by graduate students interested in design, and we always have our choice of very talented and industrious applicants. In addition to their other responsibilities, these teaching assistants have developed many of the demonstrations that we use.

Two of the courses are undergraduate level. These courses are typically taken by juniors and seniors and usually have an enrollment of 50 or more. This group meets for two 1-hour lectures a week. These lectures facilitate the introduction of new material, particularly when accompanied by demonstrations, but the size tends to discourage teacher–student interaction. The students also meet for two 1-hour recitation sections a week in groups of 25 or fewer. These sections are generally taught by faculty, although occasionally graduate students who have demonstrated both extreme familiarity with the material and excellent teaching ability teach them. New material is often introduced in recitation sections, but the format permits more interaction and question–answer type teaching.

The graduate course typically has an enrollment of 12 to 15 graduate students plus a few very gifted undergraduates. It meets an average of 4 hours a week.

One of the undergraduate courses focuses on active-circuit design. In addition to the usual introductory electrical engineering subjects, prerequisites include a moderate amount of circuit discussion. Thus students enter the active-circuit subject with a good background in semiconductor device operation, facility with models that include dependent sources, and a basic appreciation of a number of circuit topologies.

A major theme that unifies much of this coursework involves the design of both linear and switching circuits for specified dynamic performance. Emphasis is given to techniques that can be used to estimate performance while retaining insight and providing guidance for improving operation. Thus, for example, numerical methods that basically provide a "binary" answer as to whether design objectives have been met are only used as an adjunct to methods that provide greater design guidance.

The analyses of linear amplifiers is introduced with a review of the common-emitter amplifier, and its dynamics are estimated via the Miller-capacitance approximation. The development leads to the introduction of the method of open-circuit time constants. This technique is used to estimate the dynamic performance of more complex topologies and to provide design insight. Such issues as the conditions under which f_T is closely correlated with performance and the maximum bandwidth that can be achieved (assuming an unlimited number of devices are available) subject to specified constraints are explored.

This portion of the subject culminates with the students conducting a multipart design exercise. They are given specifications such as "design an amplifier with a voltage gain of 250 and a bandwidth of 5 MHz." Additional parameters such as source and output resistance and dynamic range are specified. The use of fairly docile device types and restricted supply power consumption is also specified.

These latter constraints are included primarily to reduce difficulty with spurious oscillations during the breadboard phase of the project. Most students just don't have the experience necessary to effectively use ground plane and ferrite beads in the alloted time!

The exact specifications change from year to year, but we try to maintain a constant degree of difficulty. Intentionally missing from all specifications is anything that suggests a circuit topology or limits the total number of devices used.

Students are expected to complete the following multifaceted solution to this problem.

1. Guided by anything you know, find a topology and associated component values that you think will meet the specifications.
2. While we assume that the above will be determined in part by the estimation methods suggested, show us why you believe your design will meet specs.
3. Now simulate the circuit and see if the computer confirms your optimism. (If not, decide why the simulation is wrong or redo your design.)
4. Build and test the circuit. If there are problems, iterate. (This phase is usually accomplished in protoboard form. Laboratory handouts and hints from the staff have suggested efficient layout and stressed how to include parasitics in earlier steps.)
5. Talk with your teaching assistant about the above, and convince him or her that you have done a good job.

The important difference between this assignment and many of the students' earlier experiences involves the quantity of good answers. For many students, all problems they have been given earlier have only one correct answer. (Example: What is the integral of e^x? Not too much choice on this one!) Suddenly this uniqueness disintegrates. A common characteristic of design problems is that there are an infinite number of solutions to all problems; some of these work; some work much better than others.

The reaction of students to this situation is interesting. (I feel I have enough experience to justify the following anecdotal observations.) A few students who have easily jumped through all the academic hoops previously presented to them are very uncomfortable in this situation. The subset of this group that doesn't adapt as the term progresses drops this subject and presumably lives happily ever after doing something else. Ignoring the large group in between, another fraction of the students love this sort of thing. These folks may become our kind of people!

We evaluate with only course quantization. It works or it doesn't work, with little gradation in between. This approach is appropriate for the first real design experience of the group. However, it is clear that finer value judgments are possible. One discriminator, since the designs must eventually be built by their designers, is the number of transistors used. Most designs require four to six devices. Some require more—good if the resultant performance far exceeds specifications and not so good otherwise. A few students usually design and successfully implement three-transistor solutions.

The year we assigned the gain and bandwith specifications mentioned above, a copy of the assignment found its way to Bob Pease in Silicon Valley. I'm not sure how this happened, but I suspect Jim Williams may have been involved. Bob submitted a design that met specs using two transistors. (He would have gotten a very good grade had he been taking the course.) His basic trick was to use positive feedback to reduce the effective input capacitance.

Bob's performance has become a benchmark. The teaching assistants who select the specifications each year make sure that they can meet them with two transistors and present their design to the class after the due date.

As mentioned earlier, we also look at the dynamics of switching circuits in this subject and use change-control methods for estimate in this case.

In addition to the material on circuit dynamics, which represents more than half the content and which is included every time the course is offered, we select several other topics from a menu that includes DC amplification, high-voltage-gain stages using dynamic loads, band-gap circuits, translinear circuits, noise considerations, and power handling stages. The exact mix changes from year to year, reflecting our belief that the specific choice of examples doesn't matter; the important feature is that design, rather than analysis, methods are used.

The use of real-time demonstrations to illustrate many ideas was mentioned earlier. We also use many examples drawn from available integrated circuits. For example, open-circuit time constants can be used to show that the collector-to-base capacitance of the transistors used in the 733 (an earlier linear integrated circuit) must be less than 0.1 pF. Similarly, charge-control can be used to justify the inclusion of the "Miller-killer" portion of FAST logic.

The second undergraduate course is one in classical feedback. I include a discussion of this subject in a book on analog circuit design because I believe that a thorough understanding of this topic is the single most important prerequisite for the effective design of many analog circuits. I occasionally encounter designers who know so little about feedback that they should be prohibited legally from using it. The areas in which these individuals can do effective circuit design are quite limited.

The prerequisite for this course is a good understanding of basic linear-systems ideas. Students should have a reasonable appreciation of the importance of poles and zeros, and be able to sketch Bode plots. We do not require an in-depth understanding of Laplace stuff such as partial fraction expansions and contour integration for taking inverse transforms, although many of the participants have completed that part of the EE core program. There is no prerequisite requirement linking this and the active-circuits subject, although most students feel that each provides excellent background for the other.

The emphasis for many years was on the electronic feedback systems, using primarily operational-amplifier configurations as examples. However, the discussion was at the block-diagram level, as opposed to developing the tranfer functions of the blocks from the innards of a particular amplifier. This approach was used so that systems types and computer scientists from our department, as well as students in other disciplines, could take the course without first acquiring a circuit background.

Topics covered included:

- Modeling and block-diagram representation.
- Approximating responses. Under what condition can the transient response of a system be approximated as an appropriately chosen first- or second-order system? (Answer: Almost always.)
- Stability analysis via root-locus and Nyquist diagrams.
- Analysis and design of nonlinear systems via linearization and describing functions. (Describing functions is an excellent tool when one actually *wants* to design an oscillator.)
- Compensation.

The associated laboratories were design oriented and used appropriately configured operational amplifiers as the vehicle. However, there is a difficulty associated with using wideband amplifiers that limits their educational effectiveness. One of the very satisfying features of the approximate methods that can be used to predict feedback system performance is that they yield remarkably good results with minimal effort *if the system model is accurately known.* Unfortunately, strays complicate the development of models for wideband systems. Students tend to blame their inability to predict performance accurately on the approximations inherent to the methods we suggest rather than their choice of a poor model.

We reduce the chances for this self-delusion by having this student use a pseudo op amp that has been tamed by a combination of external compensation (an LM301A is used as the building block) and a two-pole low-pass filter connected to its output. The resultant pseudo-amplifier has a highly predictable transfer function that has a unity-gain frequency low enough so that strays can safely be ignored and also has negative phase margin at its unity-gain frequency. The students design compensators for various configurations using the pseudo-amplifier and verify performance.

There were several other experimental vehicles used in this version of the feedback subject. All were rather carefully selected (and possibly tweaked) so that the students could develop accurate models for them in a reasonable period of time. They were then able to experience the positive reinforcement that resulted when their performance estimates were confirmed experimentally. It is our hope that this experience will encourage them to spend the time necessary to develop accurate models when they encounter more complex systems.

Classical feedback is taught in at least four different departments at M.I.T. Last year we decided to modify the course described above so that it might be taught to a group of students from several different disciplines. We have taught the new course once to a population about equally divided between the department of Electrical Engineering and Computer Science and the department of Aeronautics and Astronautics.

The topics covered in this joint offering are the same as described earlier for the original subject. The differences come in the examples, demonstrations, and laboratory exercises. We frequently show how identical design and analysis methods can be used in quite different systems. For example, we model a velocity servomechanism where the motor dynamics are dominated by its mechanical time constant and also model a noninverting amplifier using an operational amplifier with a single-pole open-loop transfer function. The resultant block diagrams and transfer functions are identical except for bandwith-related parameters.

The joint offering provides an excellent vehicle for expanding the horizons of both groups of students. I feel that it is particularly important for electrical engineering students to learn how to model other than electrical systems, and they seem quite willing to do this in the joint format.

We have introduced demonstrations that appeal to both groups. Additions to our collection of EE-oriented demos include a magnetic suspension system and an inverted pendulum. We are working on one that stabilizes two different-length inverted pendulums on a single platform. (The analysis of this one is delightful. It is possible to show that the maximum achievable phase margin for this system is $\sin^{-1}[(R - 1)/(R + 1)]$, where R is the ratio of the natural frequency of the shorter pendulum to that of the longer one. This result confirms the intuitive realization that the task is not possible for equal-length pendulums.)

The differences in backgrounds of the two student groups convinced us that a major modification of the laboratory was necessary. We have three different experimental systems in various stages of development. One of these is a thermal control system that maintains temperature stability to better than 1 millidegree C. The system raises the temperature of the controlled surface about 25 °C above ambient by means of a resistance heater. The feedback signal is developed by a thermistor, and thermal dynamics are dominated by a 0.1 inch thick aluminum spreader plate separating the heater from the controlled surface. The disturbance rejection of the system depends on its isolation from ambient temperature variations and its loop parameters.

Another experimental setup allows students to design several different types of servomechanisms. The mechanical portion of this system consists of a DC motor with an integral tachometer geared to a potentiometer used for position feedback. Additional inertia can be attached to the motor shaft. The electronics is designed so that a velocity or a position loop can be easily implemented, using either forward-path or feedback compensation. A wide range of compensation parameters can be selected via potentiometers and plug-in components to reduce assembly anxiety for non-EE students.

The third setup consists of a lightweight "cartoon" of an airplane. The elevator angle and the pitch angle of the aircraft are driven by positioning servomechanisms (actually the type used for model plane control). Several plug-in analog-computer–type boards simulate the pitch dynamics of different aircraft, including one that is unstable in pitch. (The mock-up is, of course, only used to give a visual indication of the response to various commands applied via a joystick.) The object here is to design an autopilot that simplifies the task of "flying" the "airplane."

The basic approach, using any of the experimental setups, is to first characterize it using appropriate measurements. For example, the dynamics of the thermal system are described by a diffusion equation and thus cannot be accurately represented by a small number of poles and zeros. Its transfer function is measured over the frequency range of interest using a Hewlett-Packard 3562A dynamic signal analyzer. Alternatively, the servomechanism can be accurately modeled after important parameters have been experimentally determined.

After characterization of the fixed elements, closed-loop performance is predicted and measured for several configurations. Finally, compensators that meet specified closed-loop objectives are designed and tested.

The laboratory work is structured as a sequence of short weekly assignments that closely parallel and reinforce classroom presentations. As in the case of the active-circuits subject, evaluation of laboratory performance is based in large part on the results of a student-teaching assistant interview.

You may wonder why I spend so much time describing a course that is basically one on classical servomechanisms in a book for analog circuit designers. I remind you of my belief that this general material is the most important single topic a circuit designer can know. It is easier to teach this material using relatively slow systems than high speed electronic ones, because the slower systems are easier to model accurately. Once the basic ideas are well understood in a servomechanism context, they are readily transferred to purely electronic systems. Finally, servomechanisms are fun to work with. (Consider the two-pendulum problem, for example.)

The graduate course, which is taught every other year, has a "family and friends" type enrollment, since I generally require that participants have taken both of the undergraduate courses and done well in them. I occasionally will waive the pre-requisites, such as in the case of a person who has had extensive experiences as a practicing circuit designer. Some of my colleagues feel that this requirement unfairly discriminates against M.I.T. graduate students who did their undergraduate work elsewhere. If this is the case, at least I came by my prejudices naturally, since I am completely inbred at M.I.T. (In actuality, most new graduate students who are interested in design opt to take the undergraduate subjects early in their graduate program.) In this way we ensure a small, very competent and enthusiastic group of students. The course enjoys essentially perfect attendance, and virtually all participants earn A's.

The subject is a joy to teach. We discuss (and "we discuss" is really a better description than "I lecture") how designers have accomplished some function during two 1½ hour sessions a week for several weeks. The specific topics vary from term to term, but the selection generally includes sample-and-holds, digital-to-analog converters, and analog-to-digital converters. There is no serious attempt at theoretical rigor in any part of this; it's not necessary since we have right on our side!

We then hand out an assignment that is effectively a spec sheet and ask the students to conduct a detailed paper design of a circuit that they think meets the specifications. There are usually two or more sets of specifications offered for each topic, often a high-accuracy set and a high-speed set, and the students may chose either. In keeping with the spirit of the subject, they may work to their own set of specifications as long as they are of comparable difficulty. The design can be either IC or discrete, and they have about two weeks to complete it, during which time a new topic is being discussed in class. Either I or the teaching assistant (who is always the most senior one working in the sequence) read each design and discuss it with its author. Because of differences in the backgrounds of the participants, ranging from seniors and early graduate students to practicing design engineers, no absolute scale is used for evaluation. We reserve any severe criticism for errors that a particular student should know enough not to make. If our students disagree with any negative comments we may make about their circuit, they can always prove us wrong by building it!

It is interesting to compare the approaches of various students. Many directly adapt some topology we have discussed. Considering the difficulty of invention, this is a fine approach if the details are filled in correctly. I'm sure most of us do much of our design by combining topologies we have seen before rather than via completely original configurations.

An occasional student will try what to him or her is a completely new approach. For example, one student designed an incredibly complex circuit using an inductor as the memory element in a sample and hold. Since he did not use superconductors, extraordinary means were necessary to achieve the required self-time constant. I think it might even have worked. He chose the approach not because of naiveté, but just to prove he could do it. Needless to say, he was the teaching assistant the next time the course was offered!

The first time I offered the course as described above, about 20 years ago, several participants mentioned that the format resulted in a rather "lumpy" work load, with a major effort required preceding each assignment due date. After some consideration, we decided that the best way to remedy this situation was not by leveling the peaks, but rather by filling in the valleys. These fillers are not directly correlated with the topic being covered in class, but do offer a way of expanding coverage to include other important material.

There is, of course, an associated laboratory. We generally do not ask students to build their designs because of the time commitment that would be required. However, we may suggest building a portion of it. For example, if a student chooses the high-speed design for the digital-to-analog converter, we expect him or her to breadboard the most and least significant bits and demonstrate setting time. This is a very worthwhile exercise for students who have limited experience with ground plane and "settle-box" circuits.

In another lab exercise, we give the students a commercially available integrated circuit (I won't divulge the type and manufacturer, but many would qualify) and ask them to find at least six lies in the data sheet.

We also hand out homework problems on a regular basis. Most of these problems were developed by the generations of teaching assistants who have been associated with the course, and generally cover more advanced active-circuit and feedback concepts than are covered in the undergraduate courses.

The teaching assistant also meets with the group for 1 to 1½ hours a week. Some of the topics discussed are related to the design problems. For example, during the discussion of sample-and-holds, emitter-follower and buffer-amplifier oscillations are discussed. (Why does a series base resistor, or an input resistor on a buffer like the LM110, work?) At times, the teaching assistant gives several talks in an area of particular expertise, possibly leading to a shorter design problem.

As you gather from this outline, the overall workload in the graduate course is awesome and probably continues to increase with time as additional teaching assistants make their contributions to the package. Since we always get enough eager students who do everything we ask of them, we don't plan to ease up!

The true and enduring joy of teaching, of course, comes from the interactions we have with our students. I have had the privilege of working with many outstanding students. I have had the further pleasure of keeping in reasonably frequent contact (occasionally professionally and often socially) with many of them after they left M.I.T.

You may have sensed by now that I feel the academic endeavors outlined above have contributed in an important way to these students' development. There is an implication that "this is the only way to do it." This feeling of omnipotence is shared by some members of professions other than teaching; I have seen analog circuit designers, CEOs, physicians, and investment counsellors, to name a few, who exhibit this failing. There may even be one or two other examples in this book.

The unbiased observer notes an inherent contradiction by observing that many of the roads to the promised land suggested by practitioners in any one area are orthogonal. In humbler moments, we in education must similarly realize that the impact we can have on our students is quite limited. Many are so remarkably talented that they will be very successful regardless of what we teach them! However, we may be able to influence their professional directions through the interest and enthusiasm we display.

11. The Process of Analog Design

••

I'm not going to draw schematics for you. They have been done and, in terms of doing or designing the next circuit, are not very interesting. I'm not going to expound on all the pitfalls that are out there—I trust that they are addressed by others. This is going to be about, if you will, the philosophy of design.

Design—A Process

The process is my focus. Not the wafer-fabrication-process (which, for an integrated circuit designer, is just as important as the engine in the race car is to the driver), but the process of "designing"—how one goes about filling the emptiness with a new and, it is hoped, useful something. The "something" can be an integrated circuit, a methodology, a machine, a process for putting up wallpaper, or whatever. The "something" is not the focus here—the "creating" of something that didn't exist before is.

The Quest

The reasons that you have for starting into a design are excuses for allowing yourself to do it. They are many and varied. In almost every case you will wind up driving yourself or feeling that the "design" is driving you, or both. It is, in some sense, a "quest."

There are many possible ways to come up with a "quest." Deliberately seeking to take advantage of a recent breakthrough or significant work (yours or not, in your field or not) is an obvious possibility. Maybe you simply find yourself inspired. Whatever the impetus, there are some things we think we know in the beginning and a "goal." The question then is "now what?"

Not Much Known

Much is considered to be known and very little actually is. When one puts one's foot on the path that, one hopes, leads to creating something new, the worst thing in the world is to "come from the place" (have the mind set) where everything is known already and all things have already been done. This is the exact opposite of what is needed—seeing the world as if for the first time.

Chaos

Chaos! Confusion! That is the cauldron that you must hurl yourself into. You may have a goal, however vague, and some resources (i.e., probably the means to implement the resultant plan). You'll probably also have some preconcieved notions about what you want to do and how to do it. I won't fault you for having these (although there are those who say that even having any preconceptions is harmful), but keep them well into the background in the beginning. You can be sure that you will be dragging them out often enough later to compare to your predictions or experiences with your "creation." Just dive in first. You have to get some movement and fermentation going.

Getting Rid of Excess Baggage

A good way to start is to look at the goal—but not necessarily the one you think you have. Your original goal is often cluttered with all the assumptions about "what can't be done" that you have chained to it or have allowed others to do so. Try starting with a clean sheet of paper. If there were no knowns, if there were no limits, what would be your goal? "But!" you say? No "buts" for now! Put them all aside— you can retrieve them later if necessary. Take some time with this step. After all, why not end up with the thing you would really like to have as opposed to something that is merely in the ballpark? The effort you will have to expend is probably not much greater and, in fact, can be considerably less as a result of having a clear goal in mind. Tell yourself "this is what I want the result to be" and be specific! Start with the big picture and qualify. Then qualify and qualify some more.

Analyze Later

Analysis—to use a tool or set of tools to predict or qualify systems—is important, as they taught us in school. There will be considerable opportunity during the design process to demonstrate that we can analyze. First, however, we need to generate something to analyze. My experience is that it is exceedingly difficult (read impossible) to be in generation and analysis mode *at the same time*. So here in the "generation of ideas" phase, let the ideas have a chance to form (if not blossom) before subjecting them to rigorous scrutiny—yours or someone else's

Don't Forget the Fixed Overhead

Fixed overhead effort is something we usually fail to take into account. If the difference in design effort is, say, 2:1 to design a really good part (we won't say great because it is considered impolite) we tend to think that is too high a price to pay. But if the design effort is only 20% of the total effort required to get the job done, then the "really good part" costs only an extra 10%. Is that worth it? We all get to answer that question for ourselves. I say yes.

Forget the Window Dressing—For Now

When you are trying to get a new idea down, don't worry about the window dressing, e.g. the grammar, the neatness, drawing within the lines. There is plenty of time for that later. It gets in the way and can impede or stop the whole process. It may even be that someone else winds up doing the "window dressing."

Synergy!

Synergy works. Get a group of people together and brainstorm, or generate ideas. The people you choose should be bright, eager, interested, and open people, but if they have even one of these attributes it will be helpful. You may find that you want to get past the idea generation phase to pull the grizzled veterans (assuming there are any) into the act—they will be much more useful in helping you check to see there wasn't something you didn't take into account when you think you have a complete formulation. Seclude your group so it can focus on generating ideas. Before you begin "the generating ideas session" with these people, make sure they all agree to the ground rules. Failure to do so will probably result in a waste of time.

Be Careful; Ideas are *Fragile!*

Making other people's decisions for them is strictly forbidden in these sessions, especially for people who can "make it possible" by doing something additional or different from what they have done before. After the session, and after you have scrutinized the approach and determined that it has a chance to fly, check with those people to see if there is a way to implement that which is needed. Yes, some of the possibilities may be unorthodox, but orthodox has already been done. "It hasn't been done that way before" is not an acceptable phrase during this time. Leave at the door all criticisms, oblique as well as candid. Encourage everyone to leapfrog, to use the other ideas as "springboards"—to hear the possibilities in what is being discussed as opposed to merely what the speaker, or the rest of the group, had in mind.

Make the Proposer Explain It

Now I can hear some of you saying "I don't need any help. I can do it myself." Maybe you already have something in mind—great. I can't tell you how many times I have witnessed or been part of somebody "explaining" how their new *what-ya-ma-call-it* worked to great benefit when, in the form it was in, it didn't work! If there was a problem, they were told about it and often solved it on the spot. At the very least, they clarified in their own mind what it was and how it worked and what the implications were by forcing themselves to "explain" it.

Now Tear It Apart

When you have finished "the session" it is time to expose the ideas to the harshest scrutiny. Now you can unleash all those analytical skills without mercy. Do a thorough job of it, and once you are convinced, call in those grizzled veterans to put it through the wringer. On the other hand, do not throw out ideas because of popular conceptions about things or processes that you have not checked out for yourself or had confirmed by those you hold in regard in that arena. You will be surprised at how often you will find that popular conceptions have no connection with reality.

Back to the Drawing Board

You may have to loop through the "generation phase" a few times, each time narrowing the piece of the universe in which you are looking as a result of what you learned previously. Do not fail to ask yourself at the end of each evaluation "what have I learned?" You will sometimes be amazed at what you have learned but not made yourself consciously aware of. Take advantage of that learning. You have put the effort in—harvest the results. That includes being willing to throw your favorite possibility out as a result of seeing yet a better one. Make yourself look for that possibility often.

Let Others Contribute!

What if you need somebody else, some other contributor, in the development or production process to do some things out of the ordinary—or at least what you "perceive" is out of the ordinary? You may find they are more than willing. What they need to do to "make it happen" may not be any big deal to them. Besides they get the chance to be involved. It goes without saying that you should tell them what the significance of their contribution will be.

Change? *Aaaargh!*

Change is hard for people, including you. You may find there is a bias in you as well as others to "throw the new thing out" at the slightest sign of its not working. "Kill it" may be the predisposed response—never mind looking to see if there really is a legitimate problem, let alone one that is caused by the new element itself. You have to be on your guard when it comes to "problems" that come up in the development. Make sure you investigate all the possibilities before concluding that the problem is a result of the new approach, let alone that it won't work.

Undoubtedly you will find yourself being intense in this process. When you notice that you have been that way for a while (usually when you find you are spinning your wheels), take a break. Do some other work, play, or whatever, as long as it's a different type of activity—let the conscious mind address other things, or nothing. It is possible that when you come back to it you will see new avenues or maybe even the solution that had been avoiding you (or the other way around).

If Not You, Who?

At some point you (or someone else) is going to declare that it is time to implement. That doesn't mean you are through! Even if other people are carrying out parts of the implementation, it doesn't mean that you can ignore them or their contribution. You are the one who has to make sure that what comes out lives up to the goal—lives up to the vision. Who else knows what that is? Does anybody know it better than you? No! It is your responsibility.

Eagle's View

Occasionally, throughout the whole design process, you are going to have to pull far enough away from the project to get some perspective on it—the eagle's view if you will. You will need to this especially after any big turns or leaps. The place from which you view the world needs to change frequently during the design process. Sometimes you need perspective, while at other times you need to be so involved with a single piece or concept that nothing else exists.

Have Fun!

Designing can be a lot of fun. It has been for me. I hope it is for you. Having fun will make your work much better and possibly result in other benefits as well.

Milton Wilcox

12. Analog Design Discipline

A Tale of Three Diodes

I would like to use a true story to illustrate what I believe is the fundamental necessity for success in analog circuit design: attention to detail. I doubt that analog design is any different in this respect from any other field of intricate endeavor— be it digital circuit design or internal combustion engine design or violin design. Beyond the need for an understanding of basic laws governing solid state circuit operation, what analog absolutely demands is meticulous attention to detail.

Analog design is about taking the time to anticipate all the possible consequences of a circuit approach, and about following up *every* quirk or anomaly you might notice while evaluating a breadboard or running a simulation. It is about devising different ways to test for the same result, and about devising test conditions that might be considered out of bounds for the circuit function, because somewhere, someday you know a customer is going to. It is about knowing more than anyone else in the world about your circuit. When a circuit is used because it was written up in a design magazine, or because it is almost like one which worked the last time, or because it is a last minute change to meet schedule, disaster is invited. This applies to analog circuits large and small, from complex to very simple. This applies to the string of three diodes in my story.

My story takes place during a brief stint at a small, aggressive, and very naive company trying to break into the analog integrated circuit business. There I experienced firsthand the consequences of not adhering to this analog design discipline of attention to detail. The three diodes were in the thermal detector circuit of a neat little chip designed to turn discrete power MOSFETs on and off. The project had been started some 18 months previously with guidance from a major power FET manufacturer. Full of confidence and with the urgency of management (who were already counting the revenues), the fledgling design team had set out to create a chip in as short a time as possible. By the time I arrived, they were on their second complete mask set and third designer. Finally, when there was no one else left to work on it, I inherited the job of cleaning up the chip for release to production.

My example circuit shown in Figure 12-1 is a fairly straightforward arrangement to achieve thermal sensing on an integrated circuit. The voltage across three series-connected diodes is compared to that of a temperature independent reference voltage. When the temperature of the diodes rises to approximately 150 °C, their voltage drops below that of the reference, and the output of the comparator signals overtemperature. In the FET driver chip, this function was desired to shut down the power FET in high ambient temperatures (see Figure 12-1).

While designing an analog circuit, I believe in using every tool available to evaluate the operation of a circuit. By using both breadboards and computer simulation tools, results can be checked against each other. I don't hesitate to use first-order hand calculations, too, which can be great for keeping SPICE or other circuit simulators honest.

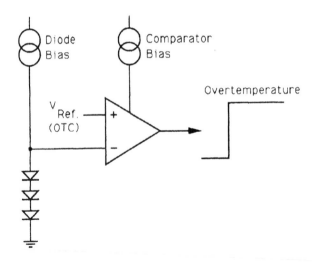

Figure 12-1.
IC thermal
detector circuit
operates by
comparing the
voltage on a
string of three
diodes to an OTC
reference.

At the risk of stating the obvious, the breadboard is not the actual circuit, and the SPICE simulation is not the actual circuit. Breadboard measurements are influenced most by parasitic capacitance—of the breadboard itself as well as oscilloscope probe loading. And in high impedance circuits, DC measurement loading can also be a problem. Simulator results are compromised by our inability to model the intricate behavior of real world devices. *The only actual circuit will be the one having the exact device types and physical placement of the final IC or printed circuit board.*

During the early stages of design, there is no such thing as going off on a tangent. I often like to just follow bias levels and wave forms through the breadboard or simulation and make sure that I can explain every last detail of what I see, even when the output appears to be correct. Anything and everything is a candidate for pausing and having a closer look. Probably four times out of five the response is easily explained and consistent with the simulation or measurement technique. But that fifth time . . .

That's what I'm really looking for: the aberrations, however small, which make no sense at all. To me, anyway. A friend of mine, Tom Frederiksen, once said, "A circuit always works exactly the way *it* is supposed to. It never disobeys any law of physics, and its behavior is exactly what you would expect if you fully understood the actual circuit you are observing." In other words, circuit behavior, no matter how weird or unexpected, can always be explained using basic network theory and device characteristics. The trick lies in understanding the circuit you really have, as opposed to the one you thought you had! By exploring these circuit aberrations, I often discover a fundamental problem or develop a whole new way of visualizing the circuit operation. But wherever they lead, I am always gaining more and more vital knowledge of my circuit.

This "search and explain" regimen can also provide a valuable indicator for a surprisingly difficult question: When is the circuit design finished? Assuming that the breadboard or simulation or both have been made to meet all of the design goals, the point at which I am no longer finding any funnies, where the response at each and every circuit node has been completely explained, is the point where I can have the confidence to call the circuit design finished.

As it happened, the thermal detector circuit had already been designed for a previous IC and was to be used again in the same configuration to save development time on the FET driver chip. The use of blocks of circuit data which can be placed on any chip made with the same fabrication process is called cell-based design. Cells are a seductive concept, and the basis for the explosion of application specific integrated circuits, or ASICs. Imagine only having to design and debug a circuit block once and then forever being able to use it to create new ICs without any further design effort! It is a concept which all but the most savvy of management have embraced wholeheartedly.

I wish it were that easy. Unfortunately, analog circuits don't always cooperate with the cell concept, as our example will illustrate. The problem is that in analog there are invariably slightly different requirements for a given functional block from ASIC to ASIC. A comparator, for example, may need a little more speed or output drive or common mode range or . . .

In our thermal detector cell the difference involved biasing. Along with signal inputs and signal outputs, analog IC cells typically have bias inputs which receive voltages or currents from centralized bias generators on the chip. This prevents needless duplication of biasing. The thermal detector circuit had originally been designed for continuous application of the bias signals; that is, the thermal detector would be biased and operating at all times that a supply was applied to the chip. However, in the FET driver chip, the bias signals were to be turned off and on by the input signal in order to minimize standby supply current. This was not a minor difference and proved to be the seed of problems to come.

So how should the analog designer handle this case of using a cell in almost but not quite its intended application? *Like a totally new design.* The smallest of changes can have a way of rippling through an analog circuit, often with dire consequences. For a thermal detector cell which is to be switched off and on, a prudent approach would be to start by evaluating the detector in the 'off' state (which in this case had not been a previous concern). Next, one would want to thoroughly investigate the detector response as bias is simultaneously applied to the the diodes, reference, and comparator. It's not obvious that there would be a problem, but then it's not obvious that there would not be. Having to modify or even completely redesign an analog cell to meet the particular requirements of the chip it is going into is always a real possibility. I assumed that the thermal detector circuit had passed this additional scrutiny since no circuit modifications were incorporated for the FET driver chip.

Readying an integrated circuit for mask layout is a special challenge since the analog circuit designer cannot be certain that the intended circuit has been created until the IC comes out. This applies equally to cell based layouts and custom layouts; the same cells reused from a previous chip may or may not yield the same results. Again, the only actual circuit is the one which has the exact device geometries and physical cell placements of the new IC itself. And the fact that the IC components all have parasitics associated with supply, ground, and each other can change the actual circuit very much indeed.

Thus again, the need for attention to detail, and for taking the time to anticipate all of the consequences of a particular device placement. Play "what if": What if I put this transistor in the same island with that resistor? What if I place this op-amp cell next to the protection cell for that pad? What if the transistor saturates? What if the pad goes below ground? Invent every kind of scenario you can think of, because I guarantee every analog circuit has the potential to behave quite differently than you expected.

Let's return now to the three thermal sensing diodes. Although I can't be certain, I would guess that somewhere near the end of layout of our FET driver chip the mask designer was running out of space on the layout. This mask designer, being a pretty creative type, saw a way to reduce the area occupied by the diodes by making a change in their geometry. This should immediately have raised a flag, if for no other reason than *last minute changes are always dangerous*. The specified diode geometries were proven to work in previous chips; the new ones *should* have worked as well, but they couldn't know absolutely, positively for sure until the IC came out. But, maybe our analog designer had something else on her or his mind that day, or just didn't feel the need to play "what if" in this case—after all, they're only *diodes*. . . So the designer gave the green light for the change and proceeded to completely forget about it. But the diodes didn't forget. They had been turned into quite different devices and had different laws to obey when they were turned on.

During the layout of an IC (or in a discrete circuit, of a printed circuit board), the analog circuit designer usually has a little free time since the design is, in theory, finished. There might be a temptation to work on a different project for fill-in; resist it. There is still plenty to focus on for the circuit in layout. First and foremost, the designer must check with the mask designer daily to make sure that the circuit being laid out is in fact the same circuit that was breadboarded and simulated. Any layout parasitics that might affect the operation of the circuit should be immediately incorporated into the simulation or breadboard to ascertain their effects. Any deviation from the specified device geometries or placement (such as in our diode string) should be thoroughly investigated. If there is any doubt, don't do it!

The most common IC layout parasitic is interconnect resistance, since metal interconnect lines can easily reach tens of ohms, and polysilicon lines thousands of ohms of resistance. Relatively small voltage drops along supply and ground lines can easily upset sensitive bipolar analog biasing, where a mere 3 mV drop can cause more than 10% change in current. If there is any doubt about the effect of a parasitic resistance, place a like-valued resistor in the breadboard or simulation and see what effect it has. Close behind in problematic effects are parasitic capacitance and inductance. And don't overlook mutual inductance between IC wirebonds or package leads. I once had a 45 MHz I.F. amplifier in which the pad arrangement was completely dictated by inductive coupling.

Also during the layout phase, more questions should be asked about the circuit (if they haven't been asked already). Questions like: What happens when an input or output is shorted to ground or supply? For an IC, what will the planned pinout do if the device is inserted backwards in the socket? Can adjacent pins be shorted to each other? In many cases, the answer to such questions may be, "it blows up." That's okay because it is still information gained, and the more the designer knows about the circuit, the better. And often, by asking some of these questions at this time, some surprisingly simple changes in layout may improve the ruggedness of the circuit.

Finally, the analog designer should also be planning ahead for evaluation of the new circuit when it comes out. While the original specifications for the project will define much of the testing required, the evaluation phase should definitely exercise the circuit over a wider range. During the evaluation phase, tests should be implemented to answer such questions as: How does the circuit "die" as the supply is reduced below the minimum operating voltage? Will the circuit survive a momentary overvoltage? What happens outside of the operating temperature range? What effect will loads other than those for which the circuit was designed have? These sorts of tests can pay big dividends by exposing a problem lurking just outside the normal operating "envelope" of the circuit.

Before I became involved in the FET driver project, the circuit had already been through the critical layout phase not once but twice, with several additional minor mask changes in between. The too brief evaluations of each new version, no doubt hastened by the urgency of management, had resulted in mask changes made in series rather than parallel. And it wasn't over yet.

Although it might seem obvious that more time invested in evaluation of a new chip could save substantial time later on, it takes a whole bunch of discipline to continue looking for problems after already identifying two or three that must be fixed. Somehow we just want to believe that what we have already found is all that could be wrong. The same advice applies here as to the original circuit design: keep looking until you can explain every facet of the circuit behavior.

One of the problems discovered in the FET driver was that the thermal detector was indicating over temperature at 115 °C instead of the desired 150 °C. I know that the cause of this 35 °C discrepancy had not been discovered, yet one of the mask changes lowered the reference voltage to bring the shutdown temperature back up. This again illustrates how vital it is to thoroughly understand the operation of analog circuits; the silicon had come out with a significant difference from the design value, and that difference had never been reconciled. If it had been, further problems might have been averted down the road.

The next problem was actually discovered by the marketing manager while he was fooling around with some of the latest prototypes in his lab at home. He had noticed that some of the parts were exhibiting a turn-on delay some of the time. He asked me to check his circuit and I quickly confirmed the problem. When a turn-on edge was applied to the input, the output would start to charge the power FET gate, only to latch low again for several microseconds before finally charging the gate completely. I quickly discovered that the delay was very temperature sensitive; as the die temperature approached thermal shutdown, the delay became hundreds of microseconds. How could this have been missed? Was it just showing up due to the mask changes, or as a result of a process problem? But most important, *exactly why was the circuit exhibiting this behavior?*

For anything more than very simple problems, the only effective way to trouble-shoot an integrated circuit is by probing the metal interconnections on the chip. As any IC designer who has spent countless hours peering through a microscope on the probe station will tell you, this is a long and tedious process. But it's the only way. After several days of probing, I kept coming back to the diodes. When an apparently clean current pulse was applied to the diode string at turn-on, the diode voltage initially rose with the current edge, then rapidly collapsed, followed by a slower recovery to the steady state voltage. It was during this collapse that the thermal detector momentarily signaled over temperature, thus inhibiting gate turn-on.

Many more days and many more experiments ruled out parasitic resistance, capacitance, or inductance effects and likewise ruled out coupling via parasitic semi-conductor structures. Finally, I realized that the reason the diodes were exhibiting such bizarre behavior was that *they weren't diodes at all.* Back when the mask designer had changed the diode geometries to save area, he or she had unknowingly created common-collector pnp transistors connected in a triple Darlington configuration! Once again the circuit had been behaving exactly as it should; it's just that up to this point I had failed to correctly identify the circuit (see Figure 12-2).

Now the pieces of this analog jigsaw puzzle started to fall neatly into place. The original shutdown temperature had come out low because the DC voltage on the "diode" string was low. The voltage was low because only the top "diode" was conducting the full bias current. In a Darlington configuration, each succeeding pnp conducts only $1/(\beta+1)$ times the current of the previous device, with the remaining current flowing to ground. As a result of ample β in these pnp geometries, the bottom pnp in the string was conducting very low current indeed! And the transient

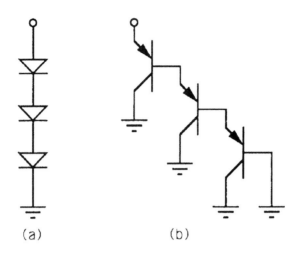

Figure 12-2.
The diode string:
(a) as shown on
the circuit
schematic;
(b) as the actual
device structures
behaved in the
thermal detector.

(a)   (b)

behavior, which had been totally unexplainable in the context of 3 diodes, was simply the natural response of a slow Darlington pnp configuration to a step change in emitter current. Basically, the long base transit time of the parasitic pnp structures causes the base-emitter junctions of the lower devices to overcharge on the initial current edge, pulling the base of the top device back down.

Could all of this have been anticipated back when the mask designer suggested the diode geometry change? Perhaps, but more likely not. This is why I have developed such a cautious approach to last minute changes in analog design. If I don't have time to completely evaluate the effects of a change, then I don't make it. In the case of the diode string this might have meant moving an edge of the die out to fit in the originally specified diodes, but as we have seen this would have been vastly preferable to the protracted evaluation and changes that followed.

Unfortunately, all of the bad news was not yet in for our power FET driver. On a subsequent processing run the chip started to draw significant supply current at elevated temperatures while in the 'off' state. The culprit? Once again the thermal detector. Since the cell had not been originally designed with the constraint of drawing no 'off' current, this was hardly a surprise. But up until that run, a balance of leakage currents had favored holding the comparator output circuitry off over temperature. Now, it was evident that the balance could be tilted in favor of having the output turn back on.

With process sensitivity added on top of the previous problems, and the chip more than a year behind schedule, we reluctantly decided to defeat the thermal detector completely. The FET driver was introduced without thermal shutdown.

The subject of this book and my example is analog circuit design. But for an integrated circuit there are two other required elements: the process and the package. There is a saying in analog IC design that if only one of the three elements is new—which is generally the circuit design—then there is a reasonable chance for success. If two out of three are new, then the chances decline dramatically. Three out of three is the kiss of death.

Practical wisdom, this. Requiring parallel developments in different technologies with no major problems is simply unrealistic and requires adding serious time to schedules. For example, any time a new process is involved, there is no history of device characteristics from which to draw, and an analog designer may unwittingly require a device to meet specs a process can't deliver.

The next project selected by management at my aspiring analog IC company? A "three out of three" on an *accelerated* schedule. That's when I departed.

13. Should Ohm's Law Be Repealed?

When I was a kid, the Stearn family lived nearby. Dr. Stearn, his wife, and two daughters had a really nice place. The house, a big old Victorian, was grand but friendly. They had a pool, shuffleboard and tennis courts, dogs, and a horse named Fred. Inside, there was lots of spirited modern art, a terrific library with a ladder that slid around the room on a rail, and great junk food. They had a wonderful collection of old surgical instruments and some great stained glass lamps. There were also pool and billiard tables, a pinball machine, and a darkroom. One daughter, my age, had cute freckles and long, chestnut hair. Once, she even baked me chocolate chip cookies and presented them in a blue box with a ribbon. They were good. I can't be sure, but I think I missed a cue. A born engineer.

For an eight-year-old boy, it should have been a really fun place. All of the attractions were of some passing interest but really weren't even distractions. Because what Dr. Stearn had, what he *really* had, was in the basement. There, sitting on something called a "Scopemobile," next to the workbench, was a Tektronix 535. That I loved this oscilloscope is an understatement. I was beyond infatuation, long past mesmerization (see Figure 13-1).

The pure, unbounded lust I spent toward this machine probably retarded the onset of my puberty, delaying sexual nascency by at least a year.[1] It also destroyed my grade school performance. I read the mainframe manual instead of doing my homework and studied the plug-in books (they were smaller and easier to hide) in Mrs. Kemp's English class. I knew every specification and all the operating modes. I lived for that 535, and I studied it. But, best of all, I used it.

Dr. Stearn, when he wasn't doctoring or being with his family, shared his electronics hobby with me. Since no amount of pleading, scheming, bamboozling, or anything else would get my father to buy one, Dr. Stearn also shared his 535 with me. Oscillators, amplifiers, flip-flops, modulators, filters, RF stages—we circuit-hacked them all with ferocious intensity. And with that 'scope you could really *see* what was going on. You knew the excitement Leeuwenhoek felt when he looked in his microscope.

In fact, the Tektronix 535 was a sublime masterpiece. In 1956, it was so vastly superior, so far ahead of everything else, that it made a mockery of the competition. The triggered sweep worked unbelievably well, and the calibrated vertical and horizontal really were calibrated. It had an astounding 15 megacycles (it was cycles then, not Hertz) of bandwidth and something called "delayed sweep." The plug-in

Versions of this chapter's text have been published by Linear Technology Corporation and *EDN Magazine*.

1. Testament to the staying power of this childhood desire is the author's current ownership of copious amounts of Beaverton hardware.

Figure 13-1.
The Tektronix 535. Introduced in 1954, this vastly superior instrument made a mockery of all competition. I knew I could use it to make my breadboards work. Or so I thought. (Photo Courtesy of Tektronix, Inc.)

vertical preamplifiers greatly increased measurement capability, and I always quickly concurred when Dr. Stearn decided to buy another one.

The 535's engineering concepts and production execution were a bumpless combination of interdisciplinary technology, inspired design, attention to detail, aesthetics, and usability. It combined solid knowledge of fundamentals, unbounded thinking, and methodical discipline to produce a superior result. The thing just radiated intellectual honesty.

Using that 'scope inspired confidence bordering on arrogance. I knew I could use it to make my breadboards work. Or so I thought.

One afternoon I was having trouble getting a circuit to work. Signals looked about right, but not really, and overall performance was shaky, with odd effects. I 'scoped everything but got nowhere. Dr. Stearn came by (after all, he lived there). He listened, looked, and thought awhile. Then he moistened two fingers, and started doing a little hand dance on the circuit board. His hand moved around lightly, touching points, as he watched the 'scope. He noticed effects and, correlating them to his hand movements, iterated toward favorable results. When things looked good, he stopped his motion. He rocked his fingers gently back and forth, watching the display respond. He paused, thought, and then soldered a small capacitor between the last two points his fingers were on. To my amazement, the display looked good, and the circuit now worked. I was dumbfounded and, propelled by frustration and embarrassment, a little angry.

He explained that the circuit had a high frequency oscillation, perhaps 100 mega-cycles, and he suspected he'd damped it by loading the right points. His finger dance had surveyed suspect points; the capacitor was his estimate of the electrical equiva-lence of the finger loading.

"That's not fair," I protested. "You can't see 100 megacycles on the 'scope."

He looked right at me and spoke slowly. "The circuit doesn't care about fair, and it doesn't know what the 'scope can't see. The 'scope doesn't lie, but it doesn't always tell the truth." He then gave me a little supplementary lecture which has served me well, except when I'm foolish or frustrated enough to ignore it.

"Don't ever get too attached to a way of solving problems. Don't confuse a tool, even a very good one, with knowing something. Concentrate on understanding the problem, not applying the tool. Use any tool that will help move your thinking along, know how those tools work, and keep their limitations in mind when you use them—it's part of the responsibility of using them. If you don't do this, if you stop thinking and asking, if you simply believe what the 'scope says, you're done for. When you do that, you're not listening to the problem, and you're no longer designing the circuit. When you substitute faith in that instrument, no matter how good it is, for your judgment, you're in trouble.

"It's a tricky trap—sometimes you don't even know you're falling into it. People are very clever at fooling themselves that way. We're all human, we all very badly want things to be simple and go smoothly. But that circuit doesn't know that and it doesn't care."

That was 34 years ago. I'm still absorbing that advice, although not progressing as rapidly as I'd like. I think Doc Stearn was right. I remember him often, usually after I've been stung by me again. My interest in tools, applying them, and human tendencies continues, and hopefully I'll get better at it all.

Lately, I've been hearing quite a bit about CAD systems, computer-based workstations, and powerful software modeling techniques. At Linear Technology, where I work, we have CAD systems and they save tremendous amounts of time. They're very powerful tools, and we're learning how and when to use them efficiently. It's a tough process, but the rewards are high and well worth the effort.

Unfortunately, I see substantive and disturbing differences between what I feel these tools are and what some of them purport to be.

There is a great deal of fanfare surrounding CAD systems today (see Figure 13-2). Promotional material, admittedly always suspect, emphasizes speed, ease of use, and elimination of mundanities and odious tasks in the design process. Unbearably attractive engineers in designer clothes reside in immaculately clean and organized work areas, effortlessly "creating." Advertising text explains the ease of generating

Figure 13-2.
CAD advertising assures high productivity with minimal hassle. Becoming the next Edison is only a keystroke away.

ICs, ASICs, board functions, and entire systems in weeks, even hours. Reading further, the precipitators of this nirvana are revealed: databases, expert systems, routers, models, simulators, environments, compilers, emulators, platforms, capturers, synthesizers, algorithms, virtualizers, engines, and a lot of other abstruse intellectual *frou-frou* Ohm and Kirchoff never got to. These pieces of technological manna ostensibly coalesce to eliminate messy labs, pesky nuts and bolts, and above all, those awful breadboards. Headaches vanish, fingers and the lab (if it hasn't been converted to the company health spa) are clean, the boss is thrilled, and you can go fishing. Before you leave, don't forget to trade in your subscription to *EDN* for one to *Travel and Leisure*. I can hear Edison *kvetching*: "It's not fair, I didn't have a CAD system." It's okay, Tom, you did pretty well, even if your lab was a mess.

Well, such silliness is all part of the marketing game, and not unknown wherever money may trade hands. *Caveat emptor* and all that. So maybe my acerbic musings are simply the cynicism-coated fears of a bench hacker confronting the Computer Age. Perhaps I'm just too invested in my soldering iron and moistened fingers, a cantankerous computer technopeasant deserving recuse. But I don't think so, because what I see doesn't stop at just fast-talking ad copy.

Some universities are enthusiastically emphasizing "software-based" design and "automatic" design procedures. I have spent time with a number of students and some professors who show me circuits they have designed on their computers. Some of the assumptions and simplifications the design software makes are interesting. Some of the resultant circuits are also interesting.

Such excessively spirited CAD advocacy isn't just found in ad copy or universities. Some industry trade journals have become similarly enamored of CAD methods, to the point of cavalierness. Articles alert readers to the ease of design using CAD; pristine little labeled boxes in color-coordinated figures are interconnected to form working circuits and systems. Sometimes, editorial copy is indistinguishable from advertising. An editorial titled "Electronic Design Is Now Computer Design" in the January 1988, issue of *Computer Design* informed me that,

"For the most part, the electronic details—the concerns of yesteryear about Ohm's law and Kirchoff's law, transconductance or other device parameters—have been worked out by a very select few and embedded in the software of a CAE workstation or buried deep within the functionality of an IC. Today's mainstream designers, whether they're designing a complex board-level product or an IC, don't need to fuss with electronics. They're mostly logic and system designers—computer designers—not electronics designers."

That's the road to intellectual bankruptcy; it's the kind of arrogance Doc Stearn warned about. Admittedly, this is an extreme case, but the loose climate surrounding it needs examination.

CAD is being oversold, and it shouldn't be. It shouldn't be, because it is one of the most powerful tools ever developed, with broad applicability to problem solving. If too many users are led astray by shuck and jive and become disappointed (and some already are), the rate of CAD purchase, usage, and acceptance will be slowed. In this sense, the irresponsible self-serving advisories of some CAD vendors and enthusiasts may be partially self-defeating. The associations being made between CAD tools and actual knowledge-based, idea generation, and iterative processes of design are specious, arrogant, and dangerous. They are dangerous because many of us are human. We will confuse, admittedly perhaps because our humanness begs us to, faith in the tool with the true lateral thinking and simple sweat that is design. We will cede the judgmental, inspirational, and even accidental processes that constitute so much of what engineering is. In the rush to design efficiency, we may eliminate time and sweat at the expense of excellence. Very often the mundanities and mental grunt work aspects of problem solving provide surprises. They can force a review process that mitigates against smugness and ossification. Most of the time this doesn't occur, but when it does the effect is catalytic and intellectual left turns often follow.

In misguided hands, a group of packaged solutions or methods looking for a problem will produce nothing at worst, an amalgam of mediocrity at best.

I also said associations between CAD tools and critical elements in the design process were arrogant. They are arrogant because in their determination to stream-line technology they simplify, and Mother Nature loves throwing a surprise party. Technologically driven arrogance is a dangerous brew, as any Titanic passenger will assure you.

Most good design is characterized by how the exceptions and imperfections are dealt with. In my field, linear circuits, just about everything is exceptions. A lot of the exceptions you know about, or think you do, and you're constantly learning about new exceptions. The tricky thing is that you can get things to work without even realizing that exceptions and imperfections are there, and that you could do better if only you knew. The linear circuit designers I admire are those most adept at recognizing and negotiating with the exceptions and imperfections. When they get into something they're often not sure of just what the specific issues will be, but they have a marvelous sense of balance. They know when to be wary, when to hand wave, when to finesse, when to hack, and when to use computers. These people will use CAD tools to more efficiently produce superior work. The others may be tricked, by themselves or by charlatan-hucksters, into using CAD to produce mediocrity more efficiently. (See Figure 13-3.)

The time has come to sum up. When reading, I enjoy this moment because I want to watch the author become more definitive without getting the foot in the mouth.

Figure 13-3.
Combining other approaches with CAD yields the best circuits.

When writing, I fear this moment for the same reason. On this outing, however, I'm not so fearful. The ground seems pretty solid.

CAD-based tools and techniques, although in their infancy, will prove to be one of the most useful electrical engineering tools ever developed. In some areas, they will become effective more quickly. They have already had significant impact in digital ICs and systems, although their usefulness in linear circuit design is currently limited. As these tools emerge, the best ways to combine them with other tools will become clearer. And they will combine with other tools, not supplant them. Right now, the best simulator we have, a "virtual model" if you will, is a breadboard. In current parlance, breadboards are full parallel, infinite state machines. They have self-checking, self-generating software and heuristically generated subroutines with infinite branching capability. If you're listening, the answer, or at least the truth, is there.

I'm reasonably certain breadboardless linear circuit design is a long way off. I suspect similar sentiments apply in most engineering disciplines. The uncertainities, both known and unknown, the surprises, and the accidents require sweat and laboratories. CAD makes nail pounding easier, but it doesn't tell how to do it, or why, or when. CAD saves time and eliminates drudgery. It increases efficiency but does not eliminate the cold realities involved in making something work and selling it to someone who wants it and remains happy after the purchase.

Where I work, we eat based on our ability to ship products that work to customers that need them. We believe in CAD as a tool, and we use it. We also use decade boxes, breadboards, oscilloscopes, pulse generators, alligator clips, screwdrivers, Ohm's law, and moistened fingers. We do like Doc Stearn said back in 1956—concentrate on solving the problem, not using the tool.

Intuitions and Insights

Every master analog designer has developed his or her own approach to design tasks and a set of mental tools to use. Such approaches and tools are the result of experience, and in this section several contributors share some of the formative events of their engineering careers.

A difficult but ultimately successful project can be a superb training ground for an analog designer, as John Addis shows in his account of his early days at Tektronix. But Bob Blauschild tells in his chapter how a failure can sometimes be a stepping stone on the path to ultimate success.

Paul Brokaw describes a process used in the design of linear integrated circuits that's equally applicable to other analog design tasks. It's possible to successfully design analog circuits in your head, says Richard Burwen, and he tells of his techniques for doing so. It's also never too late in your career to become an analog wizard, as George Erdi illustrates in his account of how he came late to the analog party and learned to like it. Of course, an early start never hurts, especially if you're nine years old and discovering the principles of feedback on your own, as Barrie Gilbert did.

Barry Hilton is one who firmly believes that good analog design has a large element of art to it. In his chapter, he shows how a mastery of basic circuit configurations is to the analog designer as a mastery of primary colors is to the artist. Phil Perkins takes a slightly different route, using "idealized" basic circuit elements to speed the design process. In his chapter, Phil shows how this approach can be used in the design of feedback loop circuits.

Insights are often where you least expect them. Jim Williams tells of how he thought he was going to spend an afternoon at the zoo, and ended up getting analog design pointers from a bunch of primates.

14. Good Engineering and Fast Vertical Amplifiers

For Brian Walker and me, the night flight from Lima to Miami was the end of a vacation, the primary objective of which was the 1986 Rio Carnival. Brian and I were sitting in the DC-10's two outside seats. A woman, a man, and two small children occupied the center four seats. I became curious why the woman repeatedly jumped up to look out the window while we were still at cruising altitude.

Very little in all of South America works exactly right. No one really expects it to. Cars retain the scars of every accident. Buses and trucks spew soot at an alarming rate. Graffiti mark even the recent buildings of Brasilia, and the pockets of abject poverty would be edifying for every American to see. I doubt there is a single jetway in all of South America. You walk across the tarmac to board your plane.

Sodium vapor lamps light up Miami like a Christmas tree, and the sight is made more impressive by contrast with the sea's total blackness. As we neared Miami, the lady once again was politely but enthusiastically looking past other passengers, straining to see out the window. My curiosity got the better of me, and I asked her if she had ever seen Miami from the air before. In broken English she explained that she and her husband were emigrating to the United States from Lima and that neither had seen their new country. Family history in the making.

She asked our aid in going through U.S. Customs, and after 2 weeks of relying on others, we were anxious to help. I tried to look at the experience from her perspective. As we left the airplane through the luxury of a carpeted jetway, I realized that she had likely never seen a jetway before. The International Terminal at Miami was new, and it was well done. The walls were pristine white, the carpets royal purple; there were lush palms in white ceramic pots. What a sight! We were guided toward a rapid transit car whose floor was flush with the building's interior. The gap between the car and the lobby was less than a quarter of an inch and did not change as 20 people walked on board. I was impressed, and in looking at this perfection through her eyes, I was even getting a little choked up!

The car doors closed automatically and the vehicle whisked us out over the tops of 747s and DC-10s into another gorgeous building. This family's first exposure to the United States was one of perfection. People here expect everything to work right. It occurred to me that only a handful of countries have this *expectation of perfection*. All of them are technologically advanced. All of them have contributed disproportionately to the sum of humankind's knowledge and comfort. *Expectation of perfection* characterizes those countries. It also characterizes good engineering.

One of the pressures that management exerts on engineers is embodied in the phrase "time to market." That phrase is held in such reverence that I have dubbed it "The Time to Market God." This sophomoric argument contends that product development time must be minimized because a product introduced one month late loses one month's sales from the beginning of its finite lifetime while the peak and tail end of the sales vs. time curve are unaffected. If a perfectly acceptable product were put in a warehouse for a month before it was introduced, I suppose that is what would happen, but no one ever does this.

When a product is delayed for a month, changes take place. It is not the same product it was a month earlier. Had the earlier product been introduced, it may have fallen on its face, with the result that sales would be lost from the both the peak and the end of a shortened product life. Another result could be long-term damage to a company's reputation. The time spent in perfecting a product *lengthens* its life and *increases* its sales, until diminishing returns ultimately set in. Of course, bad engineering can lengthen a product's design time and still result in a bad product, but that is a separable problem. Neither would I argue that no engineer had ever spent too much time in the pursuit of perfection. The Time to Market God needs to be shot, and engineers must not abdicate their responsibility to determine when a product is saleable. That is part of their job.

History of Fast Vertical Amplifiers

I have always been fascinated by oscilloscope vertical amplifiers. 'Scopes seem to be the fastest instrumentation around. For example, in 1957, Tektronix was regularly producing a 30 MHz oscilloscope (the 545), but the contemporary HP400D voltmeter was rated at 4 MHz. Furthermore, the 'scope let you see the actual wave form! Now, to be perfectly fair, the voltmeter was a more precise instrument, using an incredible 55 dB of feedback around four (count them, four) stages, rated at 2% accuracy[1] and only 5% down at 4 MHz, but the Tek 545 was rated at a passable 3% accuracy even if it was 30% down at 30 MHz. Furthermore, there was something romantic about the fact that the oscilloscope's output was not interpreted by some other electronic device—it was displayed directly on a cathode ray tube (CRT). No hiding the truth here! If the amplifier distorted, you saw it!

Early (vacuum tube) DC-coupled verticals were typically differential pairs of tubes, sometimes pentodes for their low plate to grid capacitance, and sometimes a cascode configuration to accomplish the same result, as shown in Figure 14-1. Between stages, sometimes a cathode follower served to lower the impedance level, driving the next stage without adding much capacitive loading.

Inductive peaking was an important part of these circuits. A 1.37 times bandwidth improvement could be obtained by adding inductance in series with the plate load resistor or in series with the load capacitance of the next stage.

The T coil was a combination of two inductors, usually with mutual inductance, which was capable of exactly twice the bandwidth of series peaking alone for a 2.74 times improvement over a totally unpeaked circuit. With the addition of one capacitor, the bridged T coil can present a purely resistive load at one terminal. This made it possible to terminate a transmission line and peak a load capacitance with the same circuit.

What made the 545 (and later the 545A, 585, and 585A) so fast was the distributed amplifier[2] which drove the CRT. See Figure 14-2. In the 545, six tubes were strung along two lumped element transmission lines made up of inductors and the

---

1. The 400H had a mirrored meter scale and was rated at 1% accuracy. The March 1961, *Hewlett-Packard Journal* describes the calibration of meter movements in the article by Bernard M. Oliver.
2. The basic idea of a distributed amplifier was first disclosed in a British patent specification dated July 24, 1936. The term "distributed amplification" was coined for the title of a paper by Edward Ginzton, William Hewlett, John Jasbert, and Jerre Noe in the August 1948, *Proceedings of the I.R.E.* The authors also discussed the bridged T coil.

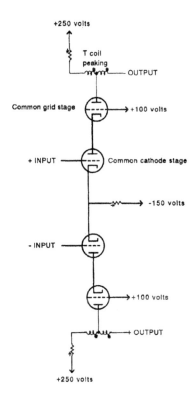

Figure 14-1.
Cascode vacuum tube amplifiers. The common cathode stage feeds the low input impedance of the common grid stage. Most of the voltage swing is at the plate of the common grid stage. The output tube's output capacitance is T coil peaked.

capacitance of each tube. An input line connected the grids together through the inductors, and an output line did the same for the plates. The total gain was that of six tubes in parallel, but the bandwidth was that of one tube operating into a load impedance too low to obtain significant gain from a single tube. The input and output capacitances were just parts of transmission lines, and did not act as if they were in parallel. All those tubes in parallel provided the large current swing neces-sary to drive the 600 Ω per side load of a distributed-deflection CRT[3]. No subtlety here; this was the brute force approach!

The first transistors available were germanium. No one in instrumentation liked them much because they exhibited substantial collector-to-base leakage at room temperature, leakage that doubled for every 10 °C increase. Silicon transistors were available, but they were frightfully expensive and not as fast.

Transistors had base resistance which made the input impedance far more "lossy" than that of vacuum tube amplifiers. That loss made transistorized distributed amplifiers impractical.

The first transistor vertical amplifiers were simple differential cascodes, a differ-ential pair of common emitter amplifiers followed by a differential common base stage. Because the transistor transconductance was so high, the collector load impedance could be much lower than the plate load for the same gain. Since tubes and transistors had comparable stray capacitances, the bandwidth for transistor amplifiers was higher than simple tube amplifiers, once the transistor F_t[4] signifi-cantly exceeded the required amplifier bandwidth. At 25 MHz, the Fairchild

---

3. The distributed-deflection structure broke up the CRT deflection plate capacitance into several sec-tions separated by inductances. As with the distributed amplifier, the deflection structure was a lumped element transmission line. The electron beam moved between the deflection plates at roughly the same velocity as the signal, so the effective deflection plate length was equivalent to only one of the sections and did not compromise the CRT's bandwidth.
4. Essentially the frequency at which the current gain drops to unity.

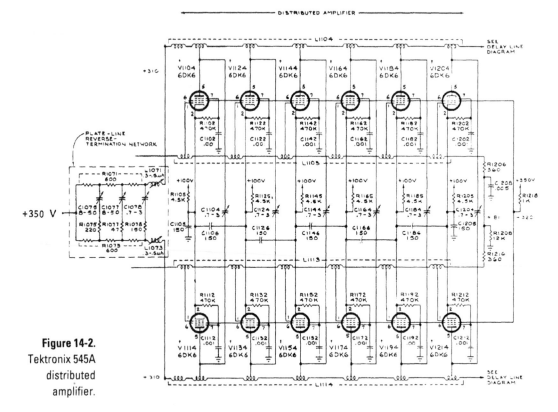

Figure 14-2.
Tektronix 545A
distributed
amplifier.

(DuMont) 766 was the first wide-band transistorized oscilloscope. Strictly speaking, the 766 was not all transistors, but only the input devices were tubes. Fairchild made silicon transistors, and Fairchild had bought out DuMont, the preeminent scope manufacturer at the end of World War II, so it was natural that Fairchild should try its hand at 'scopes.

The general philosophy was that transistors had no known failure mechanisms. Fairchild had no reason to dispute that philosophy; in fact, Fairchild was probably anxious to demonstrate its truthfulness. Consequently, the 766 had no fan and ran very hot, but Fairchild had certainly blazed (!) a trail. A transistorized high-speed, 'scope was possible.

There were some new problems with transistors. Transistor amplifiers seemed to have time constants other than those which determined bandwidth. Transient response was usually undershot by a few percent in the millisecond or even microsecond range. Tubes had funny time constants too, but they were in the 1–10 second range, and the only people who worried about them were oscilloscope designers. No one was quite sure what caused "DC shift," but large resistors and electrolytic capacitors were part of the plate loads to boost the gain below 0.5 Hz by a percent or so. Early transistor circuit designers were preoccupied with the temperature dependence of these new devices, and it was not long before they figured out that the transistors were so small that they could heat and cool in microseconds! Once that realization hit, the explanation was apparent: The signal changed the transistor's operating point, which in turn changed its power dissipation and hence its temperature. The transistor's V_{be} (base to emitter voltage), which is a function of the transistor's temperature (−1.8 mV/°C), changed accordingly. The V_{be} is effectively in series with the input signal and becomes a small error signal inseparable from the desired signal. These errors are known as "thermals," or "thermal tails."

By analogy with DC shift in tubes, the first correction tried for thermals was the addition of series RC networks between the emitters of the differential pair. The results were not very good for wideband amplifiers because these amplifiers dissipated the most heat and had the most severe thermals. Heat dissipated in silicon spreads with multiple time constants. It was a mess. Too many time constants!

No doubt several people discovered the solution to this problem independently in the early '60s.[5] By solving the simple equation for power dissipation as a function of signal and bias conditions, it becomes obvious that the correct load line biases a differential pair so that the power in both transistors decreases identically with signal. This makes the thermals into common mode signals and effectively eliminates them. Unfortunately, the required load line is totally incompatible with a wideband amplifier! To get any decent voltage across the device (e.g. 2.5 V, V_{ce}) and stand enough current to drive a capacitive load (e.g. 20 mA, I_c) requires a load line (collector plus emitter resistor) of 2.5 V/20 mA, or 125 Ω. These amplifiers were cascodes, and there was no collector resistor for the input devices, but a 125 Ω emitter resistor would have required at least a 250 Ω load in the cascode's output collector for even a gain of two. A 250 Ω collector load would have killed bandwidth, as would a resistor between the two halves of the cascode . . . unless the resistor were bypassed with a small capacitor! Yes! That was it, the bottom device was a current source, so it did not care if its load changed with frequency. This technique is known as "thermal balancing," as shown in Figure 14-3.

Thermal balancing worked reasonably well until the late '60s, by which time vertical amplifiers needed to be integrated to reduce the lead length (i.e. inductance) between the two cascode halves. There could be no thermal balance resistors because their bypass capacitors would have to be off chip, and that would destroy the advantage of integration, i.e., short lead lengths.

The usual IC manufacturers were apparently not interested in an IC process oriented around high speed analog circuits. Such a process would have to be very high speed, have good breakdown voltages, and should have low temperature coefficient resistors (e.g. thin film nichrome). These characteristics were not the same as requirements for digital ICs, which were being produced in much more interesting volumes. Furthermore, an IC manufacturer would have to support and share its technology with another company. An in house IC manufacturing facility was the key to making custom ICs for analog instrumentation.

Tektronix had a small IC facility in 1968, but the process was too slow to use in fast amplifiers.[6] A prolific Tektronix engineer named Barrie Gilbert did, however, find a clever way to write pertinent information on the CRT by analog IC methods. He also published several methods of continuously varying amplifier gain. The variable gain circuits are now collectively known as Gilbert multipliers,[7] as shown in Figure 14-4.

In August 1969, Hewlett-Packard was first to introduce an integrated circuit-based vertical amplifier in an oscilloscope, the 183A. Introduced at the same trade show was the Tektronix 7000 series, whose vertical amplifiers were designed with discrete, purchased transistors. The 183A was a 250 MHz scope with truly impressive triggering. It was small, light, and less expensive than the Tektronix 7000 series competition at 150 MHz. HP's Al DeVilbiss, Bill Farnbach, and others were

---

5. In the early 1980s, one oscilloscope company discovered it again and made an advertising claim that their engineers had discovered a way to instantly compensate for the small thermal error voltages generated in transistor junctions.

6. Two chips were in the 105 MHz 7A12 introduced in August 1969.

7. Barrie Gilbert, "A Precise Four-Quadrant Multiplier with Subnanosecond Response," *Journal of Solid State Circuits*, December 1968, p. 365.

Figure 14-3.
The technique known as "thermal balancing." R1 and R4 cause Q1 and Q3 to be biased at their maximum power points. Any signal decreases the power dissipation in both Q1 and Q2, even though one has increasing voltage across its terminals and the other has increasing current flow. Since both have the same power dissipation, no thermals are introduced. C1 and C2 prevent high frequency signals from appearing at the collectors of Q1 and Q3, preserving the bandwidth of the cascode.

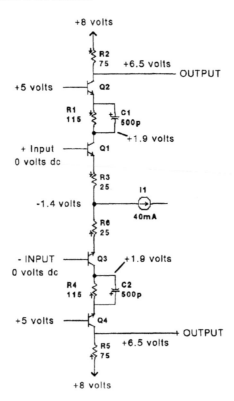

causing us considerable grief! These first ICs did not have any resistors integrated onto the IC, so lead inductance acquired on the way out to the emitter resistors was a major limitation on bandwidth.

The circuit design was still that of the differential cascode, but the integrated circuit approach not only provided higher bandwidth, it eliminated the slowest thermals. When the transistors were close together on the same piece of silicon, each transistor increasing in temperature heated a mate which was simultaneously cooling, and vice versa. The self heating still created fast thermals, but these thermals could be removed with small RC networks from emitter to emitter. Such networks were already required to make up for the delay line losses anyway.[8]

I can remember the review of the 1969 WESCON show. Management tried to paint a bright picture of what had basically been a disaster for Tektronix. Our advertisements tried to make the best of our unique feature, CRT readout. HP's ads mocked us with a picture of horns, bells, and whistles, then added a plug for their solid performance. We knew something needed to be done in a hurry, and a high speed in-house IC process was essential. George Wilson was an early contributor to Tek's first high speed IC process, SHF (Super High Frequency), and to its immediate successor, SHF2 with a 3.5 GHz F_t.

---

8. Skin effect loss, the dominant loss mechanism in vertical delay lines, requires multiple RC networks for compensation.

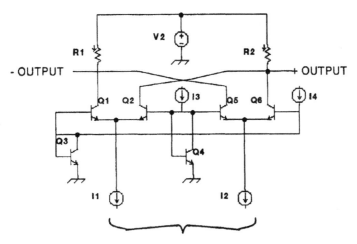

Figure 14-4.
One of the Gilbert
multiplier
configurations
used to obtain
electronically
controllable,
continuously
variable gain.

Push-pull input current

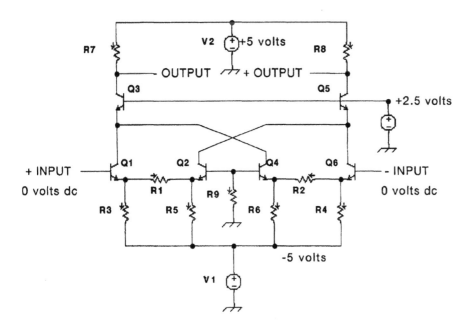

Figure 14-5.
The Battjes F_t
doubler in a
cascode
configuration.

A new circuit design, the F_t doubler shown in Figure 14-5, contributed as much to vertical amplifier bandwidth as did the faster IC process. The F_t doubler,[9] patented by Carl Battjes of Tektronix, put the inputs of two differential amplifiers in series (raising the input impedance) and put their outputs in parallel for twice the gain. At F_t, where the transistor beta is unity, the new circuit produced a current gain of nearly two.

These new amplifiers had rise times comparable to the propagation time between stages, so it became important to eliminate reflections between stages. The bridged T coil became extremely important because it allowed the input of each stage to be very well terminated. The F_t doubler's simple high frequency input impedance could be well modeled with a series RLC. Bob Ross wrote the equations for the

---

9. U.S. Patent 3,633,120

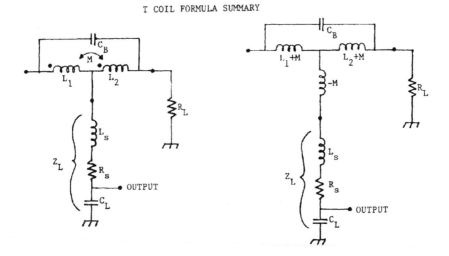

T COIL FORMULA SUMMARY

Figure 14-6.
Bob Ross's
equations for the
asymmetrical T
coil. The two
diagrams are
entirely equiva-
lent descriptions
of the T coil.

$$\theta = \tan^{-1}\sqrt{\frac{1}{\delta^2} - 1} \qquad \delta = \frac{1}{\sqrt{\tan^2\theta + 1}}$$

$$L_1 = \frac{C_L}{4}\left[1 + \frac{1}{4\delta^2}\right]\left[R_L + R_s\right]^2 - R_L R_s C_L - L_s$$

$$L_2 = \frac{C_L}{4}\left[1 + \frac{1}{4\delta^2}\right]\left[R_L + R_s\right]^2 - L_s$$

$$L_T = R_L^2 C_L$$

$$M = \frac{C_L}{4}\left[R_L^2 - R_s^2 - \frac{1}{4\delta^2}\left[R_L + R_s\right]^2\right] + L_s$$

$$C_B = \frac{C_L}{16\delta^2}\left[\frac{R_L + R_s}{R_L}\right]^2$$

$$s = \frac{-4\delta^2}{(R_L + R_s)C_L} \pm j\frac{4\delta}{(R_L + R_s)C_L}\sqrt{1 - \delta^2}$$

$$L_1 + M = \frac{R_L C_L}{2}\left[R_L - R_s\right]$$

$$L_2 + M = \frac{R_L C_L}{2}\left[R_L + R_s\right]$$

asymmetrical bridged T coil which could peak a series RLC load.[10] The derivation is quite complex, and the equations were held as a trade secret until recently. See Figure 14-6.

The F_t doubler and the bridged T coil were used extensively in the Tektronix 7904 introduced in 1972. Thor Hallen, who designed both the 7A19 plugin and the 7904's main vertical amplifier, had never worked on a high speed amplifier before. His achievement is all the more remarkable in that no other oscilloscope manufacturer has produced a real time 500 MHz oscilloscope in the intervening 19 years.[11]

---

10. The shunt resistance component of the input impedance was only important at DC.

11. A small company, B&H Electronics, did custom modifications of the 7904, and later the 7104, and achieved greater bandwidth than Tek did with the same CRT. There have been numerous direct access oscilloscopes with bandwidths limited only by the CRT.

What was not generally known is that the first 7A19 shown was a hand-built instrument, totally redesigned after the 1972 IEEE show, where it was introduced.

Tektronix still had customers, mostly the nuclear community, asking for more bandwidth. The people who built the bombs had wanted to measure high speed events with a very low repetition rate, once a month or so. If one wanted to examine predictable, high repetition rate signals, one could always use sampling techniques which availed bandwidths commonly over 10 GHz. The problem was that it was getting difficult to display a single subnanosecond event brightly enough for a camera to record it, even with 10,000 speed Polaroid film.

A CRT is a great example of engineering tradeoffs limited by physics. For a given technology, you can trade spot size, brightness, and sensitivity in any combination but never improve one without the detriment of the other two. Of course, you can change the technology and improve all three. An early example of such a technical improvement was the postdeflection accelerator tube from the mid 1950s. Added inside the classical CRT was a shallow spiral of resistive material from the deflection plate to the screen. Most of the acceleration voltage was placed across the spiral. This allowed the beam to be accelerated after it had passed through the deflection plates at a pace leisurely enough to be deflected easily in the horizontal and vertical directions. The accelerated beam was very bright, and the sensitivity was better than a simple CRT. Had the beam traveled too slowly through the deflection plates, mutual repulsion among the electrons would have caused the beam to spread and therefore increase the spot size.

Zenith had appeared at Tek with a CRT containing an image intensifier called a microchannel plate (MCP). The MCP is basically a piece of glass with tens of thousands of small holes (or channels) in it. The inside of each hole is coated with a material which, when hit by an electron, emits a cascade of additional electrons which in turn bounce into the walls farther down the hole. There is a small accelerating potential across the plate, up to 1 kV, which keeps all the electrons moving through the holes. What starts as a single electron entering a hole on one side of the plate ends up as 10,000 electrons exiting the same hole on the other side! Engineering advances usually come in prosaic numbers like 2. An advance of 5 times is quite good. A 10 times improvement is rare, and a single advance of 100 times is almost unheard of, but the microchannel plate offered the possibility of a once in a lifetime improvement of 10,000 times! This was just too good to pass up. CRTs had never been improved by 10,000 times before, and never would be again. In fact, the cumulative advances in CRT technology from their first manufacture was probably only 1000 times by 1972!

In 1973, Tek formed an engineering team to build a 1 GHz real time oscilloscope using a microchannel plate CRT. Tek designed a new CRT from scratch.[12] I had responsibility for the vertical system and the 7A29 plugin, Wink Gross had direct responsibility for the main vertical amplifier, and we shared design of all the hybrid amplifiers. Doug Ritchie had proposed a new IC process, Super High Frequency 3 (SHF3) designed specifically for analog applications. The process had 20 V BV_{cbo}, enough to drive a CRT with a 1.2 V/division deflection factor, a state-of-the-art F_t of 6.5 GHz, and 10 Ω per square thin film nichrome resistors with a 5% tolerance.[13]

The CRT took a 1000 times improvement in writing rate (brightness), a 3.5 times improvement in sensitivity, and a 3 times improvement in spot size, for its 10,000

12. Hans Springer, "Breakthroughs Throughout Push Scope to 1 GHz," *Electronic Design*, January 18, 1979; pp. 60–65.
13. SH3 has become an 8.5 GHz process with either 10 or 50 Ω per square nichrome resistors.

times total. When the 7104 was finished, I had great fun showing single shot events which occurred in "one-third of a billionth of a second" to nontechnical people. You can see this with the unaided eye in normal room light!

Three competing philosophies existed on how to design a 1 GHz amplifier. One was to use Gilbert gain cells, another Barrie Gilbert innovation, which cascaded the standing current through a series of stacked amplifiers. A second was a complicated scheme which split the signal up into several bands in the frequency domain and used microwave and transmission line techniques to recombine the signal at the output. I backed the third alternative—standard F_t doublers with T coil peaking as in the 7904. Ultimately F_t doublers won, although a great deal of effort continued on the split path scheme.

The amplifiers would not be easy.[14] The usual technique, used in the 7904, was for the output transistor collectors to drive a transmission line connected to the distributed deflection structure and a forward termination. There were no collector resistors in the amplifier. If the forward termination and transmission line closely match the CRT's impedance, very few aberrations are produced in the system. At 1 GHz, we thought that the CRT should be driven from a terminated source. The 7104 CRT had a 200 Ω push pull deflection structure and a 0.9 V/division deflection factor. Compared with the 7904's 385 Ω impedance and 3 V/division, the 7104's CRT and double termination actually required 10% more current gain from the amplifier than the 7904.

We also thought it necessary to doubly terminate the plugin-to-mainframe interface, and that required twice the gain. All together, the 7104 vertical required 2.2 times the gain of the 7904 at twice the bandwidth, or 4.4 times the gain-bandwidth product with about twice the F_t to work with!

Thermals were another problem. The 7904, as with all previous IC verticals, had used multiple series RC networks between emitters to compensate for thermals and delay line losses. The stages in the 7104 would literally be so fast (140 ps rise time) that the front corner of the transient response would be over by the time the signal left the amplifier package and got to the RC network! We figured that if we could not bring the front corner up to meet the DC response, we could bring the DC response down to meet the front corner. This technique,[15] which Wink named *feed-beside*, had the advantage that only a low-frequency correction signal is required. Operational amplifiers and high-impedance circuitry easily handled the thermal correction signal. The delay line compensation required a separate passive hybrid.

Initially I felt that we could get the bandwidth and fix the thermals, but I did not have the foggiest notion of how we would connect the ICs together without microwave packaging and coaxial connectors, which were out of the question for reasons of cost. Yet it was essential to me that we be able to remove and replace these amplifiers quickly and easily. The packaging scheme, called the Hypcon, for hybrid to printed circuit board connector,[16] allowed us to make virtually perfect connections between thin film hybrid amplifiers and an etched circuit board. It was the first of many elastomeric connectors now used in industry and the only packaging method ever used in vertical amplifiers in which there were no parasitics to limit bandwidth! Bond wires or spiral inductors formed the T coils. A typical stage had a 2.5 GHz bandwidth and a gain of 4.6. The 7104 was introduced in January 1979, after a

---

14. John Addis, "Design and Process Innovation Converge in 1 GHz Oscilloscope," *Electronics*, June 21, 1979; pp. 131–138.
15. U.S. Patent 4,132,958
16. U.S. Patent 4,150,420

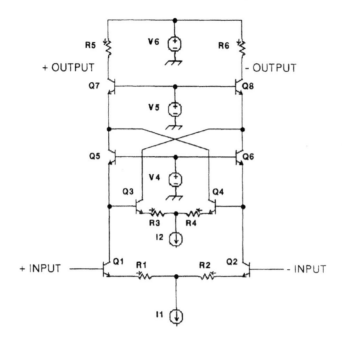

Figure 14-7.
Pat Quinn's
cascomp. The
circuit was used
extensively in the
2465 to reduce
the power
requirements
by increasing
linearity for a
given standing
current.

6 year development cycle. The fact that it still sells well after 12 years is evidence that a well-designed product can have a long life in electronics.

The 7904 and 7104 sewed up the high frequency oscilloscope market for Tektronix. It did not seem like another company was going to easily exceed the 7104 performance, and there were larger markets out there.

One of those markets was portable oscilloscopes. In 1972 Wink and I had done the vertical amplifier for the 485, a 350 MHz portable which I thought was one of the best portable 'scopes anyone had ever made. After the 7104, Wink went to work on the 2465, a portable which was destined to be the largest selling Tektronix 'scope of all time, and I stayed with the 7000 series group, which started to work on the 11000 series.

In portable 'scopes, power is at a premium. A large power supply will increase the size and weight of what is supposed to be small and light. Labor costs were gaining in importance too, and a large part of labor costs consisted of setting all those thermal adjustments. There were 32 thermal adjustments in a 7104 and its two 7A29s. Some way of eliminating these adjustments and saving power would be a big help. Such a way had been invented by Pat Quinn of Tektronix. Ian Getreu dubbed it a cascomp amplifier[17] shown in Figure 14-7.

The idea behind the cascomp was to sense the error voltages in the base-emitter junctions and cancel them out. Both thermals and amplifier nonlinearities showed up in the base-emitter junctions. Canceling the thermals could reduce the labor costs, but canceling the inherent junction nonlinearities meant that each stage could be operated closer to its maximum output with the same distortion. Alternatively, each stage could operate at lower power for the same amount of nonlinearity. The 2465 took full advantage of the cascomp, and the cascomp is widely used elsewhere at Tek. Here was a new circuit concept building on the traditional differential pair, or F_t doubler. There are actually a number of variations on the cascomp scheme. One, which we called the Traashcomp[18] after its inventor, Einar Traa of Tektronix, combined the F_t doubler with the cascomp error scheme. In some cascomp circuits,

17. U.S. Patent 4,146,844
18. U.S. Patent 4,267,51

the error amplifier was what we called a unity gain buffer (UGB) out of ignorance of any official name. The cascomp was a second generation of amplifier circuitry. The 2465 contained the last of the highspeed vertical deflection amplifiers for Tektronix. The digitizer was supplanting the CRT as an acquisition device.

The basic UGB design shown in Figure 14-8a is not new. It dates back at least to the National LM108 voltage follower in 1968 and perhaps earlier. It is, after all, an op amp whose output is connected directly to its inverting input. I remember thinking how clever it was when I first saw it in a lecture Floyd Kvame gave at Tektronix. You can look at the whole circuit as a single high transconductance, high beta transistor with just three leads, collector, base, and emitter. With that concept in mind, almost every transistor circuit you ever knew about can be redesigned by substituting this compound transistor.

In 1982, it seemed to me that accuracy was becoming a more important factor in circuit design. There was a lot of activity with 12 bit and 16 bit digital to analog (D/A) converters. More and more instruments had microprocessors to control their inner workings. Pots were being replaced with microprocessor-controlled D/A converters, and there was simply a lot more accuracy there than in any oscilloscope. I turned my attention to increased precision with the intention of not sacrificing any speed. That meant not the 1 GHz of a 7104 which, with a 50 Ω only input was not applicable to high precision, but 300–400 MHz with a 1 MΩ input.

In playing with the Traashcomp, I noticed that the UGB exhibited more bandwidth than the Traashcomp itself. Of course, the Traashcomp had a nice clean input impedance and was expected to be used with a T coil. The T coil would have given the Traashcomp the bandwidth advantage, but several UGBs could be integrated into one chip while the Traashcomp would be restricted to a single stage at a time if T coils were required between stages. The UGB was an excellent candidate for a high level of integration, and the feedback should give it higher precision and lower thermals than any of the cascomp configurations. In a week or two, I had figured out the basic way that an entire 'scope preamp could be put on one chip. There was enough versatility that all the functions in a vertical plugin (except impedance conversion from the 1 MΩ input) could be put on one chip! Here is a third generation of vertical amplifiers!

I used a variation on the UGB shown in Figure 14-8b. Tying the collector of Q1 to a positive supply voltage instead of the emitter of Q3 reduces the inherent alpha[19] of Figure 8a to that of a single transistor, but I was trying to push bandwidth of an input stage and wanted both the additional speed that came with an improved operating point for Q1 and the additional stability that came with eliminating the bootstrapped collector of Q1.[20]

The alpha loss was a problem for the IC because there was a Gilbert multiplier following this stage to obtain a continuously variable gain, and the Gilbert multiplier has an inherent alpha-dependent loss in its gain. There was an output stage too, another UGB. A total of three stages, each with a gain proportional to the transistor's alpha, would have led to a gain temperature dependence on the order of 225 ppm/°C. That was more than 1% over the ambient temperature range of the instrument, even with nominal beta, and a little too much for the precision I desired. There were other reasons too for reducing the alpha loss.

---

19. Alpha is the ratio of a transistor's collector current to its emitter current, and here it refers to the compound transistor's collector to emitter current ratio.
20. *Bootstrapping* too is a very old concept. In Figure 14-8a, Q1's collector is forced to change in voltage by exactly the same amount as the input signal, making the collector to base voltage constant. To some engineer long before my time, this was like pulling someone up by his bootstraps.

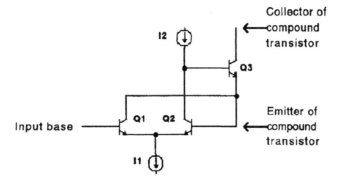

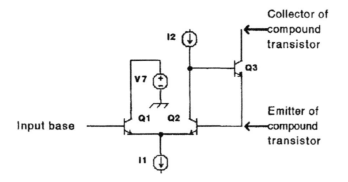

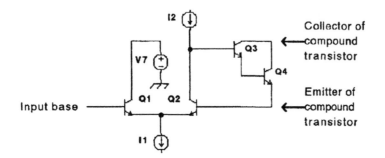

Figure 14-8.
(a) The classical unity gain buffer which may be viewed as a single transistor with very high gain.

(b) A simplified schematic of an alternative unity gain buffer having better high frequency characteristics but the alpha of just a single transistor.

(c) A unity gain buffer with good high frequency characteristics and high alpha

To get the alpha back, I substituted a Darlington for Q3 as shown in Figure 14-8c. The Darlington, of course, had its own stability problems, which necessitated some loop compensation, but at least the input was somewhat isolated from this.

A side benefit of the UGB showed up when I tried to figure out its behavior during overdrive and what could be done to improve recovery. The details are beyond the scope of this chapter,[21] but it turned out that it was possible to make this circuit recover from overdrive about 1000 times faster than anything else we had.

It is not the ability of a transistor to come out of saturation in a few nanoseconds that limits the overdrive recovery time of a linear amplifier. Once overdriven, linear amplifiers generate thermals which can last for milliseconds. It is those thermals that prevent a 'scope user from blasting the input with a signal and examining the details of the waveform with good accuracy.

---

21. John Addis, "Versatile Analogue Chip for Oscilloscope Plug-ins," *Electronic Engineering*, London, August 1988, pp. 23–28, and September 1988, pp. 37–43.

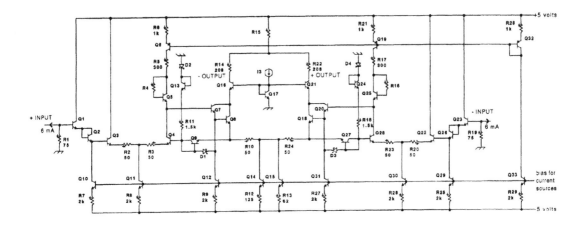

Fig. 14-9.
Input circuit presented at a design review.

The resulting IC, the M377, is the preamp in most of the 11000 series of Tektronix plugins. It has been successful enough that it is being used in several other Tektronix products.

Good Engineering and a Counter Example

Contributing to the general comfort of life on earth is one purpose of engineering. In some broad sense, we in the technologically more advanced nations live with modern conveniences because of engineering. Bad engineering has a way of correcting itself in the marketplace, but we should seek to avoid those disasters. Good engineering is elegant and no more complicated than necessary.

Engineers are sometimes asked to review other's IC designs. The more formal occasions involve as many as 20 people, including the designers and reviewers. As an enticement, lunch or refreshments are usually served during the 3 or 4 hour process. The intent is to avoid some simple mistake, something not modeled by SPICE.[22]

On one such occasion, several designers presented preliminary circuits to obtain early criticism for two very complex chips. Parts of the design were based on some concepts I used in the M377, and I was flattered to see someone make use of the same design ideas. One of the ICs would have about 1200 transistors, and that made it the largest analog IC Tektronix had ever attempted. Most of the circuitry was replicated four times, so one engineer had to design no more than 300 transistors' worth of circuitry. Yet 300 transistors is a daunting number of devices to simulate and be responsible for, so there may not have been much time spent looking at the design from a broader, systems point of view.

The complexity of analog design is still roughly proportional to the number of active devices (transistors or, at one time, tubes). When I began my professional career in 1963, a top-of-the-line Tektronix 545A oscilloscope and CA plugin had about 110 tube functions. It took an engineering team of seven to design the system.

22. Simulation Program with Integrated Circuit Emphasis (SPICE) is widely used throughout the analog integrated circuit industry to analyze circuit designs far too complex for manual analysis. The original SPICE, the outgrowth of work by Larry Nagel at the University of California at Berkeley, has spawned a dozen or so more or less compatible extensions since its inception in 1970. Tektronix has a dedicated group of about ten programmers whose constant task it is to expand and improve Tekspice and our other analog design tools.

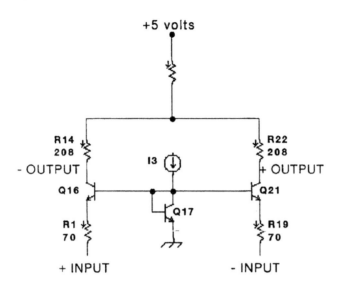

Fig. 14-10.
The circuit of Figure 9 can be simplified as shown here. R1 and R19 are reduced from 75 to 70 Ω. Standing current in Q16 and Q21 comes from the preceding stage.

Today, design tools such as SPICE allow one analog engineer to handle about 300 transistors. The engineering teams tend to number about 20 for a top-of-the-line instrument, and about half of them are software engineers. Then, as now, such a team may be designing two or three versions of the same instrument simultaneously.

For the project being reviewed, it had been suggested that designing one UGB, one which met all of the diverse chip needs, would be quicker than designing a different UGB for each need. Time to market, you know! Consequently, designing the most complicated buffer, one meeting all the possible requirements, was one of the first tasks undertaken.

The most important design goal of this project was low cost. Connection between ICs and circuit boards (generically known as "interconnect") was thought to be very expensive and unreliable, so cost was to be reduced through a high level of integration, i.e., making fewer but more complex ICs. It is my opinion that it takes at least as much talent to engineer something to be low cost as it does to make it high performance. It probably takes more. Fortunately, I have always worked on high performance projects where one could justify brute force if necessary.

Figure 14-9 shows one stage of the design as presented at the review. This is the IC's input stage with a 75 Ω termination resistor for the transmission line from the previous stage, and it uses the general purpose UGB. The preceding stage's output came from the collector of a transistor which had a 75 Ω output load resistor of its own. So the interconnection was double-terminated, and the system bandwidth was expected to be about 500 MHz (0.7 ns rise time). Even at a 1 GHz system bandwidth, one does not usually doubly terminate between two stages unless there is a good reason for it. A long and complicated path such as a plugin to mainframe interface would be a good reason, but that was not the case here. Here it was done to make the etched circuit board tolerance (which sets the impedance of the interstage path) less critical and to make the layout of that path an easier job. With so many transistors, current gain is cheap, even at 500 MHz, right?

The review made several of us uneasy because of the circuit's complexity, yet none of us had anything more concrete to contribute than an intuitively based malaise. Consider, however, the two goals of this circuit. The first is to provide a 75 Ω input termination (R1 and R19), and the second is to supply an output current into Q16 and Q21 emitters whose quiescent voltage is zero. Little current gain is required, as evidenced by the ratio of R1 to R10. One milliamp into each 75 Ω load would result in 1 1/2 mA into the emitters of Q16 and Q21. Note that there is also a 6 mA standing current from the previous stage which caused the input common mode voltage to be –0.45 V. Q1, Q2, Q23, and Q26 drop the common mode input voltage by another 1.6 V to prevent saturation of Q7 and Q20.

Figure 14-10 shows how the circuit can be simplified. The entire amplifier is eliminated! There is a 33% loss in current gain, but that seems like a small price to pay for saving 30 transistors, reducing the noise, and increasing both the linearity and bandwidth. In fact, since this circuit was replicated four times, the actual savings amounted to 120 transistors and 300 milliwatts!

There is just one hitch. A common base stage, driven from a low impedance, can generate thermals. Fortunately, there is a way around this. Eliminate the reverse termination on the previous stage and depend upon the forward termination to hold down the fairly quick reflections. The side advantage of this is that the signal lost in the reverse termination is now available to the following stage. Taking out 120 transistors, 104 resistors, and 300 mW increased the bandwidth, reduced the noise, improved the linearity, and saved money. Transistors are not free. Good engineering is elegant and simple.

15. Understanding Why Things Don't Work

In an early attempt to build an electric light, Thomas Edison used a particular construction that glowed brilliantly for a brief moment and then blew out. An assistant made a remark about the experiment being a failure, and Edison quickly corrected him. The experiment had yielded important results, for they had learned one of the ways that wouldn't work.

Learning through our mistakes doesn't apply only in the areas of dealing with IRS agents or meeting "interesting" people in bars—it's also one of the most important aspects of the creative process in engineering. A "failure" that is thoroughly investigated can often be more beneficial in the long run than success on the first try. Understanding why something didn't work yields several benefits: (1) deeper knowledge of basic principles—air in a bulb allows a filament to burn out easily. (2) faster progress due to the elimination of many alternatives—cancel all planned experiments that don't involve a filament in a vacuum. (3) solutions for other problems—maybe photographers could use a light that shines brilliantly for a brief instant to take pictures at night.

Explanation as a Design Tool

The key to achieving these benefits is that all results must be explainable. A common definition of the difference between an engineer and a scientist is that the engineer makes things work and a scientist finds out why things work. Success in designing analog integrated circuits requires a combination of both. Design tools have evolved to the point that experimentation can be done at a very rapid pace. It's unfortunate, however, that those tools are often misused. The effects of a component value change or a modified connection can be evaluated quickly with computer simulation, but the engineer is learning very little if he continues tweaking his design on the computer without asking these questions:

1. What do I expect the result to be if I make this change?
2. Was the last result different than I expected, and if so, why?

While it may be easy (and even somewhat addictive) to submit run after run on the computer, the best results for the current task, as well as individual career development, are obtained by thinking about each experiment carefully. This doesn't mean one should limit the number of experiments. The process of invention for me involves using a lot of paper. Starting from the existing circuit that comes closest to meeting my goals, I draw a deviation "almost" randomly, based on a guess as to what to change or add. Hand analysis then shows why that alternative won't work, and that knowledge is used to refine the next guess. The beauty of this approach is that it often doesn't matter how bad the initial guess is if each attempt is analyzed and adds to your knowledge base. In most cases, I don't start computer simulations until I think that I'm nearly done with a design. The computer is used for analysis and verification, not synthesis.

What if you have no idea what to change in a design to meet a particular specification? The computer can also be a wonderful tool if you don't quite know what you're doing. When people were trying to figure out what the building blocks of matter were early in this century, they shot beams of electrons and other particles at various targets. The idea was that "something may happen, and if it does, and if it's important, and if I can explain it, then I'll be famous." An example here would be the frequency compensation of a new amplifier configuration. After all the obvious techniques have been tried, put a capacitor in a nonobvious place and test for stability improvement. By definition, each trial yields new information (since you didn't know what you were doing in the first place), and it's important to take time to understand the results. If you can't understand the cause of a particular effect, find a colleague and discuss the experiment. The key points here are "try stuff," and more importantly, "explain the results."

Be especially careful in understanding improvements that go well beyond your expectations. A young engineer once proudly told me that his amplifier had a supply rejection better than −150 dB. This was more than a thousand times better than similar circuits achieved, so I questioned his results. Eventually we found that he was using a 1 F capacitor to couple his signal to the amplifier, but the simulator interpreted his "1.0 F" capacitor value as 1 fF, or a billion million times lower than it should have been. Resimulation with a "1.0" capacitor value yielded results that were less spectacular but believable.

An important rule here is that *all* discrepancies must be explainable, even good ones. One of my goals in a recent design was to build a voltage reference with moderate drift performance over temperature. The first computer analysis of the chosen configuration showed a drift of less than 0.03% over a wide temperature range, while I had only needed and expected a drift of 0.5%. The alternatives at that point were to accept my good fortune, file for a patent, and move on to the next task, or to investigate the discrepancy. Unfortunately, in the attempt to understand why the performance was so good, I found that the particular temperature compensation was dependent on variables that would not be correlated in production of the part. Actual production would show many parts performing significantly worse than the original specification. The benefit of this analysis, however, is that I was able to catch this error immediately and minimize the amount of time going down an incorrect path.

Even supposedly inconsequential discrepancies should be investigated fully. For example, if a product does something that you wouldn't expect when tested incorrectly, the first urge is to tell yourself that it's not important, since it won't be used that way in the real world. By doing that, however, you might be missing a chance to increase your knowledge base on how things work, or even be missing a clue to a hidden flaw in the product. In characterizing a recent product, I noticed that the part would work incorrectly approximately one out of ten times if plugged into a live socket. Since the end application wouldn't require such handling, there was a strong urge to put the investigation of the problem in the "we'll look at this later if we have the time" category. This is equivalent to sweeping the problem under a rug that's covering a bottomless pit. The analysis would never take place. Intermittent problems tend to be avoided anyway, as if one in ten will magically become one in twenty, then one in a hundred, then never. A good rule of thumb: Where there's smoke, there's a definite, soon to be, when you can least afford it, firestorm that will burn your house down.

The best approach is to stimulate the product in some other way to increase the frequency of failure to allow better investigation of the fundamental problem. In this particular case, it was found that the same incorrect operation occurred each time the power to the circuit was applied too quickly. In normal laboratory testing, the power supplies increase relatively slowly when turned on, and the circuit behaved normally. The part would behave incorrectly, however, each time the wire from the circuit board was connected to an already turned on supply. Live switching of circuit boards is done occasionally and would in fact be done in the system that uses this product. A second product development iteration, with all its depressing consequences, would have been necessary if the engineer characterizing the part hadn't said to me, "It may not be important, but a funny thing happens when . . ."

Basic Tools for Understanding

Learning from our mistakes is easy in areas such as eating foods that are too spicy or insulting people who are too large and violence prone. The lesson is clear, and we can move on, although perhaps with a limp. In a technical field, such as integrated circuit design, the mistake that leads to a discrepancy with your intended result is often difficult to find, and the lesson can easily be misinterpreted or ignored. Another problem is that one has to be reasonably close to a solution to make meaningful mistakes. For example, if Edison had tried to make his filament out of a ham sandwich, a failure using rye bread wouldn't have led him down a path of rapid progress. While it's certainly possible to succeed with blind luck, a solid background in fundamentals is the surest way to get close to a solution and successfully interpret the lessons from trial and error. Edison knew enough about physics, for example, to rule out filament experiments using the major food groups. In analog circuit design, a minimum technical platform is elementary device physics—how transistors, resistors, and capacitors work, and how their properties change versus temperature and fabrication variations. While experimentation can yield this information, the simplest way to get these fundamentals is from textbooks.

Armed with only this knowledge, it's possible to create a new circuit design, although it's more likely that the infinite number of monkeys will type out several episodes of "Mister Ed" first. Familiarity with functional building blocks can dramatically improve the odds. In addition to understanding the properties of building materials and the proper use of tools, a carpenter builds a structure using well known building blocks, such as "A" frames, 2×4 studded walls, and supports bolted to a foundation. Most successful analog circuit designers carry a "bag of tricks" that includes blocks such as current sources (e.g., temperature dependent, temperature independent, supply dependent, supply independent), current mirrors, differential pairs, and single transistor amplifiers. Many engineers can recognize these blocks, but the more successful designers are the ones who understand the fundamentals and know why these circuits behave as they do. For example, a simple current mirror works because two transistors have the same base-to-emitter voltage and therefore the same current, or a proportional-to-temperature current source works because the base-to-emitter voltages of two transistors change differently with temperature if they have different current densities. Just using these blocks without having the underlying knowledge of why they work is like following a cookbook recipe without knowing what the individual ingredients add to the mixture. If the taste isn't quite right, or you want to modify the mix for a slightly different result, where do you start your trials and errors? You can still learn from your mistakes, but you're liable to have a very messy kitchen before you reach your goal. A working knowledge of these analog building blocks can be obtained through textbooks, college coursework, or work experience.

Once a designer has mastered these fundamental blocks, he or she can recognize them in the work of others and see their application in problem solving. I once gave a lecture on fundamental building blocks, and someone in the group later told me that he wished the lecture had dealt with more complicated material. The next day, another meeting with the same group began with them explaining a problem that they were having in meeting a certain functional specification. The solution that I proposed used only a combination of the basic blocks that we had discussed the day before. What originally appeared to be a complicated final circuit was actually a simple collection of basic blocks.

The real trick, of course, is to know how to hook the pieces together properly. Three ways to do this are:

1. *Luck.* Try all permutations in no particular order. This is very inefficient on a large scale, but can actually be used to find the last piece of the puzzle. A technical example might be an amplifier that is meeting all of its specifications except a required gain versus temperature relationship. Changing from a temperature-independent current source to a temperature dependent current source in one section of the circuit might be enough to meet the specifications. It would still be important to investigate and understand why the solution works once you discover it.

2. *Experience.* As the years pass, I find myself saying more and more, "These specifications are very much like ones I was given a few years ago and I met those by . . ."

3. *Borrowed experience.* Read journals and research the work of others and see how they solved similar problems. While I'm not compulsive about the number of circuits I look at, I am obsessed with understanding everything that I see. Each component placed in a design has a purpose, and I can't rest until the function of each is clear. First the fundamental blocks are discerned. Perhaps they are linked in a new combination to perform an interesting function. The fun part is then trying to figure out the function of the remaining stuff.

Attend conferences and ask questions. If possible, search out the authors and see if they are willing to discuss their work. Journal papers and conference talks are nice in that they show a final solution, but they rarely discuss the approaches that didn't work. Find someone in your workplace who has more experience and ask more questions. Technical people are generally very proud of their work and are quite willing to discuss it. This is probably due to the fact that they have no other outlet to vent their pride, since their families and friends outside the IC industry have very little appreciation for the subtleties of switched capacitor filters or current mode feedback. Outside the industry, my dog is the only one who will sit and listen intently to a dissertation on tracking down circuit latch-up—at least as long as I'm holding a doughnut in my hand. The borrowed experience can be repaid by publishing your own work or helping others in their tasks.

Integrated circuit complexity has grown to the point that there are plenty of opportunities to make mistakes during the design process. If these are investigated fully, the designer can add to his or her knowledge base with each one. Success doesn't always depend on knowing what you are doing, but continued success depends on knowing what you did.

A. Paul Brokaw

16. Building Blocks for the Linear IC Designer

Linear Synthesis for Monolithic Circuits

It was difficult to title this chapter, since it appears to be a series of design examples, but I intended it to be a demonstration of a viewpoint. In solving problems, viewpoint is extremely important. Often, if a problem can be well defined, a good solution will appear obvious when the facts are seen from the proper viewpoint. An alternative approach from the "wrong" viewpoint may lead to no solution or perhaps a poor one that seems to be the best we could do under the circumstances.

In school the emphasis on the solution of difficult hypothetical problems often causes us to adopt attitudes that work against us when solving real problems. We learn to apply very complex and complete theories to problems which might yield to simpler, and in some sense more powerful, methods. Rather than use our most rigorous (and difficult) methods we should learn to look for the simplest approach that can be made to work. In this way we will save our "big guns," and our creative energies, for the problems that really require them. We will also minimize cumbersome or overconstrained designs. As an aid to developing the skill that helps select a useful viewpoint, let's examine the concepts of *analysis* and *synthesis*.

Circuit analysis is the determination of the performance or response of a circuit, given its configuration. Analysis is useful in predicting the response of a predetermined circuit configuration and quantitatively evaluating the response, if circuit parameters are given. The value of analysis as background and as a means for gaining new insight must not be minimized. Many branches of analysis are well organized and can proceed according to a predetermined method or algorithm. Extensive circuit analysis programs have been written for computers to relieve engineers of many aspects of "routine" analysis.

Circuit synthesis is, in contrast, the determination of a circuit configuration and proper component values to realize a predetermined response and level of performance. Although there are synthetic procedures to choose values for a few canonical circuit configurations, formalized methods for selecting circuit configurations are almost nonexistent. Because of the almost limitless possibilities for useful configurations, synthetic procedures are limited to a small fraction of the circuits one would like to synthesize. Since a great deal of choice and judgment is involved in synthesis, synthesis requires much of the circuit design engineer and is his or her proper occupation.

Just as there is little formalized synthesis, it is difficult to say, with any precision, what it is that the circuit designer or "synthesist" does that is different from or in addition to what the analyst does. There seems to be an intuitive process that helps the synthesist select a plausible configuration from among many possibilities for a given design.

I say *plausible* because design or synthesis differs from analysis in an important way. The physical existence of the network to be analyzed implies the existence of some response (including zero) to excitation. The analyst's problem is just to find that response. The synthesist, however, has no such guarantee. The existence of a desire for a particular response has no implicit guarantee that it is possible to devise a circuit to provide it. In fact, a large portion of the science of passive network synthesis (one of the few organized areas of synthesis) is concerned with determining whether the desired response is realizable from a network of the kind for which a synthesis (of values) exists. Or, indeed, whether the response is possible at all. Responses which implicitly call for the effected output to precede the input signal seem to be impossible, although this implication in some specified responses is not always clear at the outset.

Anyway, the synthesist is faced with the dual problem of finding "an answer" and finding whether "an answer" exists. As in many other fields, experience often helps in finding a suitable circuit "answer." It also helps in knowing when to stop, when no answer exists. It is in this last area that experience shows its negative aspect, by the way. The inexperienced analyst may continue to search for an answer long after "any sensible person" would have realized that there is no answer to the particular problem. In so doing he may discover the answer which is at the same time "impossible" and "obvious (now that I've seen it)." Granted that these marathon efforts result most often in nothing but fatigue, many breakthroughs in circuit design and other areas are made by people who just didn't know when to quit. This effect may contribute to the observed fact that many great discoveries are made by relatively youthful and inexperienced investigators. At least I prefer this theory to the supposition (generally by youthful investigators) that one's mental powers generally decline after 25 or 30 years of age.

Getting back to that intuitive aspect of the problem, which is so difficult to specify, it seems very much like a guess. That is, the synthesist guesses the configuration to be used and possibly even the circuit parameters and then uses analysis to check to see if the guess was correct. Frequently, very little analysis is required to reject the first few faulty guesses. This trial and error process is often entirely mental, with no need for drawings or written analysis. The rapid testing of many possibilities requires a ready supply of manageable, flexible concepts from which to choose.

The process may be something like putting together a jigsaw puzzle, but not an ordinary puzzle. The ordinary puzzle is a bit like analysis. The pieces are presumably all there, and it's just a matter of fitting them together. Synthesis may be a bit more like trying to create a new picture by fitting together pieces taken from many puzzles. We must be prepared to trim each piece to fit, and perhaps even to create a totally new piece when we've identified its shape but can't find the right shape or color in our supply.

The reason I find the jigsaw puzzle analogy appealing is that in both synthesis and analysis, in some sense, we must grasp the big picture, but to complete the job, the problem must be divided into subproblems of a manageable size. Rarely are design "guesses" made which are complete and detailed for a circuit of appreciable complexity. Rather, the first guess is a sort of block diagram or operating principle concept. Then, if this idea stands simple tests of feasibility, the design proceeds by use of several concepts linked by the overall principle, each of which can be further divided by linking more detailed ideas. Proceeding in this way, the puzzle is reduced to individual "pieces," which are simple combinations of basic elements, the function of which can be completely grasped conceptually.

I want to stress this emphasis on simplicity. The most elegant and pleasing solutions are those that can be easily grasped and understood. The best design configurations often appear "obvious" once we've seen them, despite the fact that many people may have tried and failed to come up with any satisfactory solution. A reason for this phenomenon is that when most of the puzzle has been properly assembled, the shape and color of the key missing pieces is often plain to see. The trick is in getting the "easy part" of the puzzle together in such a way that the key pieces exist and become "obvious."

In order to flesh out the design, many "standard" kinds of pieces will be required. The jigsaw puzzle needs flat pieces on the sides, corner pieces to bond them, probably some blue pieces near the top, and so forth. Designers need a large store of ideas from which to choose. Often, in fact, designers discover clever ways of doing something and seek out applications. For some of us, the subconscious trials of "solutions looking for problems" against current problems is a source of inspiration and new approaches.

These ideas need to be simple enough to be easily manageable conceptually. That is, we need to be able to easily and rapidly manipulate the ideas and test them for fit and "color." To keep the ideas simple, we deal with approximate models which emphasize the key parameters of the circuit and neglect the complications. Very often our synthesis depends upon these simple models, and the problem becomes one of arranging the circuit to minimize the analytical errors or the effects of the errors which result from simplifying assumptions.

P–N Junctions

A great deal has been learned about the behavior of P–N junctions and their associated semiconductor structures. Without minimizing the value of this work, I would like to ignore most of it and use the so-called diode equation, which gives the conducted current in terms of the applied voltage and a factor I_s, which is related to the area of the junction and how the junction is made. I will neglect the resistance of the semiconductor leading to the junction as well as high injection effects and frequency-dependent effects. I will then try to use the junction in such a way that the results of this use are not much affected by the neglected parameters.

Looking now at a simple P–N junction (Figure 16-1) the conduction current I is related to the applied voltage V by the equation:

$$I = I_s\left(e^{qV/kT} - 1\right)$$

where q is the electronic charge, k is Boltzmann's constant, and T is absolute temperature.

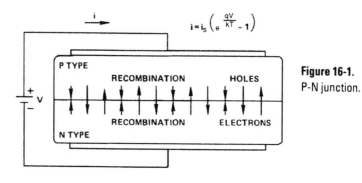

Figure 16-1.
P-N junction.

Figure 16-2.
An npn junction
transistor.

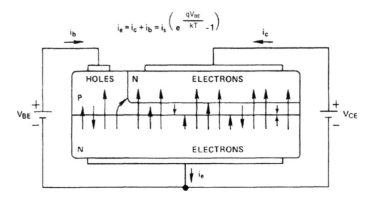

When the junction is forward biased, as shown in Figure 16-1, electrons which are the majority carriers in the N-type semiconductors cross the junction and recombine with holes in the P-type region. Holes in the P region cross into the N region and also recombine. The sum of the two resulting currents is I and is exponentially related to the applied voltage.

Junction Transistors

Assume now that we diffuse into the diode of Figure 16-1 an N-type region, as shown in Figure 16-2. Assume also that we continue to forward bias the basic P–N junction with the same voltage V (now called V_{be}) and that the added N-type diffusion is biased positive with respect to the original N diffusion.

The current crossing the original P–N junction will be approximately the same function of the voltage applied to the junction. If the new N region can be placed close enough to the original junction, however, many of the electrons which cross the junction will cross into the new region before they recombine in the P region. They will be attracted by the potential applied to the added N diffusion, and once they reach it they will be the majority carriers again and will constitute a normal current in that loop.

The three electrodes in Figure 16-2 are called emitter, collector, and base. The emitter is the original N-type region which emits (across the junction) the carriers which will be collected by the added N region, the collector. The P region sandwiched between these two N-type regions is called the base, for mechanical rather than functional reasons. Early fabrication procedures used this region as the substrate, or "base," to which the other regions were applied.

Since the forward biased junction is essentially the same as it was in the P–N diode, the current across it due to electron and hole flow is the same function of applied voltage as in the diode. This entire current flows in the emitter, so that the emitter current is given by:

$$I_e = I_s \left(e^{qV_{be}/kT} - 1 \right)$$

where V_{be}, the base-emitter voltage, is simply the voltage applied to the original junction. Some of this current will recombine in the base, particularly at the left side, where the path to the collector is very long. In the thin region of the base, however, most of the electrons crossing the junction will be swept up, and saved from recombination, by the positive collector voltage. Notice that the collector base junction is reverse biased so that very little current flows by collector base conduction. (Actually, a rigorous analysis would include a reverse biased diode term to account for the theoretical collector-base current. Since this reverse saturation current will be more than eight orders of magnitude smaller than the probable collector current, and also smaller than the "real" leakage due to "real" transistor defects, I'll neglect it). The collector current then is, in turn, governed by base voltage. The electrons which recombine in the emitter region will constitute the current flowing in the base loop. Therefore, the total emitter current will be the sum of the collector and base current, thus:

$$I_e = I_b + I_s = I_s\left(e^{qV_{bc}/kT} - 1\right)$$

The transistor of Figure 16-2 will have a large base current. If the initial P and N regions are equally doped so that they have equal densities of available electrons and holes, the hole current from base to emitter will constitute a large fraction of the emitter current. Since this current recombines and doesn't contribute to collector current, it will be part of the base current.

Since part of the base region is reserved for contacting, electrons injected into this part of the base must travel a long way to the collector. Most of these electrons will recombine with holes before reaching the collector and so will contribute only to base current.

Therefore, the base current of the transistor in Figure 16-2 may be as much as half the emitter current. Notice, however, that the collector current will be a more or less fixed fraction of the emitter current so that as base voltage increases, the collector current will increase in proportion to emitter current.

An Improved Junction Transistor

Generally, we will want to control the collector current of a transistor by means of the base voltage while providing a minimum amount of base current. One measure of quality of a transistor is β or h_{FE}, the ration of I_c to I_b. For a given emitter current, then, we would like to maximize the fraction of the current which goes to the collector and minimize the base current. A second useful measure of transistor performance is α, the common base current transfer ratio I_c/I_e.

Figure 16-3 illustrates an npn structure similar to Figure 16-2, but which has been modified to minimize base current.

Notice now that the smaller N+ region is used as the emitter, rather than as the collector. This causes most of the injected emitter electrons to cross the junction where the base is thin and minimizes the number which must travel a long distance to the collector. In practice, the gradient of P-type impurities in the base will further concentrate emission at the upper edge of the lower N+ region and minimize lateral

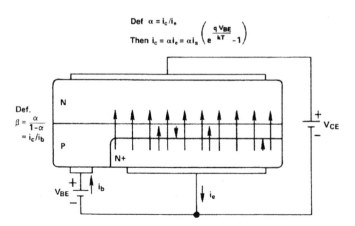

Figure 16-3. High β npn transistor.

injection from the left hand wall of this region. As a result, most of the injected electrons need only travel a short distance through the base before being rescued from recombination by the collector.

A second important step is to raise the concentration of N-type impurites in the emitter region. This high concentration is symbolized by the "+" in the N+ notation. By raising the electron donor (N dopant) concentration in the emitter, several orders of magnitude above the hole or acceptor concentration in the base, we can unbalance the ratio of electrons and holes crossing the junction. The number of electrons far exceeds the number of holes. As a result, the current of holes which recombine in the emitter can be made very small.

The transistor of Figure 16-3 is assumed to have been optimized so that base current is small. We have seen that the emitter current is determined by base voltage. Since for a given transistor the split of this current between collector and base current is more or less constant, the base current as well as the collector current is related to base-emitter voltage, V_{be}. It will most often be convenient to treat the collector current as an exponential function of base voltage and to regard the base current as a nuisance which will be neglected whenever it is practical to do so. This approach is useful in discrete transistor design and is particularly so in linear IC design, where the close matching of transistors makes it possible to treat collector currents and base voltages of several transistors using the same value for I_s.

Since we wish to relate collector current to base emitter voltage, it will be convenient to use α the collector current transfer ratio. Multiplying through by α and substituting for αI_e in the emitter current equation gives:

$$I_c = \alpha I_e = \alpha I_s \left(e^{qV_{be}/kT} - 1 \right)$$

In the discussion that follows, it will be assumed that the transistors are identically made so that for any two or more transistors the parameters I_s, α, and β, each have a single value common to all the transistors.

Simple Current Mirror

The value of I_s for a given transistor depends not only on its geometrical arrangement but also on the impurity doping levels and diffusion depths which govern base thickness. In an integrated circuit, many transistors are made at the same time in one circuit and have virtually identical doping profiles. If they are made alike geometrically they will all have the same I_s so that transistors operating at equal base

Figure 16-4.
Npn current mirror.

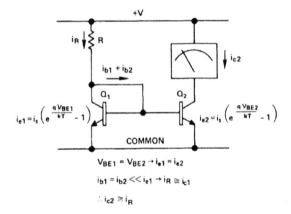

voltages will have equal collector currents. The circuit shown in Figure 16-4 takes advantage of this property to make a simple current repeater, or "current mirror."

For the sake of a simple explanation, input current to the mirror is shown to be determined by a resistor R connected to $V+$. The basic current mirror will "accept" current from any source with the proper compliance voltage range; however, the resistor input scheme of Figure 16-4 is frequently used to generate bias currents which are proportional to supply voltage in integrated circuits.

Initially I_R, the input current to the mirror, causes the voltage at the base of Q_1 to rise. As it rises, the collector current will increase and an equilibrium will be reached when the sum of the collector current of Q_1 and the base currents supplied equals I_R. Since the base and emitter of Q_2 parallels Q_1, and assuming that the collect voltage of Q_2 is prevented from going into saturation, the collector current of Q_2 will equal the collector current of Q_1.

The collector current of Q_1 will differ from I_R only by the sum of the two base currents. Assuming high β transistors are used, this difference will be small and the collector current of Q_2 will closely approximate I_R.

The simplified analysis indicates that the mirror output (collector current of Q_2) will be slightly less than I_R. In practice, the output current is frequently a little larger than the input current due to the effects of collector voltage. Diode connected Q_1 operates with its collector voltage at V_{be}. The collector voltage of Q_2 is higher and modifies the relationship between I_c and V_{be} so that the collector current is slightly increased.

A more rigorous analysis could include this effect in terms of a transistor output impedance or conductance, and some sort of collector base feedback factor. Unfortunately, these parameters are often a strong function of the wafer processing variables, so that although their effects may be small, the circuit-to-circuit variability in their effects is large. It is often important to determine the limits of these effects, but in designs where output currents must be precisely determined it is usually more practical to use a circuit which is less sensitive to them.

Improving the Current Mirror

The simple current mirror of Figure 16-4 is usually adequate for biasing and many other IC design applications. Sometimes, however, base current errors and low

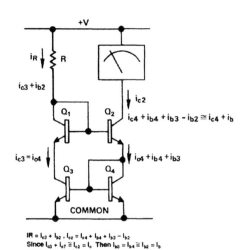

Figure 16-5.
Improved Wilson current mirror.

output impedance cannot be neglected. The circuit of Figure 16-5 substantially reduces both of these errors. The circuit uses four transistors to yield the improved function, although Q_1, which is diode connected, can be omitted when intermediate performance is required.

When voltage is first applied, the bottom of the resistor R drives the base of Q_2 positive. The emitter of Q_2 follows, driving the bases of Q_3 and Q_4 positive. As Q_3 is driven on, it begins to draw current from the resistor through the diode connected Q_1. An equilibrium is reached when Q_3 is drawing most of the input current, I_R, from the resistor.

Since Q_3 and Q_4 have common base and emitter connections, the collector current of Q_4, I_{c4}, will equal the collector current of Q_3, I_{c3}. Since Q_2 must supply the current for Q_4 when it drives the common base connection, I_{c4} will flow through Q_2 as part of the output of the current mirror. Since Q_3 and Q_4 match and supply the major portion of input and output current, respectively, the output of the mirror will have a first order match to the input current.

The total input current will divide into two components. The main one is the current I_{c3}, which flows through Q_1. A smaller current also flows to the base of Q_2. Therefore, the total input current, I_R, is given by:

$$I_R = I_{c3} + I_{b2}$$

The emitter current of Q_2 includes the base currents of Q_3 and Q_4, as well as the collector current of Q_4. This same current, less the base current of Q_2, will be the output current from Q_2.

Notice that all four transistors operate at a collector current which approximates (within an error of a single base current) the input current I_R. This means that the base currents should all be approximately equal. To the extent that they are equal, the collector current of Q_2 is I_{c4} plus two base currents, minus one base current, or just $I_{c4} + I_b$ where I_b represents the magnitude of the approximately equal base currents. Now, this current very nearly equals the input current, which is I_{c3} (which equals I_{c4}) plus one base current. The input and output currents now differ only by the difference in base currents. And, since the collector currents are so well matched, the base current errors are very small. A more rigorous analysis (which you may enjoy doing) will show the residual base current errors to be related to β^{-2} assuming that all the transistors match ideally.

Another source of error in the simple current mirror was due to the change in collector voltage at the output. Notice that in Figure 16-5 the input and output current match depends largely on I_{c3} and I_{c4} equality, since the base currents are presumably small. The diode, Q_1, insures that the collector voltage of Q_3 is at almost identically the same potential as the collector of Q_4, which connects to Q_2's emitter. This equality is nearly independent of the voltage applied to the collector of Q_2 so that the ratio of input to output current is largely unaffected by output voltage variation.

Of course, this circuit is not ideal. There is the β^{-2} residual error in base current, about 0.0002 or less for $\beta \geq 100$. there is also the modulation of the base current in Q2 by output voltage. However, if $\beta \geq 100$, base current is not more than 1% of the output current, and slight variations of it are a small fraction indeed.

The circuit in Figure 16-5 is a high performance current mirror suitable not only for biasing but also for many direct signal path applications. The configuration can be inverted to make a "positive" current mirror by the use of pnp transistors. In the case of a junction-isolated IC design application using pnp transistors, special saving are possible. The transistor pair Q_1 and Q_2 and the pair Q_3 and Q_4 can each be fabricated in a single isolation region. Lateral pnp transistors are diffused into an isolated N-type region, which forms their base. Multiple pnp transistors which share a common base can be put into the same isolation "pocket."

As was mentioned before, the diode connected transistor Q_1 serves only to equalize the collector voltage of Q_3 and Q_4. In npn current mirrors, Q_1 is often omitted (and its terminals shorted) since only a small error results. In the case of pnp current mirrors, however, the small incremental cost of Q_1 (in terms of chip area) often justifies its inclusion.

A Simple Transconductance Model

Since the collector current of the junction transistor depends, to a good approxima-tion, on the base voltage, the transconductance of a transistor is a useful concept. Transconductance is useful both in the conceptual sense that it relates it to the func-tioning of the transistor, and in the synthetic sense as well. I'm stressing the syn-thetic utility over the analytic, since there are more complete analytic models which better describe a particular transistor. The simple transconductance model is based on a property of the device which is more dependable and reproducible over the range of IC manufacture than are some of the more complete analytical descriptions.

That is, a circuit which is designed to make use of transconductance, making allowance for the approximate description, will be more dependably manufactur-able than one which depends in some critical way on more rigorous analysis, which may allow less tolerance for variation in device parameters. The rigorous analytical models are often more useful for examining the performance of a proposed synthesis at the limits of its functional parameters.

Referring now to Figure 16-6, a conventional transistor is looked at as an "ideal" transistor with "infinite" transconductance, combined with a series emitter resistor r_e. Looked at in this way, the tranconductance of a "real" transistor can be modeled as the reciprocal of r_e, the intrinsic emitter resistance.

To see how the collector current varies, incrementally, with changes in base voltage we need only to differentiate the expression for collector current. That is:

$$I_c = \alpha I_s \left(e^{qV_./kT} - 1 \right)$$

$$\frac{dI_c}{dV_{be}} = \frac{\alpha q}{kT} I_s e^{qV_./kT}$$

Using the Equation $i_c = \alpha i_s \left(e^{\frac{q V_{BE}}{kT}} - 1 \right)$

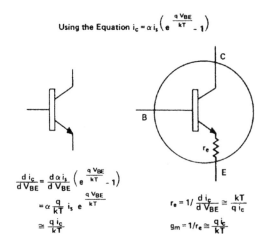

$$\frac{d\,i_c}{d\,V_{BE}} = \frac{d\,\alpha\,i_s}{d\,V_{BE}} \left(e^{\frac{q V_{BE}}{kT}} - 1 \right)$$

$$= \alpha \frac{q}{kT} i_s\, e^{\frac{q V_{BE}}{kT}}$$

$$\cong \frac{q\,i_c}{kT}$$

$$r_e = 1/\frac{d\,i_c}{d\,V_{BE}} \cong \frac{kT}{q\,i_c}$$

$$g_m = 1/r_e \cong \frac{q\,i_c}{kT}$$

Figure 16-6.
Simplified trans-conductance model for an npn transistor.

Note that for normally useful values of V_{be}, $e^{qV_{be}/kT}$ is many orders of magnitude larger than 1 so that the error in the approximation:

$$I_c = \alpha I_s e^{qV_{be}/kT}$$

is extremely small. Substituting this approximation in the derivative yields:

$$\frac{dI_c}{dV_{be}} = \frac{qI_c}{kT}$$

which is the incremental transconductance, or g_m or the transistor. Alternatively, we can look at the reciprocal of g_m, which is r_e, thus:

$$r_e = \frac{kT}{qI_c}$$

This simple model can be used as the basis of many conceptually simple circuits. Given prearranged nominal operating conditions for the transistor, we can easily predict the effect of small changes in base voltage on the collector current. For example, this model used in a simple resistively loaded gain stage yields an approximation for voltage gain which depends simply on the ratio of the load resistor to the value of r_e.

Improving the Simple Current Mirror

The transconductance model can be used to improve the simple current mirror (of Figure 16-4) by compensating for the base current error. In the simple current mirror, the two collector currents are equal, but the base current component of the input current is not "measured" and does not appear in the output. In the circuit of Figure 16-7, one of the base currents is measured and used to adjust the output current. The collector current of Q_2 is made larger than that of Q_1 by enough to compensate for the "lost" base current.

When voltage is first applied to the circuit of Figure 16-7, both transistors have their bases driven on by current through the resistor, R. As the base voltage of Q_1 increases its collector current increases until it sinks most of the input current I_R. The collector current of Q_1 differs from I_R only by the base currents of the two transistors, as in Figure 16-4. In Figure 16-7, however, the base current of Q_1 passes through an additional resistance, r. The base current develops a small voltage across this resistor. When an equilibrium has been reached, with Q_1 supplying I_R less the two base currents, the base of Q_2 will be slightly more positive than the base of Q_1 due to the voltage drop, ΔV, across r. Now, if this ΔV were zero, as in Figure 16-4,

Figure 16-7.
Improved simple
current mirror.

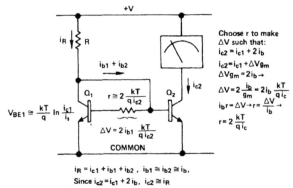

I_{c2} would be the same as I_{c1}. The incremental change in I_{c2} due to ΔV can be approximated by use of the tranconductance model. That is:

$$\Delta I = \Delta V g_m$$

where ΔI is the difference between the collector current of I_{c1} and I_{c2}. If ΔI is made equal to the two base currents $I_{b1} + I_{b2}$, then the collector current of Q_2 will equal I_R, the mirror's input current. Since the total collector current of Q_2 will differ from that of Q_1 by only a small amount (an amount $\Delta I = I_{b1} + I_{b2}$) the base currents of Q_1 and Q_2 differ only slightly. Neglecting the difference, we can set each base current equal to I_b without suffix. Then $\Delta I = 2 I_b$. The voltage ΔV is just $I_b r$ and the transconductance g_m is $q\, I_{c2}/kT)$ taken from the model of Figure 16-6. Substituting in a previous equation we have:

$$2I_b = I_b r \frac{qi_{c2}}{kT}$$

and solving for r, I_b drops out to yield:

$$r = 2 \frac{kT}{qi_{c2}}$$

This means that if the value r is selected to be $2kT/(qI_{c2})$, for the desired value of I_{c2}, the collector current of I_{c2} will be almost exactly equal to I_R, which is the input current to the mirror.

This circuit has several weaknesses. However, it serves to illustrate the use of the tranconductance model and is the starting point for slightly more involved circuits which use the base current compensation scheme. One of the shortcomings, pointed out in the discussion of the simple current mirror, is that the output current of Q_2 is affected by its collector voltage. In some applications this effect is small (when $V_{CE2} \approx V_{be}$) or may be compensated for elsewhere in the circuit.

A second shortcoming is that this mirror is properly compensated for only a particular value of I_{c2}, the output current. This is because I_{c2} is part of the information for the value of r. Moreover, the "right" value of I_{c2} is a function of temperature. If r is presumed to be temperature invariant (this assumption may be a poor one in all diffused IC technology, but works well with Analog's Thin Film on Silicon process) then I_{c2} must vary as absolute temperature T, if r is to remain optimum. As it happens, there are two major types of current bias in IC designs. One is temperature invariant current, for which this compensation scheme works poorly at wide extremes of temperature. The other is current *proportional to absolute temperature* (PTAT) for which this compensation method is ideal. This type of compensation can be used to good advantage in bias schemes involving PTAT currents.

Another Important Relationship

The transconductance model will be an important part of an interesting sort of current mirror. Before looking at that circuit we will need another expression which is related to the transconductance model but can generally be realized with more precision and reproducibility. This relationship is very frequently used in IC designs because of its analytical power and also because it can be used over wide signal or bias ranges with extremely small error.

Suppose that two identical transistors are biased to some "reasonable" collector

Figure 16-8.
The most power-
ful relationship in
IC design.

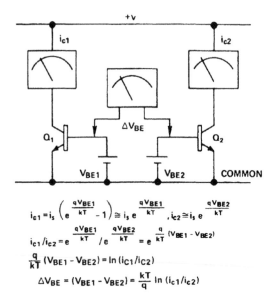

$$i_{c1} = i_s \left(e^{\frac{qV_{BE1}}{kT}} - 1 \right) \cong i_s\, e^{\frac{qV_{BE1}}{kT}}, \quad i_{c2} \cong i_s\, e^{\frac{qV_{BE2}}{kT}}$$

$$i_{c1}/i_{c2} = e^{\frac{qV_{BE1}}{kT}} / e^{\frac{qV_{BE2}}{kT}} = e^{\frac{q}{kT}(V_{BE1} - V_{BE2})}$$

$$\frac{q}{kT}(V_{BE1} - V_{BE2}) = \ln(i_{c1}/i_{c2})$$

$$\Delta V_{BE} = (V_{BE1} - V_{BE2}) = \frac{kT}{q} \ln(i_{c1}/i_{c2})$$

currents by applying fixed base voltages. This situation is illustrated in Figure 16-8. If the voltages are equal, then presumably so are the collector currents. If, however, we cause the base voltages to differ, by an amount ΔV_{be}, the collector currents will also differ. Using our expression for I_c as a function of V_{be} and again neglecting the "-1" in the second factor, we can easily relate the collector currents to ΔV_{be}.

Take the ratio of the two currents and their approximations thus,

$$I_{c1}\Big/I_{c2} = I_s e^{qV_{be1}\big/kT} \Big/ \left(I_e e^{qV_{be2}\big/kT} \right) = e^{q(V_{be1} - V_{be2})\big/kT}$$

Taking the natural logarithm:

$$\ln I_{c1}\Big/I_{c2} = \frac{q}{kT}\left(V_{be1} - V_{be2}\right) = \frac{q\Delta V_{be}}{kT}$$

or

$$\Delta V_{be} = \frac{kT}{q} \ln(I_{c1}/I_{c2})$$

$$\Delta V_{be} = \frac{kT}{q} \ln(I_{c1}/I_{c2}) \qquad \text{(repeated for emphasis)}$$

This relationship in one form or another is one of the most powerful in IC design. In practice, transistor bases aren't usually driven from a low impedance fixed voltage source; however, in many circuits it is easy to determine or to control the voltage difference among two or more V_{be}s. This expression permits us to relate the collector currents of transistors controlled in this way. Alternatively, some circuits operate to control collector currents by means of some sort of feedback to transistor bases. Again, this expression is useful in determining the resulting difference in base emitter voltage. This voltage is an extremely dependable indicator of temperature and is useful, for example, in producing temperature proportional (or PTAT) voltages and currents.

Some useful rules of thumb are that the base-emitter voltages of identical transistors operating at a 2-to-1 current ratio will differ by about 18 mV, at "room temperature." If the same transistors are operated at a 10-to-1 collector current ratio, their ΔV_{be} will be about 60 mV. In general, if the current ratio is expressed in decibel terms, the ΔV_{be} in millivolts will be numerically about three times the ratio expressed in decibels. Obviously, a more accurate and rigorous way to determine ΔV_{be} is to take the natural logarithm of the current ratio and multiply by kT/q which is about 26 mV around "room temperature."

A Current Mirror Using a Zero Gain Amplifier

In development of an integrated circuit it is important to take into account the wide variation in certain device parameters from one production lot to the next. One objective of the designer is to desensitize the design to these variations to produce a consistent product with high yield to guaranteed performance and a dependable collection of incidental (or unspecified) properties as well. Moreover, a general purpose integrated circuit is subjected to a wide range of operating conditions in a variety of applications. It is usually desirable to stabilize the circuits' performance against most of these variations.

One area of linear design which is common to nearly every integrated circuit is the establishment of internal bias level. This seemingly minor portion of the design provides some of the most challenging and interesting design problems. The bias circuitry provides the support of framework on which the functional core of the IC is built.

Many bias structures are concerned with generation of currents to operate the core of the circuit. Frequently small currents are required, and it is also desirable to minimize the current consumed in bias control circuits to limit the power required to operate the circuit. Various "high resistance" elements are available to the designer, including collector FETs. The fabrication of these collector or "pinch" FETs is difficult to control, and when small currents are desired, characteristics such as I_{DSS} may vary by factors of 4 or more among production lots.

Internal bias currents can be derived from a collector FET by use of a current mirror. However, the wide variation in input current derived from the FET and reflected in the output may conflict with dependable operation of the circuit. Alternatively, a resistor may be used to establish bias levels. The value of a normal base diffusion resistor can be more closely controlled. However, the current through it will vary as the power supply voltage. Again, this variation reflected into the bias current levels is frequently undesirable.

All of the foregoing is to introduce a current "mirror" which has an output which has been desensitized to variations in input current. To understand this circuit, it will be useful to examine the behavior of a "zero-gain amplifier," which we can construct using the transconductance model.

To begin with, look at the common emitter amplifier stage shown in Figure 16-9. Imagine that the transistor is biased "on" by the voltage E_b and a small AC signal, e_1, is added to the base drive. The voltage E_b establishes an operating point for the collector current which in turn establishes the g_m of the transistor. The signal e_1 modulates the collector current. If e_1 is sufficiently small, the operating point is not shifted drastically by the signal (which would change g_m). The modulation of col-

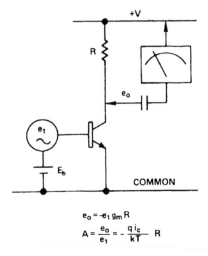

Figure 16-9.
Voltage gain of
common emitter
amplifer.

$$e_o = -e_1 g_m R$$
$$A = \frac{e_o}{e_1} = -\frac{q\, i_c}{kT}\, R$$

lector current through the load resistor R will develop an AC output signal e_o. Now, when the base goes positive, the collector current will increase, causing the collector to go negative, and on the negative base swing the collector goes positive. The output signal is inverted and amplified by this arrangement according to the relationship

$$e_o = -e_1 g_m R$$

An alternative way of looking at Figure 16-9 is to visualize r_e in series with the emitter and to note that when a signal is applied the same current (neglecting base current) flows in r_e and in the load resistor R. Therefore, the gain A, is given by the ratio of R to r_e, or

$$A = -R\Big/r_e = \frac{-qI_cR}{kT}$$

Let's modify the amplifier now by adding a signal generator in series with the load resistor shown in Figure 16-10. Since I've been neglecting the output impedance of the transistor until now, I will continue to do so and assume that the output

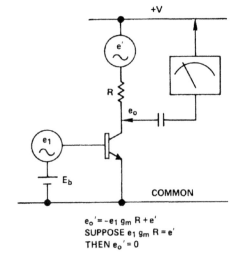

Figure 16-10.
Zero gain
amplifier.

$$e_o{}' = -e_1 g_m R + e'$$
SUPPOSE $e_1 g_m R = e'$
THEN $e_o{}' = 0$

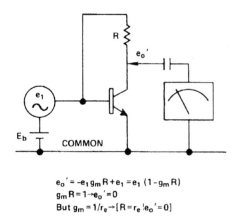

Figure 16-11.
Modified zero
gain amplifier.

$$e_o' = -e_1 g_m R + e_1 = e_1 (1 - g_m R)$$
$$g_m R = 1 \rightarrow e_o' = 0$$
$$\text{But } g_m = 1/r_e \rightarrow [R = r_e \,!e_o' = 0]$$

signal component due to the collector current modulation is unaffected by the signal generator. The new output, e_o', will consist of the original signal $-e_1 g_m R$ plus e' the output of the generator. That is, the overall output is just the linear sum of the amplifier output and whatever signal voltage is applied to the other end of the load resistor.

I suppose it's not at all clear why I might want to make the output zero (when zero signals are so readily available with less complication), but I hope to make it all clear soon. Anyway, let's suppose that the magnitude of $-e_1 g_m R$ can be adjusted (by, for example adjusting R) to equal e', or vice versa. In this case the incremental or AC output will be approximately zero. The result is only approximate due to the small variation in g_m due to the input signal which results in nonzero distortion signals in e_o'. If we now replace e' with the input signal voltage as shown in Figure 16-11 and adjust the gain properly we can again make the AC output approximately zero.

Now, with the load resistor connected to the base of the transistor as we have it in Figure 16-11, we might expect some biasing problems with the collector voltage of the transistor. Remember, however, that we are looking for a gain of only -1 so that R is very small. In fact, R will equal r_e and the voltage drop across it will be about 26 mV at the zero gain point. This slight forward bias of the base-collector junction will have almost undetectable effect on the transistor operation.

In Figure 16-11 we have a transistor biased into conduction with a collector voltage which is less than the base voltage by kT/q, and essentially a zero sensitivity to input voltage changes. The collector voltage will not depart, substantially, from

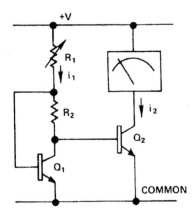

Figure 16-12.
Current stabilizer
using zero gain
amplifier.

this voltage if the input voltage changes and the transistor operating point shifts, slightly.

Transplanting the zero gain amplifier from Figure 16-11 into Figure 16-12, as Q_1 and R_2, gives us the start of a desensitizing current source. When power is first applied to this circuit, R_1 will drive the base of Q_1 positive. As Q_1 comes on it will draw collector current until the circuit stabilizes with some current I_1 flowing in R_1 and most of this current flowing into the collector of Q_1. Let's suppose that in the initial design of this circuit we have a nominal value for R_1 and $+V$ so that there is a nominal value for I_1 (in an IC, I_1 could be provided by a collector pinch FET or other structure). At this value of I_1 select R_2 so that $R_2 = r_e$ for Q1 by setting:

$$R_2 = \frac{kT}{qI}$$

or

$$i_1 R_2 = kT / q$$

If we now change I_1 slightly, it will of course change the equilibrium value of V_{be1}. In other words, changing I_1 from its nominal value changes the input voltage applied to Q_1 and R_2, the zero gain amplifier. We've already seen that with small changes in base voltage the collector voltage of this transistor is almost unaffected. The result is that the voltage appearing at the collector of Q_1 is equal to the V_{be} of Q_1 when it is operating at the nominal value of I_1 minus kT/q. For small changes in I_1 this voltage remains fixed since as I_1 increases, raising V_{be}, the voltage drop across R_2 also increases by an almost equal amount, canceling the effect of the change at the output. If I_1 falls, reducing V_{be}, the drop across R_2 will fall and once more compensate for the change.

If we now direct our attention to the collector of Q_1, we will see that its current should be, to a first approximation, unaffected by changes in I_1 around its nominal value. The base voltage of Q_2 will remain at a voltage which is kT/q less than the value of Q_1's V_{be} when it is operated at the nominal value of I_1. To find the magnitude of I_2 we substitute I_n, the nominal value of I_1, in the relation:

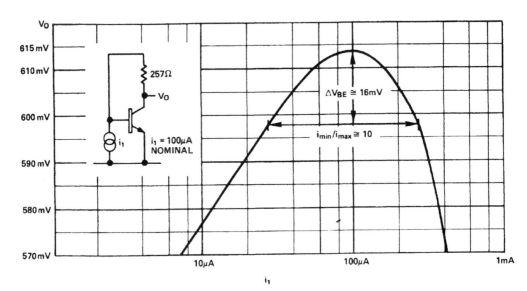

Figure 16-13.
Zero-gain output voltage versus input current.

$$\Delta V_{be} = \frac{kT}{q} \ln(I_n/I_2)$$

which can be arranged as:

$$I_2 = I_n e^{\frac{-\Delta V_{be}}{kT/q}}$$

Since $\Delta V_{be} = kT/q$, then

$$I_2 = I_n e^{-1}$$

This value for I_2 is established at the nominal value of I_1. However, the "gain" of the Q_1 circuit is "zero" for small changes in I_1. Therefore, the base voltage of Q_2 will be almost invariant, and the collector current of Q_2 will be desensitized to changes in I_1.

The output voltage of Q_1 as a function of I_1 is shown on an expanded scale in Figure 16-13. This Figure shows the top 45 mV of this 600 mV curve and illustrates the flattening of the output voltage characteristic when the input current is near the nominal value, in this case selected to be 100 μA. For small changes in I_1, changes in V_o (which determines the current in Q_2 of Figure 16-12) are very small. For ex-ample, a 2:1 change in I_1 results in about a 1.5 mV change in V_o, which corresponds to only a 6% change in the output current of Q_2. That is, the stabilizer will accom-modate a 2:1 change in input current with only 6% change in output. Even for large excursions of I_1 the stabilizer works moderately well. As shown in Figure 16-13, an input current change of 10:1 results in an output voltage change of only 16 mV, which corresponds to a factor of about 1.86 change, less than 2:1, in overall output current.

Since the voltage drop across R_2 is to be set to kT/q, the magnitude of this voltage should, then, change with temperature. If I_1 can be made to have the proper temper-ature dependence (PTAT), this will happen automatically, and the stabilizer output will be PTAT. Typically, however, variations of I_1 with temperature are one of the parameters to be rejected. In this case, R_2 can be made with a diffused base resistor or an epitaxial layer resistor both of which have a positive temperature coefficient

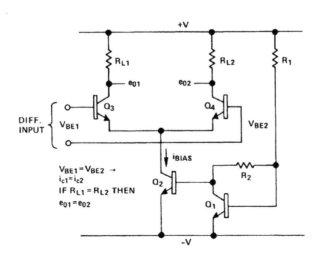

Figure 16-14.
Simple differ-ential amplifier biased by current stabilizer.

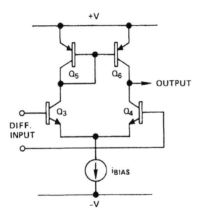

Figure 16-15.
Pnp current mirror converts differential to single ended signal.

of resistance. This compensation is usually less than ideal; however, it is often more than adequate for use in biasing applications.

Variations on the circuit of the stabilizer include connection of multiple output transistors, to the output point V_o, to provide several stabilized currents to a complex circuit. When lower currents are required, an additional resistor in series with the emitter of Q_2 will reduce the output current and improve the rejection of input current variations. Emitter resistors of several different values may be used with multiple devices in complex bias circuits.

A Simple Differential Amplifier

The current stabilizer is shown supplying bias to a differential stage in Figure 16-14. The gain of the stage will depend on the bias current from Q_2 and the load resistors. The use of the stabilizer will minimize variations in gain with supply voltage and may also permit R_1 to be implemented as a pinched epitaxial resistor or pinch FET.

The gain of the stage, as shown, is given by:

$$A = \frac{qI_{bias}}{2kT} R_L$$

where R_L is the value of one of the two equal load resistors. The transconductance of the differential transistors is determined by half the bias current, which is split between them.

In IC designs which use a differential amplifier it is often desirable to minimize the base currents drawn by the transistors. At a minimum β level, determined by process variables, the input current is proportional to the bias current. Reducing this current will reduce the input current, and the gain may be maintained by raising R_L. The current stabilizer output is less than the current in R_1 by a factor e^{-1}. The bias can be further reduced by inserting a resistor in series with the emitter of Q_2. Presumably, R_1 will be made as large as practicable when low bias currents are desired. As previously mentioned, large resistor values are difficult (or actually just expensive in terms of chip area) to fabricate. That is why it's desirable to permit the use of a pinch FET, which can provide lower currents, but with less stability or reproducibility than a resistor.

In order to maintain the voltage gain of the simple differential stage, while decreasing the emitter current, we required an increase in the collector load resistors. These resistors may also become too large to fabricate economically. In IC designs it is frequently convenient to substitute the output impedance of a transistor (which we have neglected to this point) for a load resistor. The circuit of Figure 16-15, shows a current mirror, made with pnp transistors, used as a collector load and to convert the differential signal to a single ended signal. This arrangement is one of a class which is often called active loads.

In the circuit of Figure 16-15 the input side of the pnp current mirror, on the left, is driven by one collector of the differential pair. The other side of the current mirror provides a matching current, from its high output impedance, for the Q_4's collector load. To the extent that errors in the current mirror can be neglected, the current mirror output current should exactly match the collector current of Q_4 when the differential input is shorted. If the base of Q_3 is driven slightly positive, the current mirror output will rise while Q_4's current will fall, causing the output voltage to rise. If Q_3's base goes negative, the current mirror output will fall, while Q_4's collector current will rise, and the output will swing negative. The gain of this circuit is limited by the ratio of npn transconductance to the output impedance of the combination of Q_4 and the pnp driving it. In practice, the external load impedance often dominates. As shown, the circuit of Figure 16-15 is not of much practical use as a linear amplifier (although it can be used as a comparator). Typically the gain is sufficiently high that even small errors, such as a base current error in the mirror, are amplified and cause the output to be driven to one of the limits. That is, a slight mismatch between the current mirror output current and Q_4's collector current will cause the output to saturate either Q_4 or Q_6. This circuit can be combined with a lower impedance load and used as is or it can be included in a feedback system. In a feedback amplifier the high gain becomes an asset, with the feedback loop stabilizing the circuit to eliminate limiting. The small offsets which cause limiting of the open loop amplifier manifest as small offsets multiplied only by the closed loop gain of the feedback system.

Making a Simple Op Amp

The most commonly used element in a precision feedback amplifier system is the operational amplifier. The operational amplifier has a differential input and a single-ended output. That is, change of the output voltage is proportional to change of the voltage applied between the two input terminals.

The circuit of Figure 16-16 is a simple implementation of the op-amp function. Changes in output voltage, at the emitter of Q_8, are proportional to changes in the input voltage applied between the bases of Q_3 and Q_4. The input section of this amplifier is the differential stage and current mirror active load of Figure 16-15. The single-ended output of this section drives an additional voltage gain stage, Q_7. The amplified signal at the collector of Q_7 is buffered to provide a relatively low output impedance by Q_8.

Although this circuit is quite simple (some might say primitive), it serves to illustrate a few principles widely used in IC design. Perhaps the most interesting and subtle techniques compensate for the weaknesses of the pnp current mirror.

In the circuit of Figure 16-15, the output of the current mirror from Q_6 differs from the input current in the amount of two base currents. Moreover, the current is modulated due to the variations in collector voltage of Q_6. These two effects, along with the variation in the collector voltage of Q_4, which we have previously neglected, reflect as input offset voltage. That is, the output is saturated positive or negative when the inputs are shorted, and a nonzero input voltage is required to bring the output into operating range.

In the circuit of Figure 16-16 most of the problems are compensated for. When

Figure 16-16.
Simple op amp
with full common
mode and output
voltage range.

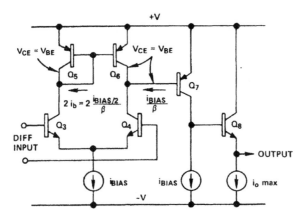

the amplifier is in its normal operating range, with the output unsaturated, Q_7 operates at about the same bias current level as do the combination of the mirror transistors Q_5 and Q_6. Assuming equal β among them, the base current of Q_7 should approximately equal the sum of the base currents of Q_5 and Q_6. This current from Q_7 will become part of the collector current of Q_4 in the same way as the base currents of Q_5 and Q_6 are part of the collector current of Q_3. As a result, the base current error due to the mirror is compensated for leaving only small residual errors due to the differences in β between the pnps and to the effects of the npn base currents.

Notice, also, that in normal operation the collector voltage of Q_6 is determined by the V_{be} of Q_7. This means that this voltage is very nearly the same as the collector voltage of Q_5, which virtually eliminates the variation in current mirror output that would result from large voltage swings at the collector of Q_6. Stabilization of this voltage also keeps the collector voltage of Q_4 equal to that of Q_3 and minimizes errors due to differences in collector base voltage of the differential pair.

The amplified signal which appears at the collector of Q_7 is buffered by the emitter follower Q_8. This transistor operates "class A" from a negative bias current slightly in excess of the largest anticipated negative load current. Assuming the β of Q_8 is sufficiently high (the base current "penalty" is low), the open loop output impedance of the amplifier will be low, and the voltage gain will be only slightly reduced by moderate loads.

The voltage gain of this circuit from the differential input to the collector of Q_7 may be quite high. Using integrated circuit transistors it should certainly be above 10,000 and might easily reach 100,000 or 100 dB. Therefore, even very small errors due to weaknesses of the configuration or simple component mismatches will lead to output saturation with zero input voltage.

However, because of the self-compensating effects of the configuration, this circuit has the potential to come to balance with only small input offset. Although this circuit lacks many of the features that are necessary in a general purpose op amp (frequency compensation, low-power operation, and positive output current limiting, to name a few), it illustrates some important principles of IC design, and it is similar to amplifiers used as internal op amps for dedicated on-chip applications. The biasing can all be arranged with current mirrors or stabilizers using multiple npn output transistors with their base-emitter junctions paralleled. This illustration shows only a few of the potential applications for the circuits described here, and these circuits are, of course, only the beginning of IC design.

The opportunities for imaginative circuit design based on the simple principles and points of view described here are almost limitless. More sophisticated and subtle techniques are available to the IC designer which provide an enormous range of possibilities. Although the types of components available for monolithic design are limited in number (the number continues to grow, however), the inherent matching and high performance available from these components and their ready availability makes possible design techniques and circuits which would be impractical or impossible to make with discrete design and manufacturing.

Richard S. Burwen

17. How to Design Analog Circuits without a Computer or a Lot of Paper

When designing a circuit some people generate pages and pages of notebook paper full of calculations. Others fill their computer screens with elaborate formulas and simulations. For over 40 years I've been designing analog and analog-to-digital circuits. While I have a small roomful of engineering notebooks, they are mostly full of schematics, prints, printed circuit layouts, and written reports on how the circuits work and not much in the way of calculations.

Not that I don't calculate; I am a pretty good computer. I design my circuits as I am drawing them and generally do the calculations in my head. By using simple approximations, I can probably come within 5–10% of the results that somebody else might generate from a notebook full of calculations.

I have nothing against using computers, calculators, manual computation, or even cookbooks, if that's the way you like to work. Probably I am the last circuit designer in the world to have acquired a computer. Right now I have two AT clones on my desk and a portable on my workbench for taking notes. I love my computers and I hate my computers. I hate them for the three months or so of my life that I wasted on program idiosyncrasies, incompatibilities, disk problems due to copy protection, IO port interferences, computer hang-ups, and a lot of little problems you have to learn about the hard way. My disks have tens of megabytes of programs I own but don't use.

My computers help me with three things. First is schematic drafting using Futurenet. I don't try to analyze or simulate my circuits, just draw them. I used to be a terrible draftsman, but now my schematics are clear and beautiful—drawn by my plotter. Second, I do a lot of writing using a word processor. That has not speeded up my writing at all. I used to dictate my reports to my secretary and ship them out with just minor revisions. Now I can change my mind as much as I want, and it is less efficient.

The third thing I use my computers for is storing tidbits of information. I use a program called MemoryMate which is very simple to use. I use one copy of it for my telephone file. I use a second copy of the same program to store my bench notes, usually done on the portable computer. The third and fourth copies of the program store information on my sound recordings and anything else I want to remember. The nice thing about this program is that if you can remember a single word of what you wrote there is no trouble finding it. Usually, just to make sure, I add a few key words at the top of the page, so if I can think of something related I can find it. Using computer memory is a lot better than stuffing notes into my engineering notebooks where I think it is too much trouble to go searching for the information.

Once in awhile I use an electronic calculator. When I was in college and during my early engineering years I used to pride myself on carrying around the most complex slide rule and a miniature version of it in my shirt pocket. Today I use a pocket calculator, a Sharp 5100. It has a zillion buttons, but what I really like about it is the 10 memory buttons that can store intermediate results. I use it mostly for calculating precision resistors and values that have to be closer than 1%.

To give you an idea of how I think about circuits, it may help to know more about my background.

My Background

I have known since I was 8 years old that I wanted to be a circuit designer. That's lucky. A lot of kids take 30 years to figure out what they want to do. My dad once sold radios back in the 1920s, and he got me started building a crystal radio. Probably most of you are too young to remember the old galena crystal with its cat whisker. The crystal was a little rock of mineral embedded in a cylinder of lead or solder about 0.4 inches in diameter and 0.2 inches high. To make it receive local AM radio signals you fished around for a sensitive spot on the crystal using a small pivoted arm with a cat whisker. A cat whisker was a little piece of springy wire. When you touched the crystal at the right spot with the tip of the wire and with the right pressure, presto—the first crystal diode or semiconductor rectifier.

With my dad's advice and finding what I could in books, I built at least three different crystal sets. They all received something. I used my bedspring for an antenna and a radiator for ground. Then I improved reception with a long wire antenna in the attic. I tried a lot of different coils, including one wound on a Quaker Oats box and different variable tuning capacitors. The third crystal set was a sleek one that I built into a redwood cigar box, which I sanded, stained, varnished, and polished like a mirror.

After the crystal radios, I graduated to a one tube radio. From the first vacuum tube set I advanced to two tubes and then three tubes. None of the circuits was exactly like what I found in the books but used the general ideas with my own improvements.

At age 11 I developed a very healthy respect for high voltage. One day when I was visiting a ham radio operator friend in his attic, I managed to get one hand across his telegraph key while holding a headphone in the other hand. There was 400 V DC between them. I couldn't let go, but fortunately he was right beside me and shut off the power quickly. Now I keep one hand in my back pocket.

When I was 13 I became a ham radio operator, but I was always more interested in building the equipment than in talking. At age 14 I took the tests and acquired commercial radio telephone and telegraph licenses. While I was still in high school, World War II started. I worked afternoons and Saturdays servicing and aligning HRO shortwave receivers used by the military. At the beginning of the war the government shut down ham radio activities, and at that point, I became interested in high fidelity music reproduction.

I realized from the beginning that frequency response was one of the primary factors in making good sound, and I have spent a lot of my life designing various kinds of analog filters, equalizers, and tone control systems related to hi-fi. The radio-phonograph I built during high school and early college days used multiple speakers, a complex crossover network, bass and treble controls, and most important, speaker equalization. It was built into large maple and walnut cabinets with transparent doors and colored lights to enhance the music.

After a half year at Harvard College, I entered the U. S. Navy and attended the electronic technician schools. That was a terrific practical education I wish every engineering school would provide. It gave an insight into the practical application of all the theory one gets in college. I often wondered how anybody could get through college physics, mathematics, electrical theory, Maxwell's equations, and such without dropping out from boredom unless he had such a practical course in electronics.

The Navy course consisted of preradio, primary radio, and secondary school totaling a year. Because I had a lot of practical experience, I was able to take the final exam at the end of my second month in primary school and graduate at the top of the third month graduating class. Students with the best grades usually got their choice of where to go to secondary school; so the Navy sent me to Treasure Island in San Francisco Bay, which was my last choice. By the time I graduated from secondary school, the war was over. The Navy made me a typist for three months, discharged me, and I went back to Harvard College.

When I wanted to take somewhat advanced electronics courses and skip one or two of the more fundamental courses, my adviser, Professor F. B Hunt asked, "Do you want to be a scientist or a hi-fi nut?" My answer was "both." I guess I've succeeded at least at the latter.

After finishing college and getting a master's degree at Harvard, I promptly retired. What I did was go down into my workshop in my parents' basement, where I tried to develop a top notch hi-fi system built into the base of a Webster record changer. My intention was to sell hi-fi systems to wealthy people. It was a great experience because I learned a lot about multiple feedback loops and equalization. I built two complete systems. About the time I finished construction I realized that the people who had the money to buy the systems didn't care about the fidelity, and the people who cared didn't have the money.

Just at that time I got a call from an electronics company and took a job in 1950 designing equipment for one of the first cable television systems. Over the next 11 years I worked for a half dozen companies designing analog circuits for laboratory instruments, hi-fi, and military equipment. Since 1961 I have been working full time as an analog circuit design consultant. Companies don't hire you for this kind of work unless they are overloaded, in a hurry, or you have a capability they don't have. So, for 30 years I have had a lot of fun involved in extremely interesting projects, including medical equipment, hi-fi, space, automotive, TV, analog function modules and ICs, power supplies, laboratory instruments, and lately, switching power amplifiers.

My first love, hobby, and part of my business has been hi-fi. My home system has 159 speakers in one room and 20,000 W. It took 25 years to build so far and will be finished around 2010.

Over the years, only a small percentage of my bread and butter work has been concerned with hi-fi. That doesn't matter; almost anything you can design in analog circuits has some bearing on hi-fi. My sound system uses about 2,000 op amps. The first op amps I designed for my hi-fi were discrete potted modules. These modules, somewhat refined, became the first products of Analog Devices. Later I developed more complex signal processing modules which helped start two high end audio companies and also companies bearing my own name.

You can see that my practical background has given me some feel for how circuits work. Although I have hardly ever directly used the mathematical theory I gained in college and graduate school, I am firmly convinced that a good mathematical background is an absolute necessity to help you make the best trade-offs in designing a circuit. Most of the tools I use in my designs are pretty simple. Here is how I go about designing analog circuits.

Breaking Down a Circuit

In the beginning there were transistors (what is a tube?). They became very cheap and you could use a lot of them. When the first op amps were developed, they were expensive and you had to conserve circuits. Now you can buy several op amps on one chip so cheaply you can use them as high quality transistors. That makes it easier to design circuits separated into simple functions.

For example, suppose you want to design a standard phono equalizer. It has a high frequency rolloff and a bass boost. You can build a single network around one op amp stage, or you can separate the high and low frequency parts of the equalization into separate op-amp circuits cascaded. Separating the circuits allows you to adjust one time constant without affecting another, and the circuit is easier to calculate.

The first thing to do is to break down a circuit into all its blocks. If each block has a very high input impedance and a near zero output impedance, one block can feed another without interaction. That's the beauty of using a lot of op amps. Noise buildup from using a lot of separate circuits can be more of a problem or less of a problem. If you keep all the signals at the highest possible level consistent with not overloading, you will probably generate less noise than trying to perform several functions in a single circuit. The more functions you perform in one circuit, the more interaction there is between them. Usually circuits toward the output of a network have to be higher in impedance than circuits near the input to reduce loading problems. The higher the impedance, the more noise it generates.

The lowest noise circuit you can make that performs a lot of different functions usually consists of a number of near-unity gain op-amp circuits. Low gain means less amplification of noise. So I use a lot of op-amp followers in my designs.

If you separate all the functions of a circuit into building blocks that don't interact, then the design job is relatively simple. Each block can be designed independently of the others, provided it can feed the load.

Equivalent Circuits

If you break a circuit apart at any point, it looks like a source feeding a load. The source has an internal impedance, and the load affects the final output. However, if you have broken your circuit into individual op-amp circuits, each with a near zero output impedance compared with its load, then you don't have to worry about the interaction. Within each individual block, you can use Thevenin or Norton equivalents to determine the gain vs. frequency.

There are two equivalents. The source can be thought of as a voltage source having an internal impedance, or the source can be thought of as a current source in parallel with its own internal impedance (see Figure 17-1). If you have a complex network, it is frequently convenient to alternate between voltage and current source equivalents. All you have to know to calculate the gains of these circuits is how to calculate the gain of a two-element voltage divider and how to parallel impedances.

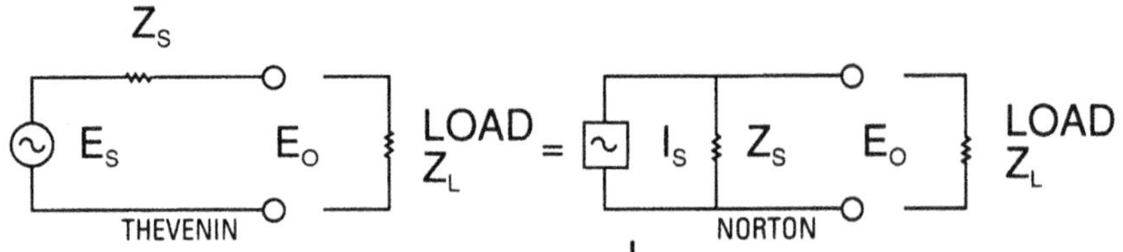

Figure 17-1.
Equivalent
sources.

$$E_o = E_s \left[\frac{Z_L}{Z_L + Z_s} \right] = I_s \left[\frac{Z_s Z_L}{Z_L + Z_s} \right] = \frac{I_s}{\frac{1}{Z_L} + \frac{1}{Z_s}} = I_s Z_L \left[\frac{Z_s}{Z_L + Z_s} \right]$$

(A)   (B)   (C)   (D)

In the case of the voltage source, formula A gives the output voltage as determined by the ratio of the load impedance to the total impedance consisting of the source and the load. In the case of the current source, the source current flows through both the source and the load in parallel. So the output voltage in formulas B and C is just the source current times the parallel impedance.

If the load is open, all the source current flows through its own internal impedance Z_s, producing an output $E_o = I_s Z_s = E_s$, the equivalent source voltage. If the load is a short circuit, all the source current flows through the load path and none through Z_s, producing zero output voltage. In between, a fraction of the source current flows through the load impedance, producing the output in formula (D), which is equal to (B).

When more than one source contributes to the output of a linear circuit you can consider the effect of each source separately. Leave all the impedances connected and short all but one voltage source. Compute the output due to that source using Thevenin equivalents. Next, short that voltage source, turn on the next, and calculate the output. After calculating the outputs due to each source, you can add them all together to get the total. If a source generates current, open it but leave its source impedance connected while calculating the effect of another source. Use whichever type makes calculation easier.

You can plug the numbers into your calculator or you can make an estimate in your head. The ratio of load to source impedance Z_L/Z_s gives you gain G.

$$G = \frac{E_o}{E_s} = \frac{1}{1 + Z_s/Z_L} \qquad (1)$$

It also gives you attenuation $1/G$.

$$\frac{1}{G} = \frac{E_s}{E_o} = 1 + \frac{Z_s}{Z_L} \qquad (2)$$

Frequent numbers appear in Table 17-1.

Table 17-1

Ratio Z_s/Z_L	Gain	Attenuation
0	1	1
1/4	4/5 = 0.8	5/4 = 1.25
1/3	3/4 = 0.75	4/3 = 1.33
1/2	2/3 = 0.667	3/2 = 1.5
1/1	1/2 = 0.5	2
1.5	0.4	2.5
2	0.333	3
3	0.25	4
4	0.2	5
9	0.1	10
100	0.0099	101

Stock Parts Values

Unless your system requires a very precise odd gain in one of its blocks, you don't have to calculate very accurately. You just have to arrive at the nearest stock resistor value. That makes calculation easy.

One percent resistors are so cheap and inserting them in boards is so expensive, there is no worthwhile savings when using 5% or 10% resistors. Your company can waste a lot of money stocking all the different values. My designs use standard 1% 0.25 W, 100 ppm/°C resistors with 12 values selected from each decade according to Table 17-2. Once in a while I need an accurate, stable resistor, and I select from a very few 0.05%, 0.1-W, 10 ppm/°C, values shown in Table 17-3.

Table 17-2

Stock Resistor Values in Ohms, 1%, 0.25 W, 100 ppm/°C

10.0	100	1000	10.0 k	100 k	1.00M
11.0	110	1100	11.0 k	110 k	1.50M
12.1	121	1210	12.1 k	121 k	2.00M
15.0	150	1500	15.0 k	150 k	3.01M
20.0	200	2000	20.0 k	200 k	4.99M
30.1	301	3010	30.1 k	301 k	
40.2	402	4020	40.2 k	402 k	
49.9	499	4990	49.9 k	499 k	
60.4	604	6040	60.4 k	604 k	
69.8	698	6980	69.8 k	698 k	
80.6	806	8060	80.6 k	806 k	
90.0	909	9090	90.9 k	909 k	

Table 17-3

Stock Precision Resistor Values in Ohms, 0.05%, 0.1 W, 10 ppm/°C

100
4990
10.00 k
49.90 k
100.0 k

Similarly, I use a limited number of capacitor values (see Tables 17-4 and 17-5).

Table 17-4

Stock Ceramic Capacitor Values in picofarads, 5%, 50 V, ±30 ppm/°C, 10 pF and Larger

5		
10	100	1000
15	150	
22	220	
33	330	
47	470	
68	680	

Table 17-5

Stock Metallized Film Capacitor Values in microfarads, 5%, 50 V, ±200 ppm/°C

	0.01	0.1	1.0
0.0015	0.015	0.15	
0.0022	0.022	0.22	
0.0033	0.033	0.33	
0.0047	0.047	0.47	
0.0068	0.068	0.68	

RC Networks

Most engineers have no feel for the relationships among reactance, time constant, and frequency response. They have to plug all the numbers into formulas and see what comes out. It's fairly simple to calculate RC circuits in your head.

First, let's look at the simple RC low-pass filter in Figure 17-2. The filter is a simple voltage divider whose gain is 0.707 or 3 dB down at the frequency where the reactance of the capacitor equals the source resistance R in ohms. The magic numbers are the –3 dB frequency, f_0 and the time constant T, simply the product of resistance and capacitance in microseconds.

All you have to remember about capacitive reactance is,

$$X_C = \frac{1,000,000}{2\pi f C} = \frac{159,155}{fC} \qquad (3)$$

$$C = \frac{1,000,000}{2\pi f X_C} = \frac{159,155}{fX_C} \qquad (4)$$

where f is frequency in hertz and C is capacitance in microfarads, is that 1 μF has a reactance of approximately 160,000 Ω at 1 Hz. You can figure out everything else from this. For example, 1 μF at 1 kHz has a reactance of 160 Ω. At 1 MHz it has a reactance of 0.16 Ω. And 1 pF at 1 MHz has a reactance of 160,000 Ω; 1 nF (0.001 μF) has a reactance of 160,000 Ω at 1 kHz.

Suppose you want to design an RC low-pass filter that attenuates –3 dB at 1 kHz (the cutoff frequency f_0). Let's start with a resistance of 4990 Ω. This is one of my frequently used stock values and is an appropriate load for an op amp. We need a capacitive reactance of 4990 Ω at 1 kHz. What is the capacitor value?

Just divide 160,000 by the frequency in hertz (1000) and then by the number of ohms (5000) as in Equation 4. The tough part is getting the decimal point right. Remember the number 160,000 is associated with ohms and microfarads. It also works with kilohms and nanofarads (1 nF = 0.001 μF) or megohms and picofarads. The simple RC low-pass filter works out to need 0.032 μF, actually 0.0318948 μF, for a 1 kHz rolloff. That's close to my stock value of 0.033 μF.

Another way of looking at the simple RC low-pass filter is to associate its time constant T = RC with its cutoff frequency f_0.

$$f_0 = \frac{160,000}{T} = \frac{160,000}{RC} \qquad (5)$$

$$T = \frac{160,000}{f_0} = RC \qquad (6)$$

T is in microseconds and f_0 is in hertz.

A filter having a 1 MΩ resistor and a 1 μF capacitor has a time constant of 1 sec

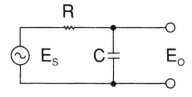

Figure 17-2.
RC low-pass filter.

or 1,000,000 µsec and cuts off −3 dB at 0.16 Hz. That is the frequency at which 1 µF has a reactance of 1 MΩ and equals the 1 MΩ resistor. You can calculate all other simple RC filters from that point.

If capacitance is in microfarads and resistance is in ohms, the time constant is in microseconds. If you know the time constant you can figure the cutoff frequency and vice versa. As examples, a time constant of 1 msec produces a cutoff frequency of 160 Hz. And 1 µsec corresponds to 160 kHz. To find the frequency, just divide 160,000 by the time constant in microseconds. To find the time constant, just divide 160,000 by the frequency in hertz.

Once you have done this calculation in your head a few times, you acquire a feel for what time constant goes with what frequency. I know that 100 µsec goes with 1600 Hz, 160 µsec goes with 1 kHz, 300 µsec goes with about 500 Hz, 10 µsec goes with 16,000 Hz, and so on.

The response of the simple RC low-pass filter at any frequency is determined by its division ratio.

$$G = E_o/E_S = \frac{-jX_c}{R - jX_c} = \frac{1}{1 + j2\pi fT} \qquad (7)$$

$$|G| = E_o/E_S = \frac{1}{\sqrt{1 + (f/f_0)^2}} \qquad (8)$$

If you know what the 3 dB cutoff frequency f_0 of a simple RC filter is, you can plot its entire response curve on semi-log graph paper or you can do it in your head. The curve is universal. You just have to move it to the right frequency. Find the ratio of the frequency of interest f to the 3 dB cutoff frequency f_0 and you can determine the response. In Table 17-6 you can see that at half the cutoff frequency the response is down 1 dB and at twice the cutoff frequency, it is down 7 dB. At 1/7th of the cutoff frequency it is down 0.1 dB. Well beyond the cutoff frequency the response goes down at 20 dB/decade

If the filter is a high-pass type instead of a low-pass type, simply interchange f_0 and f in Table 17-6.

Table 17-6

f/f_0	Decibels
1/7	− 0.1
1/2 = 0.5	− 1
1	− 3
2	− 7
10	−20
100	−40
1000	−60

Often I need to estimate the transient response of a simple RC filter or find how far the capacitor will charge in a given time for a step input. The step response of the filter is

$$\frac{E_O}{E_S} = 1 - e^{-t/T}$$

Table 17-7 is a table of useful values of output vs. time as a fraction of T, the RC time constant.

Table 17-7

t/T	E_0/E_S
0.001	0.001
0.01	0.01
0.1	0.1
0.2	0.18
0.5	0.39
1	0.63
2.3	0.9
4.6	0.99
6.9	0.999
9.2	0.9999

You can figure out most of this table in your head if you can remember that the capacitor charges up to 0.63, or 63%, in one time constant. Also, at 2.3 time constants the capacitor charge reaches within 10% of its final value, at 0.9, or 90%.

The exponential curve has the same shape in the next 2.3 time constants but starts at a point 10% away from the final value. Therefore, the value at the end of 2×2.3 = 4.6 time constants is within 1% of final value, at 0.99, or 99%. Similarly, at 3×2.3 = 6.9 time constants the capacitor charges to within 0.1% of final value, at 0.999, or 99.9%. At small fractions of a time constant the fractional charge is the same as the fractional time.

If the simple RC filter is a high-pass type instead of a low-pass type, subtract the above outputs from 1.

Stabilizing a Feedback Loop

The first rule for making a feedback loop stable is to keep it simple. Flat response feedback around a single RC rolloff or an integrator produces a low-pass filter that looks like a single RC rolloff. It goes down at 6 dB/octave.

The single RC rolloff produces a nice exponential step response with no overshoot. If the open loop response goes down at 6 dB/octave it has 90° of phase lag. When you close the loop and make a low-pass filter out of it, the phase shift at the −3 dB point is 45°. Anything else in the loop that adds phase shift tends to cause a peak in the frequency response and an overshoot in the step response.

Let's start with a simple active low-pass equalizer, Figure 17-3. The gain vs. frequency of this equalizer is $Z2/Z1$, which is simply the ratio of the feedback impedance $Z2$ to the input resistor R1. I have chosen the feedback network to be a parallel resistor R2 and capacitor C1, having a time constant of 150 μsec. This active filter has the same frequency response as the simple low-pass filter in Figure 17-2 if the time constants are the same. The difference is the active filter inverts, has a near zero output impedance, and you can design it to provide DC gain.

The −3 dB cutoff frequency associated with 150 μsec is approximately 1,000 Hz. Remember 160,000/T? You can look at this circuit as an integrator consisting of an input resistor R1 and a feedback capacitor C1. Adding resistor R2 provides an overall DC feedback path to convert the integrator into a low-pass filter.

Another way to look at this circuit is to consider A1 as a current source whose terminals are its output and its negative input. This source has an internal resistance R2. Because large feedback keeps the negative op-amp input at near-ground potential, the source current through R2 is the same as the input current through R1. This

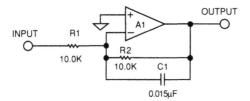

Figure 17-3.
Active
6 dB/octave
low-pass filter.

internal source resistance R2 is loaded by capacitor C1, making the equivalent circuit of the simple RC low-pass filter in Figure 17-2.

Now suppose the DC feedback, instead of coming directly from the output of op amp A1, comes from a more complex system shown in Figure 17-4. Here the integrator A1 is followed by a low-pass filter R3 and C2, buffered by a unity-gain follower amplifier A2. I have chosen the cutoff frequency of this filter at $3 \times 1,000$ Hz, or 3 kHz. Its time constant $T = RC = 160,000/3000$ Hz $= 53.3$ μsec ≈ 0.0047 μF $\times 10,000$ Ω. What happens?

The gain at 1,000 Hz is practically the same as in Figure 17-3, so the loop gain is determined almost entirely by the integrating capacitor C1 and the feedback resistor R2. The main effect of the 3000 Hz low-pass filter is an additional 6 dB/octave rolloff and a contribution to phase shift which results in about 1% overshoot in the step response.

Here is the rule. If you have feedback around an integrator and a 6 dB/octave low-pass filter, you can achieve transient response with only 1% overshoot by making the cutoff frequency of the low-pass filter 3 times the cutoff frequency of the integrator with DC feedback alone. If the cutoff frequency of the low-pass filter is lower, you get more phase shift and more overshoot. At 1,000 Hz, the low-pass filter contributes a phase lag of 18.4°. Added to the integrator phase lag of 90°, the total open loop phase shift is 108.4°, less than 110°. That's a nice number.

Remember that 110° total phase shift at the unity gain frequency gives you beautiful transient response. Unity gain means that, if you break the loop at a convenient point and connect a signal generator there, the magnitude, but not the phase, of the signal coming back is the same as that of the signal generator. If there is more than one low-pass filter in the circuit or contributor to phase lag at the unity gain frequency, you have to add up all the phase shifts. Much below the 3 dB cutoff frequency of a simple RC low-pass, the phase shift is approximately proportional to frequency based on one radian or 57° at the cutoff frequency. At frequencies higher than half the cutoff frequency, the formula is inaccurate. For example, at the cutoff frequency the phase shift is 45° not 57°.

A filter such as that in Figure 17-4 involves more than a single feedback loop. Follower A2 has its own feedback from output to input. Feeding back through R2 to A1 provides additional feedback around A2 at DC and low frequencies, producing open loop nearly a 12 dB/octave slope and 180° phase shift. A very high frequency version of this circuit might oscillate if A2 does not have favorable overload and

Figure 17-4.
Active
12 dB/octave
low-pass filter.

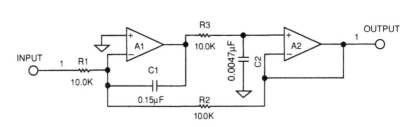

slew rate characteristics. Depending upon the method of internal stabilization, some op amps delay several microseconds in coming out of overload, effectively adding to system phase lag.

Any feedback loop involving integration or low-pass filtering in the forward path may be subject to overload recovery problems. This is because the capacitor involved becomes overcharged when the output of the system saturates. Therefore it is necessary to understand what happens to a feedback loop when various parts are driven into overload. You should know how an op amp recovers from overload before designing it into your circuit. Sometimes the problem can be avoided by limiting the input signal amplitude.

Another kind of loop you may have to stabilize is one in which the load is inductive and behaves as an integrator. This happens with many magnetic loads such as deflection circuits, magnets, and motor drives. In Figure 17-5, a low resistance shunt measures the current in the load coil, and its output is amplified by A2 to provide feedback. This loop already has nearly 90° phase shift over a wide range of frequencies due to the load. Therefore, the feedback network around A1 has to have a low phase-shift flat response region at high frequencies determined by resistor R3. We want maximum feedback at DC for accurate control of the output current, so it uses an integrating capacitor C1. Here is my simple way of stabilizing the loop without knowing anything about it.

1. Short out the integrating capacitor C1 and connect a potentiometer in place of R3.
2. Connect an oscilloscope to look at the feedback from A2 and the error signal output of A1.
3. Feed in a small signal square wave and adjust R3 to the maximum resistance value that gives you a satisfactory amount of overshoot.
4. Connect in a large value integrating capacitor C1 and then select smaller and smaller values, accepting the smallest that does not seriously degrade the good transient response you just had.

That's it. No simulation, no calculations. A great time saver. This method works for all kinds of feedback systems that can be stabilized by a simple series RC network. If the system has additional contributors to phase lag you may need to compensate by adding a phase lead network such as a capacitor in series with a resistor across the feedback resistor R2. This network can reduce the total phase shift at the unity gain frequency and thereby reduce overshoot and ringing.

Circuit Impedance

High impedance circuits are affected by small stray capacitances, and they generate more noise than low impedance circuits. When using operational amplifiers at

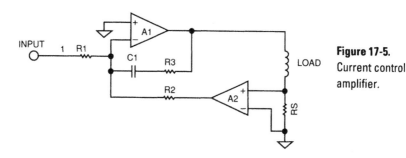

Figure 17-5.
Current control amplifier.

frequencies below 100 kHz, my rule is to use circuit impedances in the vicinity of 5000 Ω. Most BIFET op amps can feed a 5000 Ω load without distorting much at 20 kHz. If you are using an op amp as a follower, it has maximum feedback. Phase shift caused by stray capacitance can make the circuit ring in the megahertz region.

For example, suppose instead of connecting the output of the follower directly to the input it goes through a 100 kΩ resistor. Stray capacitance of only 3 pF to ground will make a 500 kHz low-pass filter at the inverting input. Its phase shift at 1.5 MHz is about 71.5°, which will cause many cycles of ringing, perhaps oscillation. If the feedback resistance is only 5000 Ω, the 3 dB point of the stray capacitance low-pass filter is 10 MHz and will not cause ringing if the unity gain frequency is only 3 MHz.

In some circuits you need a high resistance, at least at low frequencies. If the resistance is more than 5000 Ω in a circuit that produces unity feedback at high frequencies, it is necessary to bypass the output to the negative input with a capacitor. Circuits that have high closed loop gain can tolerate higher impedances because the unity loop gain frequency is lower. Remember the 3:1 cutoff frequency rule for low overshoot and less than 110° phase shift. High frequency op amps have more than 90° phase shift at high frequencies and can tolerate very little phase shift in the feedback path.

Lower impedance circuits are less susceptible to noise pickup from other circuits. On a printed circuit board, for example, where two adjacent conductors may have a capacitance between them of about 1 pF per inch, you can estimate how much crosstalk you will get by estimating the ratio of circuit impedance to coupling reactance at the frequency of interest.

New Parts

Before choosing a part you have never used before it is important to find out its characteristics, not just those on the specification sheet but important characteristics that are unspecified. For example, before using a new op amp, you should find out if the manufacturer really designed it to be stable at unity gain, or if it is on the verge of oscillation. When you overdrive either input, does the op amp go into phase reversal? Can it be driven by an op amp of the same type without phase reversal but not by a different type that delivers more voltage? Does it produce a lot of popcorn noise that may bother your system? Does it delay in coming out of overload? If you don't see the characteristic in which you are interested on the specification sheet or covered in application notes, you should assume the part performs poorly in that respect.

Many of the more serious troubles I have encountered in making my circuit designs work resulted from incomplete knowledge of parts I was using for the first time. For example, in more than one instance I have been burned by parasitic effects in an integrated circuit. One section of the chip that is supposed to be unrelated to another section affects the other when you feed current into one of the pins driving it below ground or above the supply rail. It really pays to make a few crucial experiments before designing in any part with which you are not completely familiar.

Breadboarding

If the circuit involved is closely similar to circuits I have used before and the operating frequencies are not too high, I usually skip the breadboarding phase and go straight to a printed circuit layout. Parts of the circuit that involve tricky feedback loops, unfamiliar parts, or are susceptible to wiring inductance or capacitance need to be tested.

Breadboard circuits should be carefully constructed with attention paid to grounding, shielding, and lead lengths. Use a ground-plane board. You can waste a lot of expensive engineering time finding troubles in a breadboard circuit that has been just thrown together.

Some engineers prefer computer simulation. That's okay, but the one big advantage of the experimental method is that the results agree with experiment.

Testing

I can't believe it—a technician turns on one of my circuits for the first time, feeds in an input signal, and expects the correct signal to appear at the output. I don't have that much confidence. When I test a circuit I break it into its blocks and check DC voltages, gain, frequency response, and other important characteristics of every single part of each block. It is important to know that every component you design into a circuit is really serving its purpose. If blocks of the circuit cannot be easily separated from others or test signals cannot be injected, you can measure the output signal from one block used as an input signal to the next block and then see that the next block does its job relative to that signal. Once the individual sections of a circuit are working, I check groups of blocks and finally the whole system. Even if the system seems to deliver the correct signal all the time, that does not mean every intermediate part of the circuit is really functioning correctly, optimally, or reliably.

How Much To Learn

As a consultant I have had the opportunity to work in many fields of electronics. Many times I have been surprised at how soon new circuit knowledge gained in one field became useful in an entirely different area, sometimes within a week. Efficient circuit design comprises building on what others and especially you have done before, with a bit of innovation, but not too much. While I have a stock of circuits in my computer, such as common mode rejection amplifiers, output followers, crystal oscillators, and triangular wave generators. I rarely use any circuit exactly as I did before. They keep evolving with new parts and characteristics adapted to new requirements.

Once in a while you need to take on a project involving circuits, parts, and ideas entirely new to you. Pioneering usually is not a way to make money directly. You run into too many unforeseen problems. However, it gives you knowledge which, if applied over and over again with small improvements to other projects, really puts you ahead.

Settling Time Tester

I needed a production test instrument to measure the settling time of a power amplifier used to drive the gradient coils in magnetic resonance imaging machines. In this application the output current to a load coil has to follow an input pulse and settle to within 0.1% of final value within 1.3 msec. This settling requirement applies both at the top of the pulse and following the pulse. Pulses can be either positive or negative. To avoid overloading the amplifier, the input pulse must have a controlled slope, typically lasting 1 msec, on the leading and trailing edges. For an accurate settling test, the top of the pulse has to be extremely flat and free of noise.

In addition to generating the pulse, the instrument has to provide a means of filtering out 81 kHz noise and magnifying the top of the pulse without distortion caused by poor overload recovery of an oscilloscope. I decided to build an analog signal generator and error amplifier using op amps and some HCMOS logic.

The tester consists of two sections. A wave form generator delivers the slow rising and falling pulse to the amplifier and a synchronizing signal to an oscilloscope. An error amplifier then processes the amplifier's current monitor signal for viewing on an oscilloscope. Processing consists of filtering out 80 kHz noise, offsetting the top of the pulse to zero, and then amplifying and clipping the error.

The block diagram, Figure 17-6, shows the organization of the system. The upper set of blocks is the wave form generator, and the lower set of blocks is the error amplifier. The wave form generator starts with a pulse generator block that delivers −3.3 V to +3.3 V pulses, selectable in polarity, and adjustable in width and frequency. An integrator that saturates and recovers quickly slopes the leading and trailing edges, and increases the pulse size to ±13 V.

After the integrator, two different clipping circuits select portions of the signal. One passes the portion of the integrator output signal from 0 to +10.5 V, while the other passes the negative portion of the signal from 0 to −10.5 V. After selecting the output of one or the other clipper, the operator adjusts the amplitude of the signal using a 10 turn potentiometer, and the output goes through a follower amplifier to the power amplifier under test. The diagram shows the wave forms at important points.

The error amplifier system uses a differential input buffer to get rid of ground voltage noise at the input connections. Low-pass filtering at 45 kHz attenuates 81 kHz and higher frequency noise. Then coarse and fine offset potentiometers adjust the top of the pulse to 0 V. The resulting signal is amplified 10 × in a fast recovery amplifier which clips the output at ±1 V. An oscilloscope connected to the output will display a range of ±100 mV referred to the top of the pulse. You can clearly see 0.1% of a 5-V signal as 1 cm deflection at 50 mV/cm.

Now let's run through the schematic in Figure 17-7 so you can get an idea of my thinking and how little calculation was necessary. First, I needed an oscillator adjustable from at least 4 Hz to 50 Hz. This oscillator consists of the simplest possible circuit, an HCMOS Schmitt trigger inverter U1A with negative feedback via a low-pass filter consisting of trimpot R1 and capacitor C2. The output of the HCMOS chip swings from 0 to +5 V, and its input triggers at typically +2 V and +3 V. This means that every time the output swings the capacitor charges 3 V/ 5 V = 60%. That takes about 1 time constant. Using a 200 msec network makes each half cycle last 200 msec, producing a 2.5 Hz oscillator. The 15 turn trimpot has more than a 20:1 adjustment range, so there is no problem getting to 50 Hz. I wouldn't use this circuit in a production instrument because the threshold levels of a Schmitt trigger logic device, such as U1, may vary widely from manufacturer to manufacturer and possibly from batch to batch. To construct two instruments this was no problem.

A high-pass RC network and Schmitt trigger inverter next converts the square wave to narrower pulses ranging from 3 msec to 30 msec in width for duty factor adjustment. This network consisting of potentiometer R2 and capacitor C3 converts the square wave to exponentially decaying pulses offset toward the +5 V supply. Resistor R3, which has three times the potentiometer resistance, keeps the load

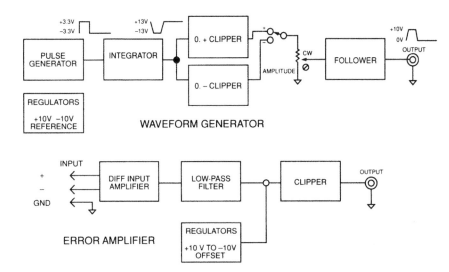

Figure 17-6.
Settling time
tester block
diagram.

impedance high when the wave form is large enough to be clipped by the input diodes of U1B. The small bypass to ground C4 prevents U1B from false pulsing on noise picked up from other circuits.

Two more gates, U1D and U1E, buffer the output to provide a synchronizing signal for the oscilloscope. Selector switch S1A selects either the narrow positive pulse output of U1B or an inverted signal from U1C. Next an inverting MOSFET driver switch U2 raises the pulse level from 0 to 5 V to 0 to +10.00 V set by a precision reference voltage.

Gate U2 feeds an integrating operational amplifier A4, which produces the leading and trailing edge slopes. To achieve equal slopes a divider, R5 and R6 connected to a precision –10.00 V reference, offsets the pulse output of U2 to ±3.3 V. I could have solved a pair of equations to determine the division ratio required for equal positive and negative swings. It was easier to try two or three different ratios in my head to converge on 2/1 for R6/R5.

Calculating the integrator part values didn't require a pencil and paper or a computer either. I wanted to adjust the leading and trailing edge slopes of a 10-V output pulse from less than 100 μsec to 3 msec using a 10 turn front panel potentiometer. Unlike the exponentially charging low-pass filter in Figure 17-2, the integrating amplifier maintains a constant charging current through the capacitor. In a time of 1 time constant, the output ramps up not 63% but 100% of the input voltage. Starting with –3.3 V input it takes 3 time constants for the output to reach +10 V.

To produce a 3 msec/10 V ramp requires a time constant of 1 msec, made up primarily of the input potentiometer R7, a 100 kΩ 10 turn type, and feedback capacitor C7, 0.01 μF. The 100 μsec/10 V ramp requires a source resistance of 100 kΩ/30 = 3300 Ω. Now here is an application for a Thevenin equivalent circuit to determine the portion of the input resistor supplied by R5 and R6. Without a load, divider R5 and R6 attenuates the 10 V output pulse from U2 to 6.67 V p-p, offset to produce ±3.3 V. Its effective source resistance is R5 and R6 in parallel, 1330 Ω. Adding R8, 1000 Ω, increases the total integrator input resistance to 2330 Ω, which meets the requirement with some safety factor. Zener diodes D1 and D2 were added across the integrating capacitor C7 to make the amplifier A4 recover quickly from saturation at ±13 V.

The next blocks in the process are a pair of clipping circuits. The circuit involving A5 and A6 clips at 0 V and +10.5 V and uses feedback to attain sharp corners. A

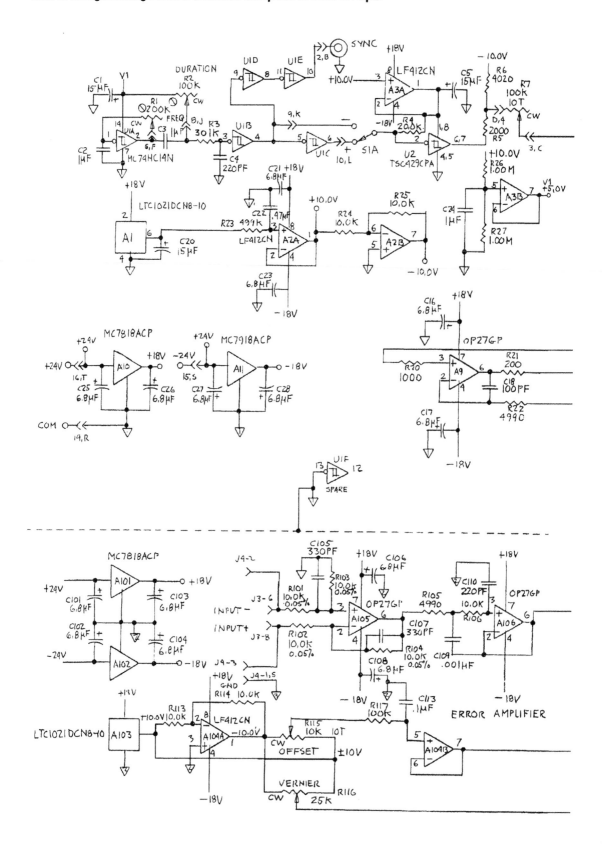

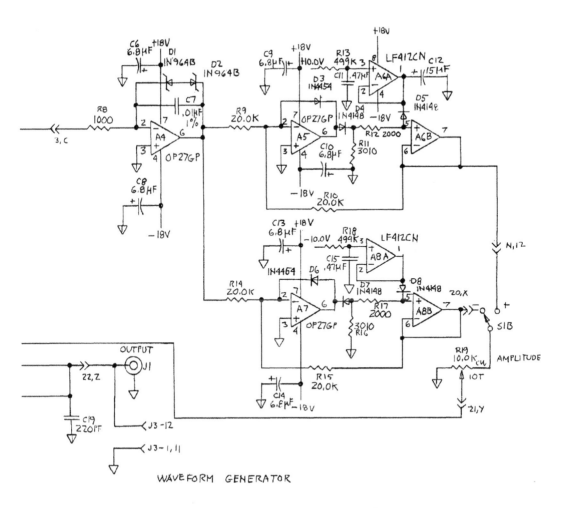

WAVEFORM GENERATOR

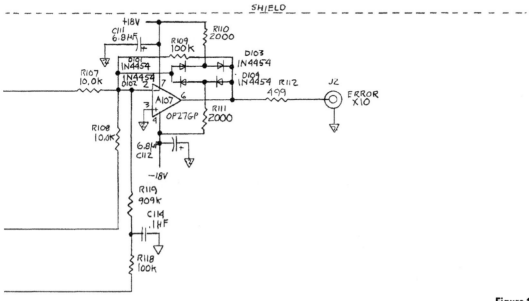

Figure 17-7.
Complete schematic of the settling time tester.

similar circuit, A7 and A8 with reversed diodes clips at 0 V and –10.5 V. The output of either circuit is selected by switch S1B for the desired polarity.

In the positive clipping circuit, A5 is a precision rectifier which clips at 0 V. Diode D4 loaded by resistor R11 conducts only positive signals while diode D3 in the feedback path prevents the output of A5 from swinging more than 0.6 V negative and causing leakage through D4. The follower A6B buffers the output of the precision rectifier circuit and delivers feedback via R10 to set the gain at –1.

The output from A6B ought to be a nice flat top pulse produced by zener diodes D1 and D2 at A4. Zeners, however, are noisy. So another clipping circuit A6A using diode D5 clamps the input to A6B at +10.5 V. A6A is a follower whose input is a 10.0 V reference. A 0.24 sec time constant low-pass filter R13 and C11 filters the reference so the top of the pulse will have very low noise. During clamping A5 goes open loop and saturates. Originally D3 was chosen as a 13 V zener diode to speed recovery by preventing saturation. It leaked noise into the pulse and had to be changed to a conventional diode.

The negative clipping circuit operates the same way, but all the diodes are reversed. When S1B selects the output from the positive clipper at A6B, switch S1A selects logic pulses that are narrow at +0 V and wide at +5 V. When selecting negative output pulses from A8B S1A selects narrow +5 V pulses.

All the reference voltages for U1, U2, D5, and D8 are derived from a 10 V reference regulator A1. Several low-pass filters and followers are used to make separate low noise reference voltages which will not interact with each other. All the followers use FET input op amps which can work with relatively high impedance low-pass filters. Convenient values for the low-pass filters were 499 kΩ and 0.47 µF, producing a time constant of 0.24 sec and cutoff frequency of 160,000/240,000 µsec = 0.67 Hz

At the arm of switch S1B we have high quality pulses either positive or negative. A 10 turn front panel potentiometer R19 adjusts the amplitude. The final block in the wave form generator is a follower amplifier which eliminates loading effects. You cannot simply connect a follower directly to a coaxial cable without the danger of oscillations. I use an isolation network at the output of the follower. The network consists of a 200 Ω resistor R21 in series with the output terminal and a 220 pF bypass capacitor to ground C19. Now load capacitances from 0 to 1000 pF will cause only a 6-to-1 capacitive load variation instead of an infinite variation.

Resistor R22 delivers DC feedback from the output terminal. High frequency feedback above 300 kHz comes directly from the amplifier output via capacitor C18. The result is a follower with zero DC output resistance and a resistive-inductive impedance at high frequencies. Up to 1000 pF loading causes a small overshoot on a step.

The error amplifier was built on a separate circuit card and has its own local regulators A101 and A102. The input circuit uses a low offset op amp A105 connected in a 0.05% resistive bridge circuit to reject common mode voltages. Bypass capacitors C105 and C108 attenuate frequencies above 48 kHz by forming 3.3 µsec time constants with the bridge resistors.

The next amplifier A106 attenuates high frequency noise at 12 dB/octave above 48 kHz using a two section RC low pass filter. To achieve about 1.5 dB corner peaking, the first section has 1.5 times the desired 3.3 µsec time constant (160,000/48 kHz), while the second section has 2/3 of that time constant.

The task of the error amplifier after cleaning up the input signal is to subtract up to ±10 V adjustable offset and then amplify the signal 10 times and clip it at ±1 V. The +10.0 V reference, A103, feeds a unity gain inverter A104A to produce –10.0 V.

Then R115, a 10 K, 10 turn front panel potentiometer spanning the two voltages produces the continuously variable ±10 V offset. A low-pass filter R117 and C113 having a time constant of 10 msec corresponding to 16 Hz cutoff (160,000/10,000 μsec) eliminates high frequency noise from the reference. The follower A104B maintains the full ±10 V offset,which is then added to the signal from A106 using resistors R107 and R108. A vernier pot R116 feeding the low-pass filter R118 and C114 adds another ±1% of variable offset in via R119. The vernier pot increases the resolution from 0.1% to about 0.01%.

Finally, the error amplifier A107 has a feedback resistor R109 chosen to provide a gain of 10 relative to the signal from A106. The biased diode feedback network provides the ±1 V clipping with near instant overload recovery. As the output feeds an oscilloscope through a short cable, the only isolation network used is a resistor R112, 499 Ω.

The instrument works. Checking its own flat top pulse shows that all contributors to a long settling tail such as op amp self-heating and reference tilt amount to less than 0.01%.

Now you can see that by using enough amplifiers and followers to isolate the functions of a circuit, the design became quite simple. The only place I needed to figure out a Thevenin equivalent circuit was between U2 and A4 in the wave form generator. Remembering $f_0 = 160,000/T$ was essential. It would have been convenient to have a table of low-pass filter values available for use in designing the error amplifier in the event I wanted to choose a precise amount of noise filter overshoot. The circuit could have been built with fewer op amps, but there would have been interaction between portions of the circuit. It would have been much more complicated to figure out and perhaps would not have worked quite as well.

One final note. Part of your circuit design job should be writing a description of how it works before you build it. If you want to find the flaws in your design, there is nothing like trying to describe it to someone else.

18. Starting to Like Electronics in Your Twenties

This book brings together a collection of talents in the analog electronics field. Jim Williams started playing with oscilloscopes at age eight in his neighbor's basement. He nagged his father until he got his birthday wish: a $250 vacuum-tube Philbrick operational amplifier. Another one of my co-authors gets an idea in his sleep and rushes to work at 3 A.M. to try out his concept. Yet another, after working a full day and more on electronics at his job, spends his recreational time designing and installing a unique electronic security system in his house. These people are born and addicted electronics engineers!

A reader of this book, in his or her sophomore year in college, considering the possibility of majoring in electronics, may despair. "How can I succeed in this field? I did not start early enough!" Well, there is still hope. I, personally, am definitely not a "born engineer." When I arrived in Canada as a refugee from Hungary after the 1956 revolution, I was interested in journalism and political science. However, without speaking any English, a career in these fields did not seem a realistic goal. Washing dishes and packing ladies' clothing was more easily attainable. After a couple of years, I enrolled in the Engineering Department of McGill University, assuming that command of the language was not a necessity in a scientific field. In my freshman year, however, the first course I had to take was English Literature. Fortunately, the first topic we covered was Geoffrey Chaucer's Canterbury Tales, written in Middle English. None of my classmates understood a word of it either. Thus, I became an electronics engineer almost by default.

When I arrived at the University of California in Berkeley in 1965 to study for a Master's degree in Electrical Engineering, I considered integrated circuits as an area of specialization. Berkeley already had a tiny wafer fabrication facility in 1965. Then I discovered that it would take a minimum of two years to conclude a project in ICs for my thesis. Therefore, I selected statistical communication theory for my thesis topic—something that could be completed in ten months. I had a couple of IC courses at Berkeley, but when I began my first job at Fairchild Semiconductor's Research and Development Laboratories in Palo Alto in 1966, I was basically a novice. To quote an old joke, I did not know an integrated circuit from a segregated one. But neither did anybody else, the situation being strikingly similar to my first English class at McGill.

In those early years of ICs in the mid-1960s, experienced design engineers had great difficulty abandoning the well-established design concepts of discrete or even vacuum tube circuits. The idea that a transistor was cheaper than a resistor was revolutionary. Everything depended on matching between resistors and transistors, not on absolute tolerances. Inductors and capacitors were unavailable.

I, as a beginner, did not have any preconceived notions. IC design was just as easy, or difficult, as discrete design. My timing was right. I also have to confess that

I am still not addicted to electronics. I work an eight-hour day and go home to a house completely devoid of electronic instruments, with the possible exception of a soldering iron (if you want to call that an instrument). I only touch my son's computer to play the occasional game or to type this chapter of our book. To the constant consternation of my relatives and friends, I do not repair TV sets or VCRs. After this short summary of my life story and philosophy, let us move to some circuit examples.

Simple Yet Powerful Circuits with Five or Fewer Transistors

What has always appealed to me is how a handful of transistors can make a gigantic difference in the performance of a circuit. All my examples will have five or fewer transistors. This will serve two purposes: all the circuits should be easy to follow, yet simultaneously, the power of simplicity can also be demonstrated.

Start-up Circuits and Current Sources

Start-up circuits provide my first examples. Our goal is to develop precise operating currents. All these configurations turn on with a poorly controlled start current; the most common implementation of this is the epitaxial FET transistor J1, in Figure 18-1. The output I_o1 should be independent of power supply voltage V+ and the current in J1. What can be simpler than the circuit of Figure 18-1? There is only one equation to write:

$$V_{be}Q_1 = I1R1 + V_{be}Q2 \qquad (1)$$

Therefore,

$$I_o1 = I1^{-I1R1/_{kT/q}} \qquad (2)$$

where kT/q = 26 mV at room temperature.

If for the nominal value of $I1$ (I_n), resistor R1 is selected to make I_nR1 = 26 mV, then a plot of I_o1 versus $I1$ is as shown in Figure 18-2. A 1.5 to 1 up and down variation in $I1$ is reduced to a less than 5% variation around the midpoint value of 0.352 I_n. An order of magnitude improvement from just two transistors!

This circuit is called the peaking current source [1] because of the shape of the

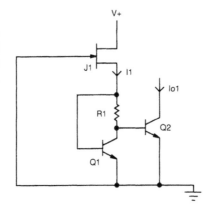

Figure 18-1.
The peaking current source start up circuit.

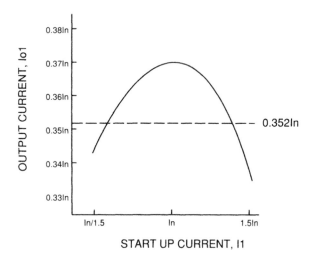

Figure 18-2.
Output current versus nominal input current. $I_n R1 = 26$ mV.

plot of Figure 18-2. The emitter areas of Q1 and Q2 were assumed to be equal in the above calculations. If we want a typical output current different from $0.352\,I_n$, the emitter sizes of Q1 and Q2 can be scaled. Another powerful tool!

The circuit of Figure 18-3 uses four transistors [2]. Initially, let us assume that R3 = 0. By inspection:

$$V_{be}Q11_{I1} + V_{be}Q12_{I_o2} + I_o2R2 = V_{be}Q13_{I_o2} + V_{be}Q14_{I1} \qquad (3)$$

The beauty of this expression is that we find transistors on both sides of the equation operating at the poorly controlled start current $I1$.

Therefore, the $I1$ dependence cancels out, and Eq. (3) reduces to:

$$I_o2 = (kT/qR_2)\ln(A11A12/A13A14) \qquad (4)$$

where A11 to A14 are emitter areas of the respective transistors. I_o2 is not a function of $I1$. Another interesting feature of this circuit is that the output impedance of current source I_o2 is negative! This can be shown intuitively by examining Eq. (3). As the collector voltage of Q13 rises, its base emitter voltage decreases because of the Early effect. The only way to maintain the equality of Eq. (3) is by a reduction in I_o2. This current source can be used to load pnp gain stages. Theoretically at

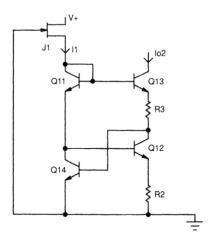

Figure 18-3.
A negative output resistance start circuit.

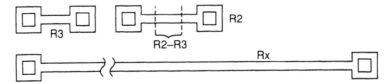

Figure 18-4.
(R2-R3) matches
long resistor Rx.

least, if the pnp output resistance equals in magnitude I_o2's output resistance, a gain
stage approaching infinity can be built.

What if R3 does not equal zero? Equation (4) now modifies to:

$$I_o2 = \{kT/q(R2 - R3)\}\ln(A11A12/A13A14) \qquad (5)$$

The end effects in the clubheads of R2 and R3 are cancelled (Figure 18-4). This
resolves one of the common problems of linear IC design: how to match a small
resistor (with dominant end effects) to a large resistor with negligible end effects.

Figure 18-5 shows another start-up circuit [3]. Here the J1 current ($I1$) is still
poorly controlled, but it is matched by an identical epitaxial FET, J2, which also
generates a current equal to $I1$. By inspection:

$$V_{be}Q22_{I1/A} + I1R4 = V_{be}Q23_{I2/nA} + I2R5 + I_o3R5 \qquad (6)$$

If $I1 = I2$ and R4 = R5, the FET current dependence is eliminated and Eq. (6) reduces to:

$$I_o3 = (kT/qR5)\ln n \qquad (7)$$

The emitter areas of Q22 and Q23 are A and nA, respectively.

There are three significant advantages to this circuit. It can work on a supply
voltage as low as 700 mV, which is only slightly above a diode voltage drop. The
output resistance of I_o3 is extremely high. As the collector voltage of Q21 rises and
its base emitter voltage decreases, the base of Q21 will simply move down a few
millivolts without any change in I_o3. As a third benefit: the voltage compliance of
I_o3 is excellent. It will provide an accurate, high output impedance current source
down to a Q21 collector voltage of 100 mV.

Figure 18-5.
Low voltage,
high output
impedance
current source.

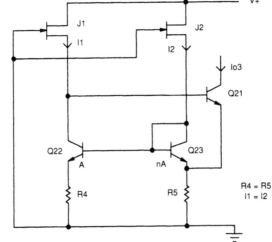

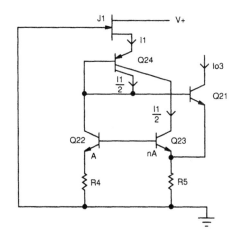

Figure 18-6.
Another version of the wide compliance current source.

If we do not want to use two epitaxial FETs, and another diode voltage is acceptable, the circuit of Figure 18-6 works equally well: Q24 splits the J1 current into two equal segments.

The Triple Function Magic Diode

The next circuit example is the input stage used on the popular, industry standard OP-07 precision operational amplifier [4], including input bias current cancellation, as shown in Figure 18-7. I know this diagram violates the promised maximum of five transistors. But, being a differential pair, the even numbered transistors are mere repetitions of the odd numbered ones; consequently, they should not count against my self-imposed limit.

The bias current cancellation functions as follows. Q33 operates at the same current and matches the current gain of input transistor Q31. The base current of Q33 is mirrored by the split collector pnp, Q35. The nominal 25 nA input current of Q31 is cancelled to less than 1 nA .

The fascinating part of this circuit, however, is the triple role played by diode Q37 (and Q38). The obvious function of these back-to-back diodes is to protect the input transistor pair. Without diodes Q37 and Q38, a differential voltage of more than 7 V would avalanche one of the input transistors, causing permanent damage.

Looking at the circuit diagram of Figure 18-7, we see that no other role for Q37 is apparent. But in ICs circuit schematics often tell only part of the story. What happens when the negative supply is lost, or when the positive supply turns on before the negative one? Assume the input at the base of Q31 is grounded. With no negative

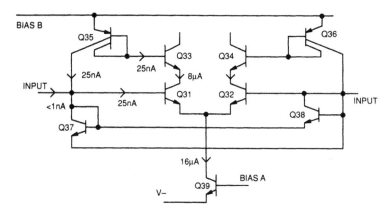

Figure 18-7.
Circuit diagram shows only one of the roles of Q37.

supply, bias A floats up to the positive supply, the collector base junction of Q39 forward biases. The emitter of Q31 is pulled close to the positive supply, avalanching—and damaging—the emitter base junction of the input transistor. This is what would happen without the presence of our hero, diode connected transistor, Q37. Like any other npn transistor, Q37 has a fourth junction: its collector substrate diode—not shown on circuit diagrams. The substrate is always tied to the negative supply. Therefore, as the substrate tries to float up, the substrate collector diode of Q37 will turn on, clamping the negative supply a diode voltage above ground, protecting the precious input transistor, Q31.

The third function of Q37 is again not apparent by looking at Figure 18-7. At elevated temperatures, say at 125 °C, the leakage from the epitaxial layer to the substrate can be as high as 10 nA. The leakage current is indistinguishable from Q33's base current and will be mirrored by Q35. Therefore, an excess current of 10 nA will be pumped into the input by Q35. Again, Q37 to the rescue! By making the isolation area of Q37 the same size as Q35's, the collector substrate leakage of Q37 will be the same 10 nA as generated by Q35, cancelling the leakage current.

Layout Considerations

The three most important factors in real estate are location, location, and location—to quote an old joke. The same thing can be said about the layout of precision analog ICs. The location of a few critical transistors—probably again five or fewer—can create major changes in performance.

A quantum jump in precision operational amplifier performance was achieved with the advent of the common-centroid (or quad) connection of input transistors, and the thermally symmetrical layout [5]. The differential input pair, such as Q31 and Q32 in Figure 18-7, is actually formed from two pairs of cross connected transistors. The effective centroid of both Q31 and Q32 is common at point X (Figure 18-8).

The heat generated by the power dissipating devices is completely cancelled. Although the temperature coefficient of transistor base emitter voltages is –2 mV/°C, the net differential effect on the input transistors in a well layed out precision op

Figure 18-8. Thermally symmetrical layout and quad connection of input transistors.

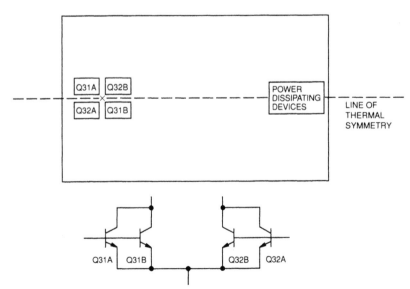

amp can be as low as 1 or 2 μV. The quad connection also improves matching. Just as thermal gradients are neutralized, processing gradients across a wafer, which translate into minute differences between adjacent devices, are cancelled.

Conventional Wisdom and Its Pitfalls

I have had some success by not always accepting the conventional way of doing things. One of the truisms in the IC business is that integrating more and more functions on a chip always represents progress. This is certainly true in digital ICs, and most of the time for analog ICs. For example, most quad operational amplifier designs are monolithic, although the performance of quads is not as good as duals, and duals are not as good as singles. One problem with analog ICs is that the business is fragmented. Very few designs sell in large numbers, the economies of scale, so prevalent in lowering digital IC costs, is seldom present. Thinking about these issues gave me the simple idea: why not design a dual op amp chip, and lay it out in such a way that two dual op amps in a package will make a quad with the standard pin configuration (Figures 18-9 and 18-10).

Look at the advantages of this approach:

1. There is only one product in wafer fabrication—the dual—not two. Assuming equal sales for the dual and quad op amps, chip volume will triple—economies of scale in action, lowering costs.
2. The cost of the quad is further lowered because the wafer sort yield of the dual will be significantly higher, since its chip size is half of a quad made the monolithic way.
3. The performance of the quad will be that of a dual—after all it uses dual chips.

All this can be achieved with the only constraint of two V+ bonding pads, which are shorted with metal on the chip. The layout follows the rules of thermal symmetry [5], easily achieved on a dual but very complicated on a monolithic quad.

The purist will say "but your quad is a hybrid"—with the usual connotations of hybrid versus monolithics, i.e. the hybrid is more expensive and less reliable. We have already shown that our hybrid is less expensive than their monolithic. As far as reliability is concerned, the two extra bonding wires will not make any measurable difference.

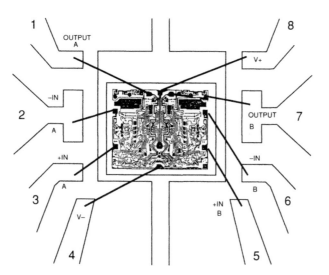

Figure 18-9.
Dual precision
op amp bonding
diagram.

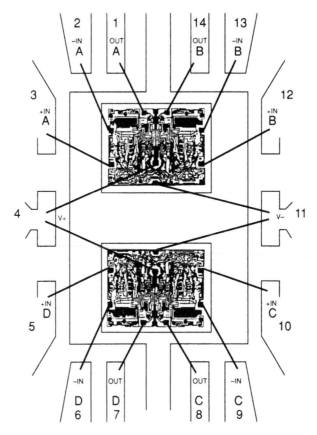

Figure 18-10.
Quad bonding
diagram.

Conclusion

What can be said in conclusion? I hope you have enjoyed my eclectic set of circuit examples. A few years ago one of Linear Technology's advertisements called me an "elegant designer." Needless to say, I was mercilessly teased for years by my peers for that description. Yet, apparently, I am still trying to live up to that ad. With my examples I was striving for elegant simplicity.

References

1. K. Fukahori, Y. Nishikawa, A.R. Hamade, "A High Precision Micropower Operational Amplifier" *IEEE Journal of Solid-State Circuits,* vol. SC-14, Dec. 1979.

2. R. Dobkin, "Input Supply Independent Circuit," U.S. Patent 3, 930, 172, Dec. 30, 1975.

3. G. Erdi, "Low Voltage Current Source/Start-up Circuit," U.S. Patent 4, 837, 496, June 6, 1989.

4. G. Erdi, "Instrumentation Operational Amplifier with Low Noise, Drift, Bias Current," in *Northeast Res. Eng. Meeting Rec. Tech. Papers,* Oct 1972.

5. G. Erdi, "A Low Drift, Low Noise Monolithic Operational Amplifier for Low Level Signal Processing," Fairchild Semiconductor, Application Brief 136, July 1969.

19. Where Do Little Circuits Come From?

When World World II ended, I was 9 years old and fatherless, living in England not far from the Isle of Wight, where Marconi made his historic transatlantic transmissions. Radio was a now worldwide reality, but "electronics" was still in its infancy. I had built a couple of crystal sets, and my dad had left one (a "Gecophone," which I still have), but I was constantly disappointed by their limited performance. They used a jagged lump of grayish silver galena and a wire whisker which had to be wiggled around until you found the sweet spot. Little did I realize that this primitive one-way conducting junction was to be so painstakingly refined in the decades to come or that it embodied the genetic material from which the translinear principle would eventually emerge. Although by no means representative of the state of the art, we had in our house—and routinely used—a beautiful TRF receiver, all ebony and brass. The baseboard-mounted components were connected by bare 16-gauge wire, carefully laid out Manhattan-style. Its few tubes (I believe there were four) had black bakelite bases and some had top caps for the grid connection. The RF and detector stages were independently tuned by black dials, precisely engraved 0–180°, like protractors. It used regeneration, positive feedback around the detector stage, to increase the selectivity and sensitivity, so consequently often used to scream and whistle. The speaker had a stiff cardboard cone that went to a point, where the mechanical drive from the moving magnet drive unit was connected. It was a far cry from hi-fi, but it provided a treasured window on the medium-wave world and brought an immense amount of cerebral nourishment to my young mind, particularly through the BBC, which to this day stands as a testimony to what radio can be, as a medium to inform, enlighten, educate, and delight its thankful listeners.

My older brother used to make shortwave "sets," basic two- and three-tube affairs built on softwood bases (the true breadboard days!), with another, thinner board nailed on at right-angles for the controls. I used to fire them up while he was at work and often would make a few little "improvements" here and there, taking care that any changes I made were pretty much back in order before he came home. Shortwave was different: the excitement of pulling in those weak signals from all across the globe far exceeded their actual content. All the receivers in our small home were battery powered, since we had no "mains" electricity: lighting was by gas lamps and heating was by coal fireplaces. Later, I began to build some receivers of my own but stubbornly refused to use the circuits published in the top magazines of the day, *Practical Wireless* and *Wireless World*. Whether they worked as well or not, they had to be "originals," otherwise, where was the satisfaction? I learned by my mistakes but grew to trust what I had acquired in this way: it was 100% mine, not a replication or mere validation of someone else's inventiveness.

My mother used to work cleaning houses, and on one lucky day the lady owner, observing my interest in radio, invited me to take whatever I wanted of equipment left by her husband, who was killed in the War. It was a garage-full of all kinds of basic components. Most valuable were reels upon reels of enameled copper wire, from the thinnest gauges, like spider webs, to unyielding rods of 14 gauge. These were later to make innumerable tuning coils, solenoids, and transformers. There were multigang tuning capacitors and variable-coupling RF transformers called "goniometers," resistors about 3 inches long, looking like fuses, that clipped into holders which were screwed to the breadboard base, as well as "spaghetti resistors" (in the form of insulated wires, to save space, I suppose), big blocky oil-filled capacitors, cylindrical electrolytic capacitors filled with a mysterious corrosive fluid (I discovered!), power transformers and E and I core stampings, wonderful old tubes, and much more—a garage-sized starter kit! This arcane stuff really got me excited and provided the means to carry out endless experiments.

With the cessation of hostilities, and for about a decade thereafter, the back pages of *Practical Wireless* and *Wireless World* were joyously bursting with advertisements for the most inspiring and magical pieces of "ex-government" equipment— real serious electronic systems. Now in my teens, and with a morning and evening paper route to provide some income, I began to mail-order these extraordinary bargains. Most memorable was an RAF Communications Receiver Type R1355, a wonderful synthesis of steel and copper and aluminum and carbon and glass, for which I paid about 30 shillings, a few dollars. It was amazingly sensitive. At that time, I was particularly interested in anything which had a CRT display in it. The Indicator Type 62 had no less than two of them, both electrostatically deflected: a 6-in. VCR517 for PPI and the other, a 3½ in. VCR138, for A-scan, as well as innumerable high-precision mechanical drives, dozens of vacuum tubes, and one or two crystals. It arrived at my home brand new in the original packing crate, and I couldn't wait to probe its secrets. The VCR517, distinguished by its unusual white phosphor, later became the soul of my first TV receiver, and the VCR138 my first oscilloscope. In time, I had earned enough to have "the electric" put into our home so could do some high-power things now, like build 20-meter PA's with push-pull 807's, dangerously operating their anodes red-hot at 400 V. It was now also possible to make Tesla-coil schemes to generate foot-long arcs and impress the other kids.

There was no end to the wondrous "government surplus" trinkets to be had for a song in those halcyon days. Many were still mysterious to me in their function and purpose, like klystrons and magnetrons. Most enigmatic was an IFF (Identification Friend or Foe) receiver: two back-to-back steel chassis, one side containing about twenty tubes, mostly pentodes (red Sylvania EF50s, I recall) but with a few double-diodes (dumpy little metal-envelope 6H6s); the other chassis included a DC–DC rotary converter and a carbon-pile voltage regulator. Between these was a space into which a detonator fitted; clearly, the secret of the IFF circuitry was something to protect at all costs, and that made it all the more interesting to trace it out. Alas, try as I might, it made no sense at all. Everything seemed to be connected to everything else, forming an impenetrable electronic labyrinth. To this day, I wonder how that stirring electronic mind made its crucial decision between friend and foe.

As long as I can remember, I've been fascinated by finding new circuit forms. It has always been a highly heuristic process. On one occasion, during the battery and gaslight era, I had just built a two-stage triode amplifier, like that shown in Figure 19-1. It worked okay, I guess: placing a finger on the grid of the first tube made a healthy hum in the headphones. But I wasn't satisfied; there *must* be something else you could do with two tubes. I idly connected the output back to the input. It so happened that there was a medium-wave domestic receiver turned on in the room, and I immediately noticed that the BBC station it was tuned to was swamped by a signal obviously emanating from my erstwhile amplifier. Tuning the receiver over the band, "my" signal was popping up all over the place, at regular intervals across

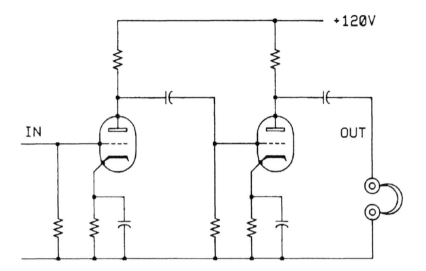

Figure 19-1.
Barrie Gilbert's
early two-stage
triode amplifier.

the dial. I had invented—or rather, stumbled upon—the astable multivibrator (Figure 19-2) and had my first encounter with Fourier series, although it was only later, after having built my first oscilloscope, that I actually saw the squarewave. Ever since then (and probably before) the "what if?" approach has played an important role in reaching new forms.

I get the feeling that the development of new circuit topologies is viewed by the newcomer to circuit design as something akin to magic. I'm not speaking here of architectures of increasing scales of grandeur—LSI, VLSI, ULSI—those best expressed on flickering VDUs as annotated rectangles linked by one-way causal arrows or described by the liturgy of disciplined algorithms, syllogism upon syllogism. Rather, I'm thinking about those happy little tunes that weave just three or four active elements together in some memorable relationship, the themes, rich in harmonic possibilities, from which countless variations unfold. In these deceptively innocent and simple systems, cause and effect are inextricably bound; we are at the quantum level of electronic structure. How many distinctly different and really useful circuits can be made with two transistors, anyway? (Answer: about twenty-four). What heady heights of functional complexity can be attained with three, or even four, transistors? And, heaven forgive us for being so gluttonous, but what

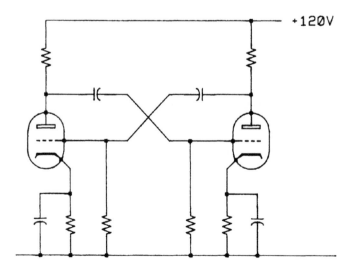

Figure 19-2.
The "accidental"
astable multi-
vibrator.

might one do with *eight* transistors? This is not something to approach as a challenge in combinatorial analysis but as an adventure in the maximum utilization of means, the distillation to essentials, the pursuit of elegance.

Discovering (or inventing—is there a difference?) new uses for just a handful of transistors seems to be difficult for many young engineers entering the field of electronics today. Perhaps this is because they are being taught that in confronting this Brave New Digital World they would be ill-advised to waste their precious college hours on such bygone and primitive notions as Kirchoff's and Ohm's laws, or be concerned about such rudimentary concepts as the conservation of energy and charge, or bother to become adept in Laplace and Fourier analysis. The enlightened contemporary view seems to be that everything of importance will be done in DSP sooner or later. Sadly, there is evidence to suggest that this message is increasingly being accepted. It is precisely the lack of these foundation skills, or even an awareness of the history of electronics, which makes it so hard for many new graduates to cope with component-level circuit innovation, analog or digital.

Several years ago Paul Brokaw was interviewing a young candidate for a job opening in circuit design. They had been discussing the merits of the μA741 operational amplifier, and at one point in the interview Paul asked, "Where do you think the designer got the idea for this particular configuration?" The candidate, without hesitation, replied: "Oh! From a book!" "Hmmm," said Paul, "Well, where d'ya suppose the author of the *book* got the idea?" After pondering this for only the briefest moment, the would-be designer confidently asserted, "Why, from another book!" Many fine textbooks have been written about analog circuit design, but few seem to address the matter of building circuits from first principles. For an amplifier, this might mean starting by fully characterizing the nature of the signal source, its impedance, the range of signal levels, any special features that need to be reckoned with, then doing the same thing for the load. The inner function—its gain, bandwidth, distortion, group delay, transient response, and numerous other matters —likewise need careful quantification before a suitable form can be chosen. It probably doesn't hurt to examine some preexisting forms to begin this process, but it is more satisfying to pursue the design with just the requirements as a starting point. The final choice may turn out looking like the μA741 all over again, but hopefully for the right reasons. In fact, in today's demanding world, it's less likely that an amplifier requirement could be adequately met by a traditional op-amp approach (with its alluring promise of "infinite open-loop gain" and other deceptions), but that's another story.

Authors of books for use in colleges frequently seem to believe that all the popular and widely used topologies which they discuss have an independent existence, and the most the reader can ever hope to do is understand how these circuits work and how they might perhaps be slightly modified to the special requirements of this or that application. The treatment of these circuits as separate forms, all neatly classified into distinct strata, is contrary to the way the skilled designer *feels* about circuits, who is more likely to perceive apparently disparate forms as somehow all part of a larger continuum, all sharing some common all-pervasive notions.

In my own experience, I've met with suspicion and even incredulity at my admission that every time I undertake a new monolithic design project I start out with just the Four Basic Truths:

1. Like elements match well
2. $V = IR$ (ignore this at your peril)
3. $dV/dT = I/C$, or its integral form $CV = IT$
4. a. $I_C = I_S{}^{(Vbe/V_T)}$ for bipolar out of saturation;
   b. $I_{DS} = K(V_{GS} - V_{th})^2$ for MOS in its saturation region

Of course, it helps to have a bit of experience to get things going in the right direction, and now and then one needs to call on some other notions or employ some analytical tools. But it is surprising how far one can go with just these foundations when working out fundamental new forms. A new circuit idea may not work well in practice for some detailed reason, but if it doesn't work at all when the above basics are applied, there is little point in pursuing it further. I'm a great believer in this idea of "foundation design," and almost always begin to explore a new circuit form with totally idealized transistors. Considerable insight can be gained in this way. The nonideal behavior of real devices is only a distraction. For example, the finite current-gain of a bipolar transistor is not of much interest at the foundation level, although it may become the only thing of importance at the practical level, say, in a power amplifier. In a similar way, the ohmic resistances and capacitances of a transistor will eventually become very important in limiting performance, but they are usually only a nuisance during the conceptual phase.

Have you noticed that many little circuits are named after their originator? Thus we have the Widlar and Wilson current mirrors, the Bowes astable, the Eccles-Jordan flip-flop, the Schmitt trigger, the Brokaw bandgap, and so on. Even ultra-simple concepts, little more than common-sense fragments, such as the "Darlington" configuration, are recognized by the names of their originators. Certainly, being the first to realize the utility of these basic forms deserves recognition, but textbooks which present them as timeless entities flowing from the pens of a few Goliaths can create the impression to newcomers that all the important circuit configurations have already been developed and immortalized in print. Circuit naming is useful for purposes of identification, but it tends to transfer *ownership* of these key concepts out of the hands of the user.

This idea of ownership is *very* important. Seasoned designers may be well aware that some ubiquitous form has been widely described and analyzed—perhaps even patented—by somebody else, but they nevertheless have wholeheartedly adopted it and have come to regard it as a personal possession, ripe for further specialized adaptation or augmentation to improve performance or to extend its utility and functionality in numerous new directions. Alternatively, they may perceive the challenge in developing the circuit with a view to *reduction*, to achieve the same or better performance with even fewer components. Indeed, one of the most appealing challenges of analog design is the ongoing search for ever more elegant and potent forms in which every device is indispensable.

This discussion is not about circuit synthesis in any formal sense but about the more organic process of circuit conception, gestation, and birth. It is my experience that the forging of basic new circuit topologies or semiconductor device structures rarely takes the form of a linear, step-by-step progression, from need to solution. Many thick books and numerous papers in the professional journals have been written about circuit synthesis, and this is a perfectly appropriate—even necessary—approach to design in certain cases. For example, it would be virtually impossible to conjure up a useful 7th-order filter without a solid procedural basis. And as mixed-signal ASICs built from standard cells become more commonplace, and product life cycles continue to shorten, a more streamlined approach to "design" will become essential. But where will the "little circuits"—the cells—come from? I happen to be of the opinion that few ever came out of a synthesizing procedure, or ever will.

Few, if any, texts about monolithic design tell the reader that most of the important circuit forms found over and over again in contemporary analog ICs were in all likelihood not the result a formalized approach but arose out of a "what-if?" attitude, or just a dogged determination to *force* a result, sometimes using an "output first" approach, plus a bit of guesswork, all of this seasoned with a generous dash of serendipity and unquenchable optimism. It seems only a matter of basic honesty to tell the reader that, rather than leaving the impression that it's all done by some hard-won esoteric skill. I can recall on several occasions having to console some neophyte designer, whose first stab at a design wasn't working, that it may be necessary to go down 99 paths before finding the 100th approach that satisfies all the conflicting requirements. Maybe this "explorer" attitude toward design cannot be taught, but it should certainly be admitted by those of us that are reckoned to be particularly capable of coming up with new forms, and it should be encouraged as a legitimate—and perhaps the only—productive methodology.

While not academically too respectable, this "prod and poke" approach to cell design is more likely to yield an interesting and valuable new pathway than a formal synthesis procedure ever could. The end result of such fuzzy mental processes may often not be the object of the original search at all, but instead the outcome is to open up some wild new terrain, to be explored and mapped for future journeys of conquest. Creative design is frequently an intensely personal pilgrimage, often quite lonely. It may be hard to justify to the casual onlooker some of the odd transient ideas jotted down in lab notebooks, or to adequately explain the destination planned for a small fragile idea in process. It is likely to have little to do with the most urgent project on the fiscal year's objectives.

It may be useful to illustrate this point of view with a couple of examples from my own experience. I've frequently been asked, "Where did you get the idea for the translinear thing?" Of course, the multiplier cells have proven the most useful, but in fact numerous circuits, all sharing the same principle, were invented following the basic idea, conceived in 1968. I was working at Tektronix on the 7000-series of oscilloscopes, where a pervasive problem was that of altering the gain of a wideband amplifier in a vertical deflection plug-in by about 10 dB without changing the transient response. This was invariably accomplished using a mechanically alterable attenuator, in which a variable resistor was mounted near the signal path and controlled by a connecting shaft to the front panel of the plug-in unit. The problem was that this shaft often had to run from the rear of the plug-in to the panel and was a bit awkward. Various electronic gain-control methods had been tried, having the advantage of allowing the remote location of the controlling potentiometer, but failed to meet all the requirements. The search for an electronically controllable gain-cell arose in that environment, although the solution was not really the result of meeting that need but more a matter of curiosity coupled to the awareness of this potential utility.

In a bipolar long-tailed pair (Figure 19-3), of the sort sometimes found in scope amplifiers (which are invariably differential from start to finish, although usually with emitter degeneration), I noted that if one wanted to end up with a linear signal in the collector, shown in this figure by the modulation index, X, the differential voltage at the bases is forced to be nonlinear, very much so as X approaches its limit values of -1 or $+1$. The exact form ("exact" in the sense of a foundation design based on ideal devices) is simply

$$V_{BB} = V_{T} \log \frac{(1 + X)}{(1 - X)}$$

The noteworthy thing about this is that the "tail" current I_E does not appear in the expression for input voltage. It's also worth noting that the required drive voltage must be proportional to absolute temperature (PTAT) since $V_T = kT/q$. Now, if we turned this circuit around and somehow forced X to a desired value, V_{BB} would have to have exactly the same form. A simple way to do this would be to *current-*

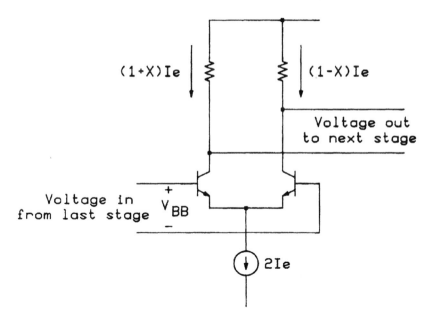

Figure 19-3.
Bipolar long-tailed pair.

drive a similar pair of junctions, from the preceding stage of the amplifier, which we'll assume to be another differential transconductance cell. Although not the only approach, Figure 19-4 shows a suitable way of arranging this. It's a small step to put the two pieces together (Figure 19-5) to make a cell with some amazing properties: the gain is now (1) entirely current-mode, the voltage swings being now only of incidental interest, (2) totally linear right up to the limit values of $X = \pm 1$, (3) totally independent of temperature (the V_T's cancel) and device geometry (the saturation current $I_S(T)$ does not appear in the expression for V_{BB}), and (4) current-controllable, being precisely the ratio I_{E2}/I_{E1}. From this quickly followed the four-quadrant multiplier cell and numerous other translinear circuits. It became something of an obsession at that time to believe that "current-mode" operation was going to be of universal applicability. One reason for this optimism was that since all voltage swings were now reduced to their fundamentally smallest value (typi-

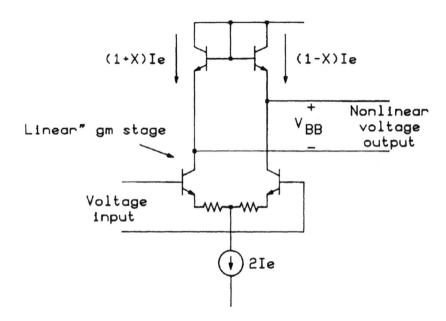

Figure 19-4.
Current-driven junction pair.

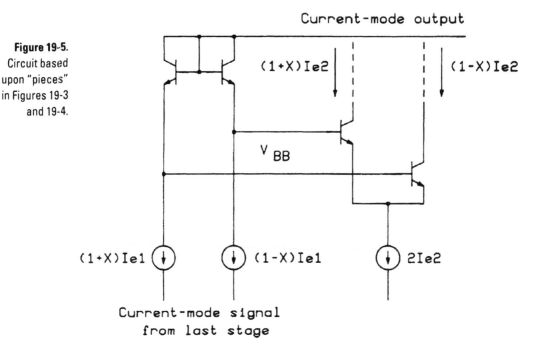

Figure 19-5.
Circuit based
upon "pieces"
in Figures 19-3
and 19-4.

cally, the full-scale value of V_{BB} is only 50 mV), displacement currents in parasitic capacitances were likewise minimized. This is just another way of saying that the impedance levels of translinear circuits are fundamentally minimal. In a wideband monolithic design, this could be quite advantageous; in particular, the pole formed at the load by the large collector-substrate capacitance need no longer limit bandwidth. In fact, bandwidths close to the f_t of the transistor were soon demonstrated in translinear amplifiers and multipliers. Another reason for being excited about these circuits was that their essential function was independent of the actual bias currents: values from a few nanoamps up to tens of milliamps could be used, depending on the speed requirement.

I've described the genesis of this particular cell because of its popularity and utility. In fact, I'm not even sure it happened in such a methodical way. I had also been doodling with the idea of making current-mode amplifiers using current mirrors, employing devices of unequal emitter area. Two were needed to handle a differential signal, and the gain was fixed by the choice of device geometry. Putting these two mirrors side by side and then breaking apart the emitter commons (Figure 19-6), leaving the outer pair grounded and supplying the inner pair with an independent current, resulted in a circuit that could still provide current-mode gain, but the gain was now dependent on the tail current I_{E2}, and not at all on the emitter area ratio. However it may have happened, there was a lot of "what if" about it, and all the translinear circuits that followed. And, if space permitted, I could go on to recall how that was the case for other circuit innovations, too.

All this is not to cast doubt on the value of a structured and disciplined approach to design, sometimes using formal methods, in order to be productive. Modern analog ICs often use dozens of cells. Maintaining focus in design requires paying close attention not only to basic principles but the judicious application of numerous well-known analysis procedures to prevent wasted effort and speed up the optimization process. But the fact remains that the conception of fundamental new topologies will always be a highly heuristic process, which relies far more on

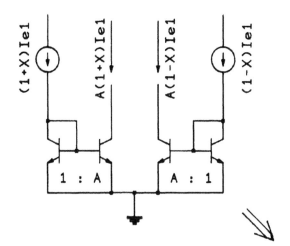

Figure 19-6.
Current-mode
amplifier using
current mirrors.

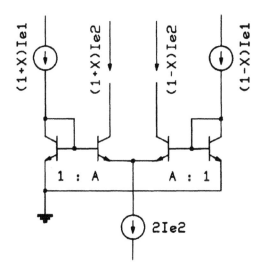

lateral thinking (I'm a fan of Edward de Bono) than on a mechanized approach, which is in the spirit of, and amenable to subsequent conversion into, a computer program. Once the creative breakthrough has been achieved, it is a simple matter to convert the design process into a machine algorithm, if "design" is now taken to mean selecting optimal transistor geometries, choosing the bias points, and deter- mining the value of all the other components in the circuit. This part of design needs a lot of help from formal methods, particularly with optimization as the objective.

Some people are busily writing computer programs for the "automatic design of circuits." I confess to doing some pioneering in this regard, in the early 1960s, using a creaky old Elliott 803 computer, painfully programmed in machine code using paper-tape I/O. Software of this sort serve a useful purpose in allowing one to rapidly parametrize and perhaps optimize a given topology, but I am concerned that as they become more widely available in the modern world of SUNs and Macintosh's they may sound the death-knell to original design. It's bad enough that hundreds of people are already "designing" CMOS VLSI without any significant knowledge of silicon devices and circuits and sometimes without much idea of the physics of hardware in the broader sense. As electronic systems become increasing complex, this type of design will inevitably dominate, certainly for large-scale digital systems. But I wonder how many potentially useful ideas in the meadow-lands of analog circuits will never be discovered because the world of the twenty-first century was taught that analog is dead?

20. The Process of Analog Design

The intent of this chapter is to shed some light onto the way one would approach the definition, design, and manufacture of high performance digitally enhanced analog instrumentation, particularly if the requirements point to a new paradigm in technique and a higher than average project risk. While most helpful and necessary, all of the books in libraries and whatever design tools may exist will not ask or answer the important questions. In such cases, even the design and manufacturing issues will be so interdependent that the whole process will have to be invented, designed, and debugged before the performance of the first working design can be proven and some of the risk of abject project failure removed. Initially there is no light at the end of the proverbial tunnel, no tunnel; no one even knows in which direction or how to dig, and the popular vote is it's pretty silly to think about going straight through solid rock in the first place. Challenges such as this are what some designers find stimulating and call fun, and if successful, many others with later involvement will, in retrospect, remember sharing that same view.

The choice of Digiphase as a vehicle to discuss analog techniques was made partly because the analog techniques are still relevant and yet no longer proprietary. One of the problems with writing about the subtleties in analog design is that many of the stories cannot be told because there are patents or information involved which might later be considered a "trade secret." The nature of things in analog art that are learned or passed along by human experience seldom turn up in print. The discussion of some of the historical and technical aspects of the design will, it is hoped, be a vehicle to convey the process of analog design, and in that sense the particular details of the design are less important, although to date the performance described would be competitive.

Historical Background

Having decided to diversify and expand its business, Dana Labs investigated many possibilties and decided to enter Frequency and Time precision intrumentation. The plan was to lever the experience gained in precision DC instrumentation (DC amplifiers and digital voltmeters) as a basis to build a future F&T division. Hewlett-Packard dominated the market in digital frequency counters and analog signal generators. After hitting the road and talking with many customers, Dana Labs felt it was clear there was a great market opportunity for a synthesized digital signal generator with the performance and price of the existing Hewlett-Packard analog signal generator products. What was required was a high resolution, low phase noise, digitally programmable, and low cost synthesizer.

At that time, direct synthesizers providing high speed and resolution were becoming available (HP 5100) but were enormously complex and cost several times more than analog signal generators. Indirect phase-lock loop (divide-by-N) synthesizers were slow and had inadequate resolution and poor phase noise. DSP frequency generators were also known but suffered from poor spurious and limited output frequency of available DACs. The obvious fact that no one had solved the problem indicated there was an opportunity.

What clearly was called for was a technique with the low cost of indirect synthesis which avoided the resolution and phase noise bandwidth limitations of the Divide-by-N technique. Noel Braymer, a company founder and resident inventor, came up with a proposal for a much higher sampling rate phase-lock loop, which on paper looked like it could work. It was later to be named Digiphase.

At that time, I had been involved in most of the analog circuits used in the DVM product line and was flattered to be called in by the company managment to lead a new development project. It didn't seem to matter at the time that no one in the company had the vaguest idea how a synthesizer worked and, to my knowledge, had never even seen one. The company president on several occasions seemed to have trouble pronouncing the word, which in retrospect I later observed to be a bad sign. Because of priorities in the ongoing business, new engineers and support personnel would have to be recruited. In fact, even space had to be found outside the plant. A small room was sublet and equipped at a nearby warehouse of the Ballard Backup Alarm Company (we were surprised with end-of-the-month alarm quality control testing). Having worked in start-up situations before, even this seemed to be a normal part of getting something bootstrapped, although starting over again from scratch and losing contact with the rest of the organization, I realized with some remorse, was initially an isolated and unrecognized organizational endeavor which would have to provide its own rewards. One of these was uninterrupted focus on the project and time to do some investigation.

The Gathering Phase

A literature search was launched, developing a tree of references from the accumulated bibliographies. The Government Printing Office turned out to be a treasure of old historical data such as *Frequency Control Symposium Proceedings* and other reports and documents. The February 1966, frequency control issue of the *IEEE Journal* was a reference, as were collected copies of patents on various schemes for synthesis. Collecting *Frequency* magazine issues also was informative. After that there were individual articles in *Electronics* and information obtained at the NBS Boulder conference every year on frequency and time standards. In all of this research nothing really did shed much light on the project; in fact, the technique initially appeared somewhat bizarre by traditional standards.

I decided to talk with as many people as possible to gather what was possible from those working in the field. I traveled to visit with Eric Aupperle at the University of Michigan (maximal sequence shift registers), Lee Blackowicz (ECI, Florida, 500 MHz divide-by N), Watkins Johnson (microwave systems), and Bureau of the Navy (shipboard requirements), and I talked with Professor Weaver (Montana State, Montronics founder) and Vic VanDuzer (Hewlett-Packard 5100 section leader, friend from college days). Locally we talked with Dick Sydnor at the Jet Propulsion Laboratory and ultimately hired Floyd Gardner as a consultant to see if we could convince and teach him the technique. This effort verified that we were completely on our own, which is what was originally anticipated.

In recruiting it was apparent that looking for experience in the field was not a practical goal, so a bright engineer working on a graduate degree in mathematics with no design experience was hired to do the logic, and an aerospace engineer with radar design experience was hired to do the fixed frequency divider/multiplier. A skilled PC designer who was not getting along with his supervisor was added from the main plant, although his intransigence vanished when he was challenged with the responsibility of working on something radically new. Within the first 6 months the research and hiring phase was complete, and while the commercial design experience was limited, the team was extremely excited about and dedicated to designing the synthesizer.

One of the first important decisions was to select a digital technology for the 50 MHz phase counter and equality circuit. It must be mentioned that at that particular time almost all logic was constucted of bipolar TTL technology. High speed TTL was reported to be capable of perhaps 25 MHz, and Sylvania was reporting a SUHL line for military customers that was reported to be able to toggle flip-flops at 50 MHz. This apparently was a scaled shrink of their standard line, and since the bond pads also were part of the mask they had been shrunk as well, contributing to bond reliability problems. The price on this logic was as high as any military contract would pay, and commercial support and applications were not available. The high transient current associated with this type of saturated logic and the variable propagation delay associated with a low yield selection process were of great concern, although it was clear IC technology of some form would be necessary due to the prohibitive cost of using discrete transistor hybrid logic, which was the alternative used by Blachowicz.

Motorola had announced MECL I, an ECL SSI logic family which was rated at 25 MHz. It too was very expensive and used round metal-covered ceramic packages to handle the heat dissipation. Conventional wisdom among logic designers of the day was that it was impossible to have reliable operation on logic with only 0.8 V swing, since fast TTL required more voltage than that for noise margin alone. Since the ECL power dissipation per gate was much larger, MSI functions were also not available. ECL did have a much more constant delay, and the differential inputs had a much lower switching threshold uncertainty. Transient currents in ECL were also balanced to a first order. To interface ECL with TTL logic was unnatural, since the output swing was too small, and the inputs were biased from a reference level inside the device.

After some investigation it was determined the risk in pushing TTL to 50 MHz would be unacceptable, both from a vendor and an EMI point of view. This left the 25 MHz ECL, and we decided to give it a try by building two identical 25 MHz ECL sections of the high speed logic and then multiplexing each between even and odd cycles to achieve 50 MHz operation. For 5 MHz and below, TTL seemed a good choice, since it offered multiple vendors and MSI functions. Rather than operate ECL between ground and −5.2 V it was decided to use +5.2 V and ground, in order to allow TTL to drive the ECL inputs. While this created the possibility of noise between the references of the two types of logic, it was not necessary for the ECL logic to drive the TTL logic, so the problem which everyone warned about was not serious. This created a design in which a high speed carry propagate had to be developed in TTL and then reclocked in ECL to reduce the timing delay uncertainty for 50 MHz operation. Once at ECL logic levels, the output could stay in ECL until it was used to drive the phase detector switch drives.

When working on something new with new people, it is very worthwhile to work toward a goal of building a model as soon as possible, since this will tend to focus effort and drive people to ask the right kinds of practical questions. Otherwise, it is possible that enthusiasm will be replaced by frustration as paralysis by undirected analysis sets in. To test the Digiphase concept, a breadboard was constructed using a sheet metal box to shield the VCO, with the logic spread all over hand-wired prototype boards and a hand-wired connector frame. To minimize the effort, the residue compensation and reference section were omitted, and 100 kHz resolution was all that would be possible without fractional frequency phase noise. Laboratory power supplies were connected with long lengths of small diameter wire. This test was the first reality check for the project. Although excitement was very high when the loop was finally locked, it served to point out how incredibly far away we were from the goal of shipping a product:

- To measure phase noise, the Tykulsky autocorrelation method (*IEEE Journal*, February, 1966) was used, and the gain for sidebands close to the carrier was observed to be inadequate. It was obvious that a new setup to measure the phase noise, using a tuned voltmeter and an external frequency synthesizer reference, would have to be designed and the instruments purchased.
- The VCO mechanical stability was such that one could readily observe planes taking off at nearby Orange County (now John Wayne) airport! The setup served as an ultrasensitive microphone for any nearby sounds, such as low level conversations. It was clear that a successful VCO would require a construction that was rigid and magnetically shielded in addition to being electrostatically shielded, since any mechanical modulation of the fields around the coil would create phase noise many orders of magnitude too large to be acceptable. Books and articles never mentioned a problem of this magnitude, since phase-lock loops were not typically used in high performance applications, and it was not clear that a suitable solution could be found. All of the work done in maximizing Q in the VCO inductor had served to make the mechanical problem worse.
- Any physical motion within a wide range of the logic circuitry created large amounts of phase change. Moving a cable or wire by hand made it clear that EMI was out of control. The logic was transmitting into everything, and the loop amplifier was essentially floating in the middle of numerous antennas tied to power supplies. From this experiment it was clear that only the most complete EMI shielding would have a possibility of working.
- The initial loop dynamics were so bad that in order to achieve lock the bandwidth had to be lowered dramatically and the VCO tuned to achieve lock. It was clear that reliable operation would only be obtained when there was careful work done on the whole loop, in particular the problem of VCO gain variation with frequency. The nonlinearity of the available varactors and early design required too much range in the gain of the VCO in megahertz per volt. If this gain change could not be reduced, the loop would not be unconditionally stable for all output frequencies. It was also apparent that the loop would have to be locked in a stable condition simply to measure a low phase noise with the available instrumentation, and an unstable loop would prevent work from even progressing.
- None of the critical electrical real signals was observable directly. The AC voltage at the varactor required to create a milliradian of VCO phase modulation at close-in frequencies was only nanovolts. The time jitter caused by a milliradian of low frequency PM at the 50 MHz output was totally unobservable by the best oscilloscopes and most spectrum analyzers. All experiments had to be done with indirect measurements which were carefully constructed so that the results were unambiguous. As the performance improved with

time, it became impossible to devise experiments with any unambiguous results, since all that could be observed was the VCO output phase noise in a locked loop. This made progress impossible until a completely improved working design was available, somewhat of a Catch-22 situation.

- The odd–even ECL logic created enormous amounts of phase components in the VCO output phase at the first VCO subharmonic. Various schemes to reclock the output edge were unsuccessful. It became clear that all points driven by the VCO output would have to be independently buffered to prevent even the slightest amount of loading of the VCO output phase. These buffers would have to be highly shielded, have high forward gain, and be overdriven at high level to limit AM/PM conversion from the power supplies.

- In addition, the odd–even logic scheme was abandoned when experiments were performed to measure the rejection of data input jitter obtained by re-clocking a flip-flop or the amount of output–input reverse transfer rejection that could be obtained by an ECL circuit at 50 MHz. Typical results indicated only 15 dB improvement, due to pin-pin capacity and package bond induc-tance. In addition, it was determined that the maximum amount of ripple at line harmonics on the ECL power supplies would have to be approximately 1 μV to prevent phase modulation of the final phase detector edges.

- The pulling caused by mutual coupling of the fixed frequency portions of the circuitry and the 1 Hz resolution VCO circuitry indicated the need for almost perfect shielding and isolation of each section of the design which contained different frequencies, otherwise the phase detector output would not achieve the almost perfect linearity required for compensation. Each section would have to have separate power supplies which entered totally shielded modules through EMI filters at every point of entry. Any 50 MHz current leaking out of the modules would surely find a sympathetic semiconductor junction that would result in pulling. All module grounds would have to be low inductance for RF energy bypassing but share no common return path with power supply currents from other modules. In addition, the need for an output attenuator required that the actual output connector common not have ground current flowing in it, since the output attenuation would have to be accurate at micro-volt levels. To solve this, a large piece of metal was required, with no ground loops from DC to 50 MHz, and the output ground of the attenuator floated from local grounds.

At this point all of the original experimental work had to be scrapped, and further progress could only be made when at least all of the above known deficiencies were removed. While enthusiasm was high, these were serious problems, the solution to which could not be confined to a predictable length of time. While Dana manage-ment continued to be supportive, they were also now quite remote and preoccupied with the short term demands of running a successful and growing company. From the discovery rate of problems in the first experiment, there were certain to be more problems in the future, which meant that several multi-quarter iterations would be minimally necessary before release of a production-worthy design. All of the litera-ture and consulting indicated this technique to be new and with little or no hope for building on prior art.

At this juncture, a commitment was necessary, since success would not be possible with a nagging timidity nor by ignoring the magnitude of the problems. This has to be a "right brain" or "no brain" process, since the data processing and consciously analytical "left brain" can't help in seeing the future. Today, business schools are now beginning to recognize that in these circumstances an intuitive decision is not (necessarily) a sign of weakness, although a deterministic situation is still given great preference. To decide between fight or flight, managers must be knowledgable about the technical aspects of the project, be self-confident and experienced enough in that knowledge to tolerate some uncertainty and personal risk, and in the absence of conclusive scientific data, be willing to move ahead by trusting in one's perceived truths, to be willing to go beyond deductive or quantitative analysis, which typically fails in these areas for lack of definitive information or data on future events.

The result of the commitment was it took 3 years to achieve a routinely shippable product, and during more than 2 of those years there was no confirmation that the product would ever meet competitive specifications, yet I don't remember anyone looking back until it was finished. Ultimately, the product became part of a new Frequency and Time Division, which lasted until the recession of the 1970–1971 period, at which time the instrumentation industry consolidated into its present form today, and Dana Labs began a decline which soon forced it to abandon the synthesized signal generator business. In all, about a hundred units were produced.

21. The Art of Good Analog Circuit Design: Some Basic Problems and Possible Solutions

In my opinion, good analog design is an art and not a science. However, a great deal of science is required to master the knowledge needed for analog design. Before any good artist can produce a great work of art, he or she has to have a thorough understanding of the materials he or she is using, as well as the performance and limitations of the tools. Then the artist must use these materials and tools to create a work of art or, in our case, a circuit. My reasoning why good analog design is an art and not a science is that many engineers who have a good technical understanding are still unable to translate their knowledge into good circuit designs. Many excellent pianists have truly mastered playing the piano but lack the creativity to compose music.

The simpler the circuit, the more I like it, and in general, the better it will perform —although it might not be as impressive to an outsider viewing it. In fact, a large circuit can be simple, with each circuit block elegantly designed to perform its function well.

The basic knowledge required for good analog design is a complete understanding of the circuit elements. This understanding must be so ingrained that one knows automatically what element to use in order to achieve the characteristics necessary for good circuit performance.

For bipolar designs, the equivalent of the three primary colors for the artist are the three configurations of the bipolar transistor: grounded base, grounded emitter, and emitter follower. A true mastery of these configurations must be achieved. This will include the DC, AC, and transient performance, as well as knowing all the port impedances, including the substrate in IC designs.

Circuit Element Problems and Possible Solutions

When designing analog integrated circuits, the single most important characteristic is the excellent matching of devices on the same die. This matching will allow many circuit imperfections to be corrected by adding the same imperfection in an appropriate location, with the result that one will cancel the other. This technique can be used not only for cancelling V_{be} errors (which is done regularly) but also for cancelling base current errors as well as Early voltage errors.

The following are analog circuit design problems which I solved by using cancellation techniques.

Base Current Errors

I was designing a simple pnp buffer for part of a chip on the AT&T CBIC U process which had to have a bandwidth of 200 MHz and a low offset voltage temperature coefficient. The pnp transistors can have a worst-case current gain of 10 at low

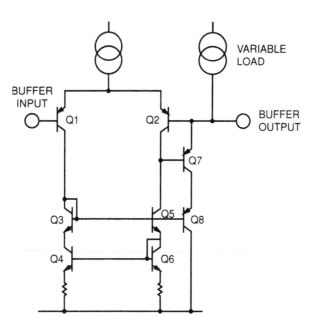

Figure 21-1.
Base current
cancellation.

temperatures, which was causing a problem. The base current variation of Q7 (Figure 21-1), the output emitter follower, with both load current and temperature was causing an offset voltage error.

By adding Q8, running at almost the same current as Q7, the base current error is cancelled. Because the base current from Q8 is added to the input (Q3) of the Wilson current mirror, and Q7's base current takes away current from the output (Q5) of the mirror, the only differential current error seen by the differential pair, Q1 and Q2, is the difference in the base currents of Q7 and Q8. Now the input differential pair, Q1 and Q2, may be run at a much lower current since they no longer have to supply the base current of Q7. As a result, the buffer has a lower input bias current. This is a far better solution than making Q7 a Darlington with its terrible dynamic characteristics.

Early Voltage Errors

When designing a transimpedance amplifier on the Analog Devices complimentary bipolar process, I needed a very high (>100 MΩ) impedance on the high impedance node, as this was the only node that gave the amplifier gain. The input to the high impedance node was a grounded-base stage, Q5 (Figure 21-2), which gave the highest output impedance that I could get from a single transistor. This was still not high enough due to its Early voltage. The collector-to-emitter voltage variation changed the current gain of the transistor and thereby lowered the grounded-base output impedance. The only solution was to cancel the Early voltage error with another device.

Transistors Q1, Q2, Q3, and Q4 form a Wilson current mirror with all base currents compensated. The output from the mirror, Q4, supplies the output grounded-base stage, Q5. The output buffer has a total of three current gain stages (emitter followers), the first being Q6. The three stages of current gain are required so that any output load on the amplifier will not load the high impedance node and lower the amplifier gain. The collector of Q6 is connected to the output so that the voltage across it will not change and thereby modulate its current gain. The constant collector-to-emitter voltage makes the input impedance of Q6 very high (>10 GΩ). The

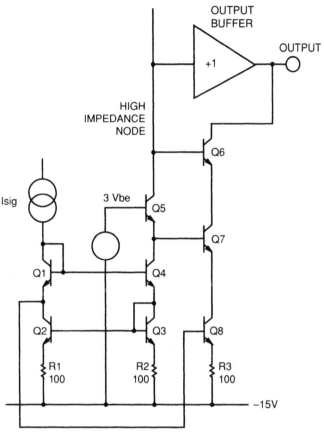

Figure 21-2.
Early voltage
cancellation.

current at which Q6 runs is the same as the rest of the circuit and is supplied by Q7, which in turn is supplied by the current source, Q8. The voltage across Q7 is the same as Q5. The base of Q7 is connected to the emitter of Q5. The base current lost by Q5 is compensated for by adding the base current of Q7, including the modulation of base current due to Early voltage. The only error left is the base current component of Q5 of the base current of Q7, which is a very small $1/\beta^2$ term. Now Q7 has compensated for the Early voltage error of Q5. The impedance on the high impedance node has now gone from 100 MΩ to 4 GΩ in simulation! In silicon, it runs from 200 MΩ to 1 GΩ, which includes the impedance of the pnp side.

The current source, Q8's base, is connected to the input side of the Wilson current mirror so that the base current compensates for the base current required by Q6. As a result, the high impedance node current is exactly equal to the input signal current with all base current errors compensated for. Even at –55°C, the currents are matched to within 0.06%. In silicon, this current matching will be in error due to h_{fe} mismatch of transistors in the circuit (see Computer Simulation, page 195).

Emitter Follower Problems

A simple emitter follower is a great circuit element and will be found in most circuits. If care is not taken to run it at sufficient current, however, one can find it causes terrible distortion on transients due to capacitance on the emitter (Figure 21-3). This capacitance may be the collector-to-substrate capacitance of the current source supplying the emitter current. If sufficient voltage headroom on the current source is available, it is quite often a good idea to insert a resistor in series in order to isolate

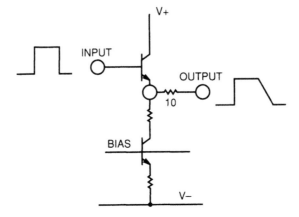

Figure 21-3.
Emitter follower.

the collector capacitance of the current source. This resistor will not stop the emitter follower from switching off on large negative edges but will help current spiking on positive edges.

In addition, emitter followers are known to oscillate very easily if capacitively loaded. This is due to the output impedance looking like a negative inductance as the ft rolls off. This inductance forms a tuned circuit with the capacitive load. To stop the oscillation, a small (10 Ω) resistor in series with the output will normally be sufficient to damp the tuned circuit and stop the oscillation.

Recently I designed a very high speed buffer (>5 V/nsec slew) and arranged that during the high slew conditions, the current sources are increased dramatically so that the emitter followers do not switch off. Power dissipation, which is a constant problem in very high speed circuits, is greatly reduced by only increasing the current sources when needed.

Transient Problems

The really difficult design problems are usually getting a circuit to settle fast. Often I design a circuit with excellent DC and AC performance and then perform a transient simulation with results that bear no resemblance to what I would have expected from the AC transfer curve. In my design work, I probably spend 90% of my time sorting out transient problems. I have just finished designing a very high speed (>5 V/nsec) closed-loop buffer (Figure 21-4) on the AT&T CBIC U process. I had the AC performance looking great with no peaking, a −3 dB point of 600 MHz and a nice clean −6 dB/octave rolloff. I looked at the transient response to a 5 V pulse and had a 0.5 V overshoot. It was the usual problem of a stage cutting off under high slew conditions. When designing closed loop systems, try to avoid stages that switch off under high slew conditions in order to eliminate long settling times as these stages slowly recover. In my case, the problem resulted from the current mirrors in the input stage switching off as the input differential pair switched hard to one side. To stop these stages switching off, I added two extra current sources, Q10 and Q11, to continually supply 1 mA of current to the mirrors.

The circuit diagram (Figure 21-4) is not complete and will not slew at 5 V/nsec as shown. What happens when the current mirrors switch off is that the base nodes of Q3 and Q4, and Q5 and Q6, collapse, discharging the base-stored charge in the devices. When the stages are required to switch back on again, all the charge has to be restored, which takes time. During this time, the output continues in the direction it was going, since no current is available to stop it, the result being overshoot.

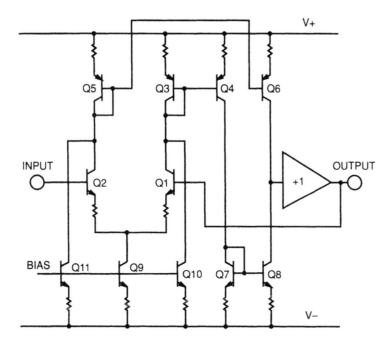

Figure 21-4.
High speed
closed loop
buffer.

Adding the current sources, Q10 and Q11, the mirrors never switch off and the base emitter voltages of the devices only have to change by about 30 mV between the two conditions. This allows the current mirrors to recover very fast and stop the overshoot.

Computer Simulation

Computer circuit simulation is a great learning tool inasmuch as one can examine new circuit ideas and learn from the results. Computer simulation is a great help in designing circuits but will not design the circuit for you. Also, more expensive computer workstations will not design better circuits.

I have used computer simulation for circuit analysis since the late 1960s. At that time, I was designing digital voltmeters for Solartron Ltd. in England. I used a 110-baud teletype machine with punch paper tape. The analysis was purely AC, and each individual transistor had to be inserted as its hybrid-P equivalent. Since the introduction of the IBM PC, I have upgraded my computer about every 2½ years and realize a speed improvement of between 4 and 5 times. Computer speed is a real factor in deciding on what simulations to perform. I will think twice about starting a run that I know is going to take an hour, such as a high-resolution transient run with process variations. Currently I am using a 33 MHz 486 machine which I bought in December 1990. The spice software I use is PSPICE, which is excellent. For chip layout I use ICED from I.C. Editor, especially helpful now with its DRC (design rule checker) for checking the layout.

With the new simulation software and good spice models that cover production process variations, it is now possible to check product production specification spreads. Using the cancellation techniques I have described, the performance of the circuit really depends on device matching. By simulating statistical variations of h_{fe} and V_{be} on the chip, it is possible to determine the production spreads of the cancellation techniques.

Design Approaches

When starting to design a circuit to meet a required specification, several different approaches should be tried. You usually will find that a particular topology will fall close to your requirements. If you stick to your first circuit idea to solve the problem, you will probably find that you need to add a lot of extra circuit fixes to get around circuit imperfections. When the circuit fixes are bigger than the main circuit, you know you have taken the wrong approach.

In simulating different design approaches, you will learn what the problems are with each particular design. Then this knowledge will guide you in designing a new approach that avoids the previous problems.

Future Designs

The new semiconductor processes and those under development are truly remarkable and enable us designers (artists) to create wonderful new circuits. Now it is possible to design new integrated circuits that perform functions that would not have been thought possible a few years ago (100MSPS Track and Hold). It is unfortunate that the one-time engineering costs for making integrated circuits are so *high*.

Today's analog circuits certainly would not have been possible without the remarkable advances in digital electronics—from the IC processes generated for high density digital chips to the use of microcomputers for simulation and chip lay-out. The real world is always going to be analog. The demand for analog circuits is only going to grow as electronics becomes more and more a part of society's daily life. The challenges for analog circuit designers are getting greater as the demand for higher speed and accuracy continues to increase.

22. My Approach to Feedback Loop Design

I like designing feedback loops. I have been designing and building feedback controlled systems for audio and low frequency control since high school. My interest in high fidelity audio started in the late 1950s. Transistors were scarce and not very good, so I worked with vacuum tube circuits. I learned that negative feedback would improve just about every characteristic of an audio amplifier. I built Heathkits and modified them to suit my own preferences. I experienced oscillation when there was too much feedback.

For a freshman project at M.I.T., I learned how negative feedback could transform an unstable device into a stable one. I built a device to suspend a steel ball a fraction of an inch below an electromagnet driven by a tube amplifier. The position of the ball was sensed by a light shining on a photocell. The ball would partially interrupt the light as it was pulled higher. The photocell output was fed back to the amplifier input to control the magnet. After I got the hookup right, the first thing the circuit did was oscillate. I tried out my newly acquired capacitor substitution box and discovered a network that would tame the oscillation. I later learned that it was called a lead-lag network. I was developing an intuitive feel for what to do to make feedback stable.

During my studies at M.I.T., I learned about circuit theory, circuit analysis, and feedback circuit analysis. M.I.T. taught methods for analyzing a circuit "by inspection" as well as the usual loop and node equations and mathematical analysis. I learned the theory of analyzing circuits and transforming between the time domain and the frequency domain. Then I could relate my early experiences to the theory. Along with learning the theory, I really appreciated learning methods of analyzing a circuit by inspection to get approximate results.

Much of the feedback loop design work I do is satisfied during the design phase with only rough approximations of performance. Actually, there are so many variables and effects you cannot consider during the design of a circuit, it is often useless to analyze with great precision. (You don't need a micrometer to measure your feet for a new pair of shoes.)

Since graduating from M.I.T., I have worked in the semiconductor Automatic Test Equipment (ATE) field, designing instrumentation and other parts of ATE systems. First I worked for Teradyne and now I work for LTX. At Teradyne and LTX I have designed several programmable power sources. These programmable sources make heavy use of feedback loops. I have developed a method for design and analysis, which I would like to describe here. I work and communicate better when I use a specific example to illustrate what I am designing or describing. The example I will use here is a programmable voltage/current source I designed for LTX. The drawings are based on sketches I made in my notebook during the development of that product.

My Approach to Design

First, I need a specification of the instrument I am going to design. Then I make a block diagram of the circuit. Also, I will draw a "front panel" of the instrument to show its functions and how they are controlled. This "front panel" has knobs and switches and dials, even though the finished product may be software controlled. The "front panel" helps to evaluate the functions that were specified and to investigate interactions between functions. In other words, does it do what you wanted, the way you want it to?

After I have a block diagram, I like to start the circuit design with a greatly simplified schematic made with a few basic building blocks of ideal characteristics. These blocks are simplified models of the real circuit elements I have to work with. I prefer to design the final circuit with blocks that work similarly to these basic blocks. The basic circuit blocks I like to use include:

- Amplifier with flat frequency response
- Ideal op amp
- Ideal diode
- Ideal zener (voltage clamp)
- Voltage output DAC

To analyze a specific aspect of a design, I make a new schematic that eliminates anything that won't significantly affect the results. I want to be able to analyze by inspection and sketches. After reaching a conclusion, I might check my assumptions on a more complete schematic, or I might build a prototype and make measurements.

I use a notebook to record, develop, and analyze my designs. I draw block diagrams, schematics (from simple to detailed), and sketches of wave forms and frequency responses. I keep the notebook in chronological order and draw new drawings or sketches when there is a significant change to consider. During the preliminary design phase of a project, I might draw dozens of similar sketches as I develop my ideas and the design. I draw with pencil so I can make small changes or corrections without having to redraw the whole thing. I date most pages in my notebook, and I usually redraw a diagram if I invent a change on a later day. I also record the results of experiments and other measurements.

I have my notebooks back to the beginning of LTX. They have been a great source of history and ideas. Sometimes a question comes up that can be answered by going back to my notebooks rather than by making new calculations or experiments. There is real value in having diagrams, sketches, notes, and test results all in that one place. Recently, though, I have been using various CAD and CAE systems to record some of my design developments. Sometimes, the precision that the computer insists upon has been helpful, and other times it's a hindrance. With CAE results, I now have two or three places where parts of the design process are documented. I need to develop a new system for keeping all the historical data in one place. Even with CAE, I don't expect to ever give up hand writing a substantial part of my design development notes.

What Is a V/I Source?

Integrated circuits need to be tested at several stages of their manufacture. Electrical testing is done with automatic test equipment (ATE). One of the instruments of an ATE system is the programmable voltage source. It is used to apply power or bias voltage to a pin on the device under test (DUT) or to a point in the DUT's test circuit. Programmable voltage sources usually can measure the current drawn at the output, and sometimes include the capability of forcing current instead of voltage. In that case, the instrument is called a V/I source.

A V/I source I designed at LTX is called the DPS (device power source). It is part of the Synchromaster line of linear and mixed-signal test systems. The DPS can force voltage or current and measure voltage or current. It's output capability is ±16 V at 1 A to ±64 V at 0.25 A. The current measure ranges are 62 µA full-scale to 1 A full-scale. There is great opportunity for analog design sophistication and tricks in designing a V/I source. Some typical performance requirements are:

- 0.1% forcing and measuring accuracy (at the end of 20 ft of cable)
- 100 µsec settling time after value programming or load change
- Minimum overshoot and ringing (Sometimes 1 V overshoot is acceptable, other times overshoot must be zero.)

There are many interesting aspects to the design of a V/I source. Perhaps the most challenging is to design a V/I source that can force voltage with acceptable performance into a wide range of capacitive loads. Why capacitive loads? In many (perhaps most) of the situations in which a voltage source is connected to the DUT or its test circuit, there needs to be a bypass capacitor to ground. This is the same bypass capacitor you need when actually using the DUT in a real end-use circuit. Occasionally, bypass capacitors need to be gigantic, as in the case of an audio power amplifier IC that requires 1000 µF.

At the same time the V/I source is forcing voltage, the test may require measuring the current drawn by the DUT. The current measure function usually requires a selection of resistors in series with the source output. The current is measured as a voltage across the resistor.

An Ideal V/I Source

It would be ideal if the capacitive loading on the V/I source output had no affect on the force voltage response to programming changes. However, one response effect we should accept is a reduction of the output voltage slew rate as the capacitive load increases, due to the limited amount of current available from the source. Since the V/I source must have some slew rate limit even at no load, the ideal voltage wave form would look like the one in Figure 22-1.

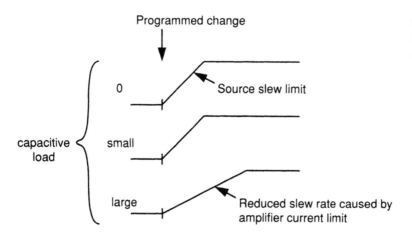

Programmed change

Source slew limit

capacitive load

small

large

Reduced slew rate caused by amplifier current limit

Figure 22-1.
Ideal voltage wave form.

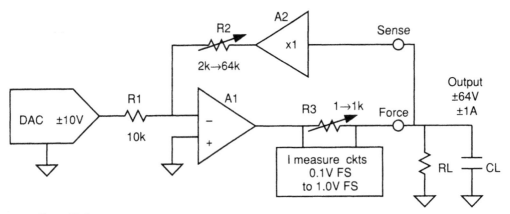

Figure 22-2.
First model of the
programmable
voltage source.

In this ideal case, the voltage will settle to the programmed value just as soon as it is finished slewing. In practice, there is probably no way to achieve the ideal case. The design of a V/I source involves lots of compromises.

Designing a V/I Source

To illustrate my approach to feedback loop design, I will describe a circuit that I developed for the LTX V/I source called DPS. A feature of the DPS is compensation for capacitive loading. With capacitive load compensation, the DPS can drive loads of any reasonable amount of capacitance with good stability and without overshoot.

The figures in this sections are very close to the kind of drawings I do in my notebook. Here is a first model, shown in Figure 22-2, of the circuit to consider. I have simplified it to concentrate on the force voltage mode and the response when driving capacitive loads. This model meets the performance requirements listed above.

I picked 10 kΩ for R1 because that value works well in real circuits and calculations would be easy. I like to have a real value in mind just to simplify the analysis. R2 sets the output voltage full-scale ranges of 2 V to 64 V. R3 varies from 1 Ω at 1 A full-scale to 1 kΩ at 100 μA full-scale. The ×1 amplifier A2 eliminates the loading effect of R2 on the current measure function. With CL possibly as large as 1000 μF, I expect that the rolloff caused by R3 and CL will be a major source of stability problems.

To investigate the effects of capacitive load on this model, I will simplify it further (see Figure 22-3). One of the objectives of a model is to reduce the problem to the basics and make approximate performance estimates easy (or at least possible). Figure 22-3 shows the simplified version.

I have given the simplified model a single voltage and current range. I have left out the load resistor RL, retaining only the capacitor load. I have retained the ×1 amplifier to remind me that R2 does not load the output. If the compensation works in the simple case, then I expect it can be expanded to work on the complete DPS.

I like to design a feedback loop with just one integrator or dominant pole. The other stages are preferably flat frequency response. With this approach in the model, the analysis is easy. I want to design the real circuit the same way. Settling time is related to the lowest frequency pole in the loop. Therefore, extra poles

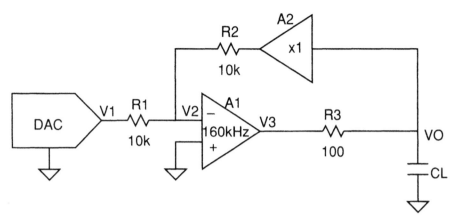

Figure 22-3.
Simplified model
to investigate
capacitive
loading effects.

below the loop unity gain frequency (UGF) make the settling time longer. That's bad in ATE applications, where fast settling is very important. Poles below the loop UGF may be needed, but they should be carefully considered.

I have given op-amp A1 a UGF of 160 kHz. Why 160 kHz? My past experience with programmable sources has shown 100 kHz to be the maximum practical loop bandwidth. More than 100 kHz leads to oscillation due to phase shift from the many amplifiers that will ultimately be in the loop and from the output cable. The output cable? The output cable may be 20 ft long; the cable inductance has a significant effect.

As with other parameters I choose, A1's UGF is a convenient approximation. 160 kHz corresponds to a time constant of 1 μsec. We often need to switch between time domain analysis and frequency domain analysis. I like to remember a simple conversion to eliminate the need to consult a table or calculator while designing. Just remember that 1 μsec corresponds to 160 kHz; you can extrapolate from there— 2 μsec → 80 kHz, 100 μsec → 1.6 kHz, etc. Conversely, 1 MHz corresponds to 160 nsec.

Now we need to analyze the model. Bode plots for the individual stages and the complete loop will give us an idea of the loop stability and bandwidth. Let's start with CL = 0, as shown in Figure 22-4:

The loop gain is unity at 80 kHz and the phase shift is 90°, the "ideal" case for stability.

What happens when CL is increased? First, make new Bode plots as shown in Figure 22-5.

Figure 22-4.
Bode plots for
individual stages
and complete loop.

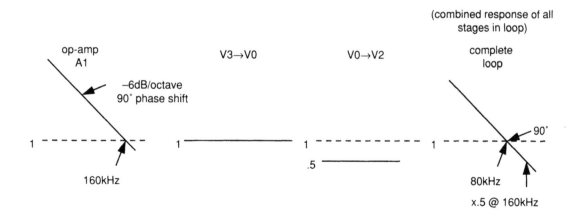

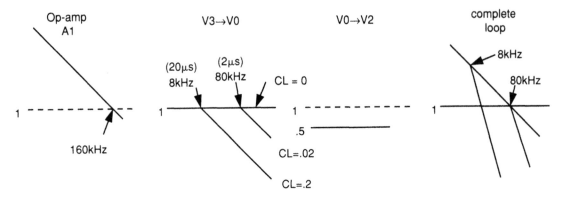

Figure 22-5.
Revised Bode plots.

For CL = 0.02 μF the R3CL rolloff starts at 80 kHz adding 45° phase shift at the loop UGF. For CL = 0.2 μF the added phase shift at 80 kHz is nearly 90°. From experience and by inspection, I would estimate the step response of this circuit to be as shown in Figure 22-6.

For CL = 0, the only time constant in the loop comes from the op amp A1. The step response will be a simple decaying exponential of time constant 2 μsec (based on loop UGF of 80 kHz). At CL = 0.02, there is 45° phase shift added to the loop at the UGF; I estimate a small amount of ringing, perhaps 1 cycle. At CL = 0.2, I know the added phase shift is higher (nearly 90°) and expect more ringing.

I don't need to estimate closer than ±30% or so because this circuit is a greatly simplified model. The main thing we need is to get within a factor of 2 or so of the real performance.

Capacitive Load Compensation

In a previous attempt at cap load compensation, I added an RC network as shown here in Figure 22-7. This technique is similar to adding a resistor and capacitor from force to sense on the output of a power supply. That is one way to stop oscillation

Figure 22-6.
Simplified step
model response.

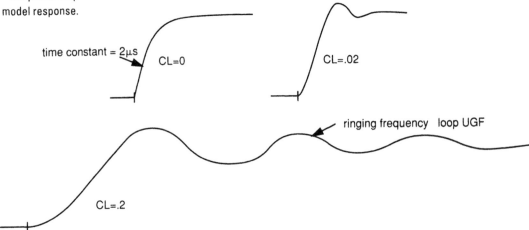

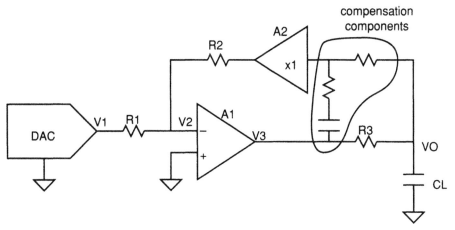

Figure 22-7.
Previous attempt
at capacitive
loading
compensation.

when there is a long cable to a capacitive load. This V/I source has a programmable
choice of four RC networks for compensation. The RC network can be chosen to
eliminate ringing for a range of capacitive loads, at the price of increasing settling
time for no load. The overshoot is not reduced, however. Overshoot has become a
serious issue in a few cases, so when I designed the DPS I wanted a better compensation technique that would reduce ringing and overshoot.

What Causes the Overshoot?

It will be easier to investigate the cause of overshoot with a better model. The above
model is completely linear. It behaves the same for small or large signals. The
amplifier in the model is an ideal op amp. A real amplifier will be limited to some
maximum slew rate and will be limited in its current output. Here we can do a
thought experiment. Assume instant settling time and perfect stability. Make a
feedback loop from an amplifier with a slew rate limit and a current limit. Then the
step response would look like the ideal, as shown in Figure 22-8.

These wave forms show the ideal effect of capacitive load on a voltage source:
no effect until the capacitor current would exceed the amplifier current limit, then
the slew rate is decreased according to the current limit. Let's see how close we can
get to this ideal.

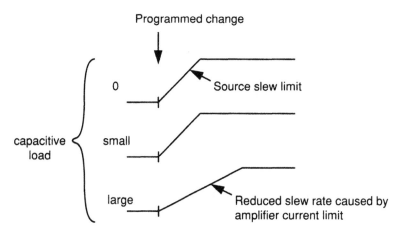

Figure 22-8.
Ideal step
response.

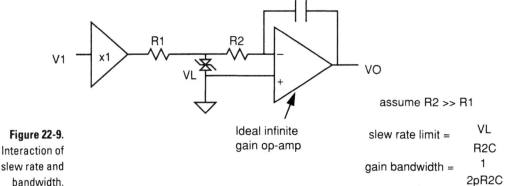

Figure 22-9.
Interaction of
slew rate and
bandwidth.

Ideal infinite
gain op-amp

assume R2 >> R1

$$\text{slew rate limit} = \frac{VL}{R2C}$$

$$\text{gain bandwidth} = \frac{1}{2pR2C}$$

Improving the Model

To add slew rate limiting to the circuit, we could simply redefine the characteristics of the op amp A1 to include a slew rate limit. However, that would add several parameters to remember when analyzing the circuit. I prefer to add a few ideal components to the circuit that would add the slew rate limit and make it easier to visualize the effect.

To model a slew rate limited op amp, I take an ideal op amp with very high bandwidth and put components around it. The components show the effect and interaction of slew rate and bandwidth, as shown in Figure 22-9.

The double anode zener represents an ideal bidirectional clamp that limits the voltage to VL. I have simplified the model to be an inverting amplifier only, since that is what our circuit needs. A "better" model with differential inputs would just make the analysis by inspection harder.

Note that slew rate and gain-bandwidth (GBW) can be set independently of each other by changing VL. Real op amps show this interaction of parameters. Compare a bipolar op amp and a FET op amp of similar GBW. The FET op amp has higher slew rate but needs a larger input voltage than the bipolar to get to its maximum slew rate. A feedback loop built from this model will be linear when V1 < VL and will be in slew rate limit when V1 > VL.

Model to Investigate Overshoot

I have made a new circuit model to include the slew rate of a real amplifier. I simplified the op amp model a bit to make the circuit easier to analyze, as can be seen in Figure 22-10. In this case the simplification should have no effect on the accuracy of the analysis. This op amp model is good here because it is easy to see at which point the circuit becomes nonlinear. When V2 < VL the circuit is linear. When V2 is clamped at VL, the op amp is at slew rate limit (see Figure 22-10).

With this model, I can estimate the time response. When CL = 0, the output will be at slew rate limit until V2 becomes smaller than VL. From that time on, the circuit is linear and the output is a simple time constant response.

This circuit *must* overshoot with capacitive load. That's because CL causes a delay between V3 and VO and there is also a delay from VO back to the amplifier output V3. When the output is programmed from zero to a positive value, V2 starts negative and stays negative until VO gets all the way to the programmed value. At that point, V3 is above the programmed value and positive current is still flowing into CL. In order to make V3 go down, V2 has to be positive. VO has to go above

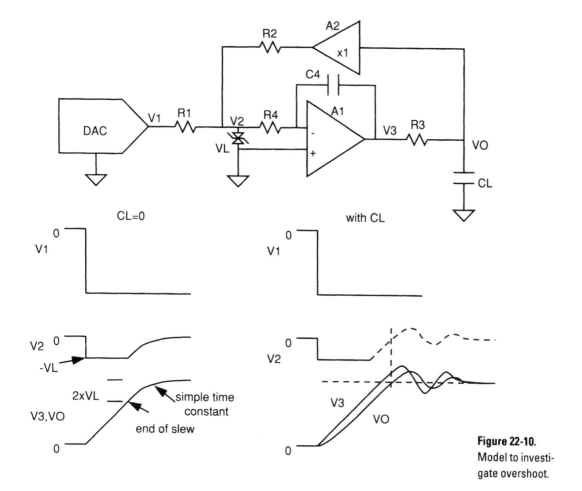

Figure 22-10.
Model to investigate overshoot.

the programmed value in order for V2 to go positive, and that's overshoot. It's a little like trying to stop a boat the instant you get to the dock. There is no brake—all you can do is row backward, which takes time to take effect. Likewise, it's too late to stop VO from overshooting if you don't reverse the current in CL until after VO gets to the value you want to stop at.

How do you stop the boat at the dock? Start slowing down before you get there. In this model, we need an anticipator to reduce the slew rate of V3 before VO gets to the programmed value. One way to do this, as shown in Figure 22-11, is to add a capacitor C2 across the feedback resistor R2. The faster the output is slewing, the more current C2 adds to the null point (V2) and turns down the amplifier input in anticipation of VO getting to its programmed value (see Figure 22-11).

During the time that VO is slewing at a constant rate, the current in C2 is constant and provides a fixed offset current into the loop null point at V2. Without C2, VO started to slow down only when it was 2VL away from the final value. With C2, VO starts to slow down when it is an additional (slew rate × R2C2) away from the final value. That's the anticipation. What happens with different values for C2? If C2 is too small, then VO will overshoot. If C2 is too large, then VO will be too slow to settle. Since ATE applications need minimum settling time, C2 needs to be set just right. Therefore, the value of C2 would have to be programmable according to the load capacitance.

In addition to affecting the transient response, C2 changes the loop gain. At high frequencies where C2 is low impedance, the loop gain is increased by a factor of 2.

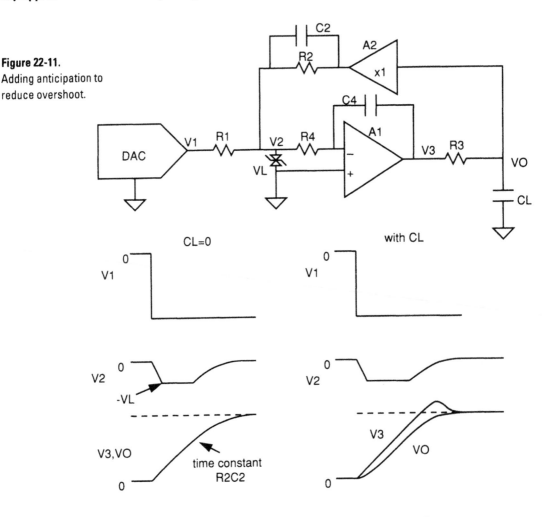

Figure 22-11.
Adding anticipation to
reduce overshoot.

Looking ahead to the real DPS, the closed loop gain from DAC output to source output will be programmable. With other values for R1, C2 would increase loop gain by a different amount, even though the anticipation was the same. I want to make the frequency response independent of range, so I will put C2 in an inverting op-amp circuit. This way, the compensation will behave more nearly ideal. My goal is to have simple, independent circuit blocks that can be implemented in the final circuit (see Figure 22-12).

I have added R6 in series with C2 because the real A3 circuit would have a gain limit at high frequency. In the model, R6 allows us to use an ideal infinite gain op amp for A3 and still set a realistic maximum gain for this stage. For frequencies up to several times the loop UGF, I want the gain of each stage to be set by the passive components and not limited by amplifier gain. That way, the frequency response is sufficiently predictable and the amplifier could be changed to another type without serious effect.

Back To The Frequency Domain

At this point, we have an idea that improves the time response of the circuit, but we have ignored loop stability. To get an idea of the small-signal stability, we need to

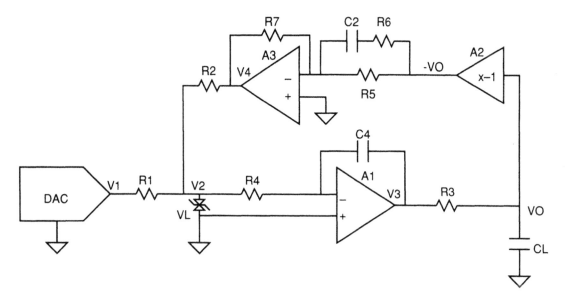

Figure 22-12.
C2 moved to be in
series with R6.

make Bode plots of the new model, as shown in Figure 22-13. From a Bode plot, I
can estimate phase shifts according to the sketch in Figure 22-14.

My criteria for stability of this circuit are:

1. Loop UGF shall be 80 kHz or less.

    Although the Bode plots don't show it, the real circuit will have many poles
    above 100 kHz. If the loop UGF were higher than 80 kHz, there would be lots
    of phase shift added by these poles causing serious instability.

2. Phase shift at UGF is no more than 135° to 150°.

    At 135° the step response has a small amount of ringing. Much more phase
    shift would increase the ringing unacceptably.

Applying the stability criteria above to the drawing of compensated Bode plots,
I conclude:

When CL is zero the UGF is too high.
When CL is small both the phase shift and UGF are too high.
When CL is just right, the loop will be stable.
When CL is too large the phase shift is too high (this case is not shown in
    drawing, but similar to uncompensated).

Is this a problem? I'm not surprised that making CL too large would cause instability. This compensation scheme has a limited range. The big problem is the excessive UGF for small CL. Overcompensation is expected to result in slower than
optimum time response but should not cause oscillation!

To reduce the UGF when the loop is compensated, but CL is small or zero, we
need to reduce the loop gain. This gain reduction has to be done in the forward part
of the loop at A1. Changing the gain in the feedback patch would change the closed
loop gain which sets the full-scale range. Increasing R4 is the best way to reduce the
loop gain. Changing C4 would also change loop gain, but using solid state switches
to switch capacitors is likely produce side effects. Increasing R4 to reduce the loop
gain by a factor of 4 shifts the Bode plot down, decreasing the UGF to an acceptable
80 kHz.

By reducing the gain of the compensated loop we have achieved stability when
CL = 0. However, the settling time is still seriously affected by the compensation

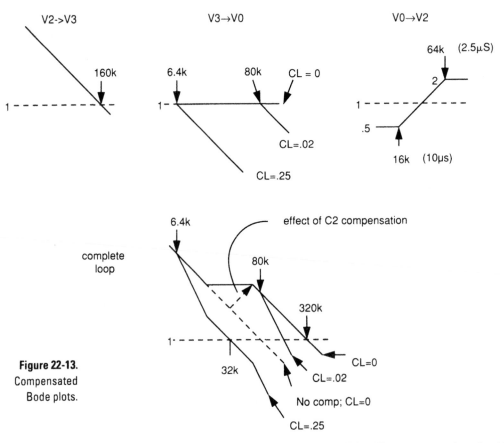

Figure 22-13.
Compensated
Bode plots.

because there is now a pole well below the loop UGF. The compensation should be programmed or switched to an appropriate value according to the load capacitance. In Figure 22-15 I have drawn a circuit to show switching one value of compensation.

S1 adds the anticipation or frequency compensation and S2 reduces the loop gain. To achieve both stability and the elimination of overshoot, these functions must be located in the different parts of the loop as shown.

Range of Compensation Required

The basic stability problem of capacitive loading is caused by the time constant of R3 and CL. R3 is expected to range from 1 Ω to 1 kΩ. CL could reasonably range

Figure 22-14.
Bode plot phase
shift estimates.

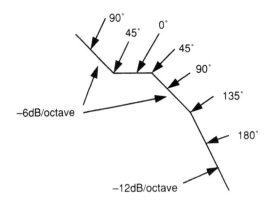

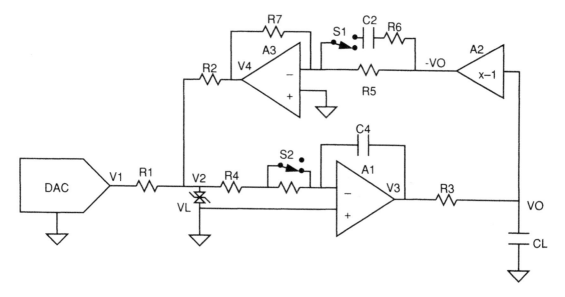

Figure 22-15.
Switchable
compensation.

from 1 nF to 1000 μF. That is a time constant range of 1 nsec to 1 sec. Time constants up to about 1 μsec will have little effect since the loop time constant is 2 μsec. Time constants above 100 msec are probably too long to be useful in an ATE system. Therefore, we have a time constant range of 100,000:1 to compensate for. The circuit above has only a narrow range of good compensation. There would have to be too many programmable choices of compensation networks to cover this wide range. Wouldn't it be neat if there were a way to cover a wider range with one compensation network?

Phase Margin Approach to Loop Compensation

We have been looking at the frequency domain. Now let's consider the capacitive load problem from a phase margin point of view. If the loop frequency response is a single pole, then the loop phase shift is 90° at UGF. A capacitor load adds, depending on frequency, from 0° to 90° additional phase shift to the loop. Phase shift approaching 180° is a problem. What if the no-load loop frequency response were –3 dB/octave instead of –6 dB/octave? That would give the loop a 45° phase shift at no load. Adding a capacitive load would increase the phase shift from 45° to a maximum of 135°. Anywhere in that phase range, stability would be acceptable. This

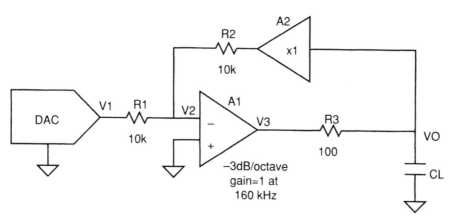

Figure 22-16.
–3 dB/octave
loop gain.

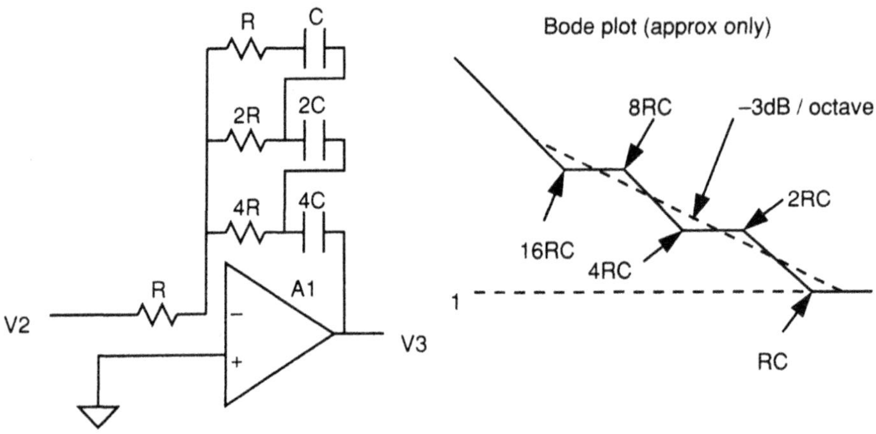

Figure 22-17.
−3 dB/octave
amplifier.

idea sounds like it would work for any value of capacitive load without any switching.

How would we get a loop gain of −3 dB/octave? One way would be to give the A1 stage −3 dB/octave response and the feedback path a flat response, as shown in Figure 22-16.

Replacing C in the inverting op-amp model with a series-parallel RC network comes close to −3 dB/octave and 45° phase shift over a range of frequencies, as shown in Figure 22-17.

This approach to a −3 dB/octave loop does not fix the overshoot problem since it does not allow for the anticipation network we need in the feedback path. A better approach to −3 dB/octave loop response is to leave the forward gain (A1) at −6 dB/octave and make the anticipator circuit +3 dB/octave. Adding more series RC's to R6 C2 will give the anticipator the desired +3 dB/octave over a range of frequencies. The more RC's, the wider the range. However, each RC adds a pole below the loop UGF, and these poles increase settling time. The compensation network does cover a wide range, but it still needs to be switchable to minimize settling time.

At this point, analysis by inspection became less reliable for me, so I used SPICE to simulate the multiple RC compensation network. I ran a SPICE analysis of a triple RC network. Phase analysis is easy with simulation but hard to do by inspection or on the bench. I found it tough to get close to a 45° phase lead out of a small number of components. I decided to shoot for 30° phase lead. SPICE showed me I could get 30° from a double RC network over a 50:1 frequency range. That covers a sufficiently wide range of capacitive load without making settling time too bad.

LTX Device Power Source (DPS) Performance

The LTX DPS turned out pretty good. The LTX Cadence programming language includes a DPS statement to specify the expected load capacitance and a choice of modes: minimum risetime, minimum settling time, or minimum overshoot. The operating system takes into consideration the voltage range and the current measure range, then selects one of four compensation networks (or no compensation) and the corresponding loop gain settings. The three modes are just different degrees of compensation. To minimize overshoot takes the greatest compensation and results in the longest settling time. The compensation works well for all practical values of load capacitance. Overcompensation (specifying a larger capacitance than the actual) makes the settling time longer but causes no stability problem.

The DPS contains other functions, more stages, and many more details than I have outlined here. Each of the analog functions I designed using the same method. All the stages can be grouped together into blocks which match very closely the simplified blocks I based the design upon. The DPS behaves very much like the block diagram and simplified models would predict.

Summary of My Method

By way of an example, I have shown the method I like to use for analog design, especially for feedback loops. Here's an outline:

(Simplify!)

1. Draw a "front panel" of the instrument to be designed. "Try out" its functions.
2. Make a simple circuit model for one function or aspect of the instrument. The model should emphasize that one aspect and deemphasize other aspects.
3. Make simplifying assumptions and analyze the circuit by inspection where possible. Go back and forth between time domain and frequency domain analysis. Check your assumptions.
4. Change the model and analyze again until the results are acceptable.
5. Repeat steps 1–3 for other aspects of the instrument.
6. Design the full circuit with circuit blocks that behave like the ideal blocks in the models.
7. Test a prototype of the instrument to see if it behaves like the models.

Simple, isn't it?

23. The Zoo Circuit

••

History, Mistakes, and Some Monkeys Design a Circuit

•••••••••••••••••••

This chapter is dedicated to the memory of Professor Jerrold R. Zacharias, who saved my ass.

A couple of years ago, I was asked to design a circuit for a customer. The requirements were not trivial, and the customer was having difficulty. I worked on this problem for some time and was asked to present my solution in a formal design review at the customer's location.

When I say "formal," I mean it! I came expecting to talk circuits with a few guys over a pizza. Upon arrival, I was taken to a large and very grand room, reminiscent of a movie theater. About 150 engineers were in attendence. There was every audio-visual machine known to humanity at the ready, and I was almost embarrassed to report that I had no slides, overheads, charts, or whatever (although a piece of chalk would be nice). A "senior technical management panel," positioned in a boxed-off section adjacent to the lectern, was to present a prepared list of questions. A video camera duly recorded the proceedings. The whole thing was chaired by somebody who introduced himself as "Dr. So-and-So, senior vice-president of engineering." Everybody in the place talked in whispers and nodded his head a lot. I found myself alternating between intimidation and amusement.

I gave a fairly stiff presentation, clutching my dear little piece of chalk the whole time. Things seemed to go okay, but not great, and then the panel began with their prepared list of questions. The first question went something like, "Can you explain, precisely, where the ideas for this and that piece of the circuit came from? Can you detail what design procedures, programs, and methodologies were helpful?"

I considered various acceptable answers, but decided to simply tell the truth: "Most of the ideas came from history, making mistakes, and the best source of help was some monkeys at the San Francisco Zoo."

You could have heard a pin before it dropped. There was absolute silence for a bit, and then some guy stood up and asked me to elaborate "a little." Everybody cracked up, the mood shifted, and we finally began to really *talk* about the circuit.

This customer originally came to me with a need for a "CMOS voltage-to-frequency converter." The performance requirements were as follows:

Output frequency	0–10 kHz
Input voltage	0–5 V
Linearity	0.04%
Drift	100 ppm/°C
PSRR	100 ppm/V
Temperature range	0°–55°C
Step response	< 5 cycles of output frequency
Output pulse	5 V CMOS-compatible
Power supply	Single 9 V battery (6.5–10 V)
Power consumption	200 μA maximum
Cost	< $6.00/100,000 pieces

Figure 23-1.
The customer's
circuit, which
was deemed
unsatisfactory.
Despite all-CMOS
construction,
performance was
poor and power
consumption too
high.

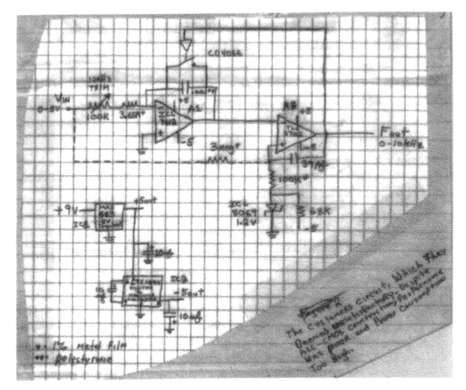

Figure 23-1. The customer's circuit, which was deemed unsatisfactory. Despite all-CMOS construction, performance was poor and power consumption too high.

These people had been working on a design for several months. It functioned, but was described as wholly unsatisfactory. I asked why they needed CMOS and was assured that "the low power requirement is nonnegotiable." Without further comment, I asked them to send me their breadboard. It arrived the next morning, and looked like Figure 23-1.

This is probably the most obvious way to design a V/F converter. The 9 V battery is regulated to 5 V by IC1 and a –5 V rail is derived by IC2. The input voltage causes current flow into A1's summing point. A1 responds by integrating negative, as shown in Figure 23-2, trace A. When A1's output goes low enough, A2 trips high (see trace B in Figure 23-2), turning on the CD4066 switch and resetting the integrator. Local positive feedback around A2 (A2's positive input is trace C) "hangs up" the reset, ensuring a complete integrator discharge. When the positive feedback decays, A1 begins to ramp again. The ramp slope, and hence the repetition frequency, depends upon the input voltage-dependent current into A1's summing point.

As soon as I saw the schematic, I knew I couldn't salvage any portion of this design. A serious drawback to this approach is A1's integrator reset time. This time, "lost" in the integration, results in significant linearity error as the operating frequency approaches it. The circuit's 6 µsec reset (see Figure 23-2, traces A and B) interval introduces a 0.6% error at 1 kHz, rising to 6% at 10 kHz. Also, variations in the reset time contribute additional errors. I added the 3 M resistor (shown in dashed lines) in a half-hearted attempt to improve these figures. This resistor causes A2's trip point to vary slightly with input, partially compensating for the integrator's "lost" reset time. This Band-Aid did improve linearity by more than an order of magnitude, to about 0.4%, but it ain't the way to go.

There are other problems. Quiescent current consumption of this entirely CMOS circuit is 190 µA, rising to a monstrous 700 µA at 10 kHz. Additionally, the polystyrene capacitor's drift alone is –120 ppm/°C, eating up the entire budget. The 1.2

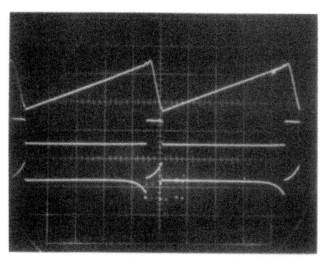

A = 0.5 V/Div.
B = 10 V/Div.
C = 10 V/Div.
Horiz. = 10 μsec/Div.

Figure 23-2.
Wave forms for Figure 23-1's circuit. Finite reset time prevents good linearity performance.

V reference and the input resistor-trimmer could easily double this figure. There are a host of other problems, but what is really needed is an approach with inherently better linearity and lower power consumption.

There are many ways to convert a voltage to a frequency. The "best" approach in an application varies with desired precision, speed, response time, dynamic range, and other considerations.

Figure 23-3's concept potentially achieves high linearity by enclosing Figure 23-1's integrator in a charge-dispensing loop.

In this approach, C2 charges to $-V_{ref}$ during the integrator's ramping time. When the comparator trips, C2 is discharged into A1's summing point, forcing its output high. After C2's discharge, A1 begins to ramp and the cycle repeats. Because the loop acts to force the average summing currents to zero, the integrator time constant and reset time do not affect frequency. Gain drift terms are V_{ref}, C2, and the input resistor. This approach yields high linearity (typically 0.01%) into the megahertz range.

Figure 23-4 is conceptually similar, except that it uses feedback current instead of charge to maintain the op amp's summing point. Each time the op amp's output trips the comparator, the current sink pulls current from the summing point. Current is pulled from the summing point for the timing reference's duration, forcing the integrator positive. At the end of the current sink's period, the integrators output again heads negative. The frequency of this action is input related.

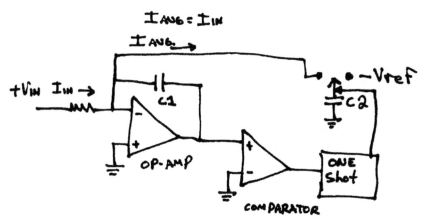

Figure 23-3.
Conceptual charge-dispensing type voltage-to-frequency converter.

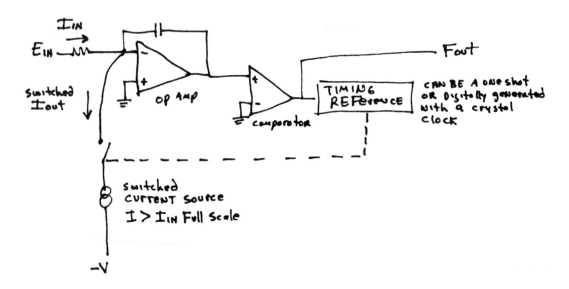

Figure 23-4.
Current balance
voltage-to-
frequency
converter.

Figure 23-5 uses DC loop correction. This arrangement offers all the advantages of charge and current balancing except that response time is slower. Additionally, it can achieve exceptionally high linearity (0.001%), output speeds exceeding 100 MHz, and very wide dynamic range (160 dB). The DC amplifier controls a relatively crude V/F converter. This V/F converter is designed for high speed and wide dynamic range at the expense of linearity and thermal stability. The circuit's output switches a charge pump whose output, integrated to DC, is compared to the input voltage.

The DC amplifier forces the V/F converter operating frequency to be a direct function of input voltage. The DC amplifier's frequency compensation capacitor, required because of loop delays, limits response time. Figure 23-6 is similar, except that the charge pump is replaced by digital counters, a quartz time base, and a DAC. Although it is not immediately obvious, this circuit's resolution is not restricted by the DAC's quantizing limitations. The loop forces the DAC's LSB to oscillate around the ideal value. These oscillations are integrated to DC in the loop compensation capacitor. Hence, the circuit will track input shifts much smaller than a DAC LSB. Typically, a 12-bit DAC (4096 steps) will yield one part on 50,000 resolution. Circuit linearity, however, is set by the DAC's specification.

Figure 23-5.
Loop-charge
pump voltage-to-
frequency
converter.

If you examine these options, Figure 23-3 looks like the winner for the customer's application. The specifications call for step response inside 5 cycles of output fre-

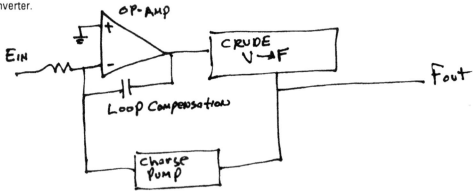

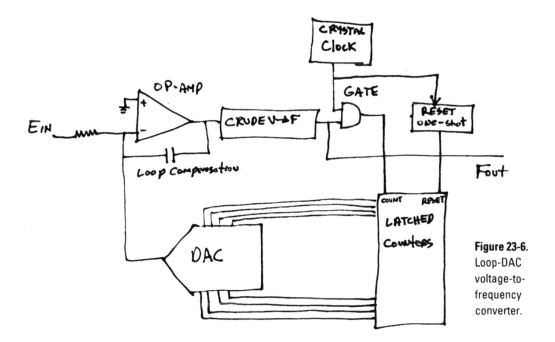

Figure 23-6. Loop-DAC voltage-to-frequency converter.

quency. This eliminates the circuits in Figures 23-4, 23-5, and 23-6 with their DC amplifiers' response time lag. Figure 23-4 requires a timing reference and a precision switched current source, implying some degree of complexity. In theory, Figure 23-3's approach can meet all the specifications without undue complexity.

This technique is not new. I first saw it back in 1964 in a copy of the *GE Transistor Manual*. T. P. Sylvan used a discrete op amp and a unijunction transistor to form the loop. Hewlett-Packard built rack-mounting V/F converters in the early 1960s which also relied on this approach. In 1972, R.A. Pease developed a commercially produced modular version (Teledyne-Philbrick Model 4701) using a single op amp which routinely achieved 0.01% linearity with commensurate drift performance. Pease's circuit is particularly relevant, and a version of it is shown in Figure 23-7.

Assume C1 sits at a small negative potential. A1's negative input is below its zero-biased positive input, and its output is high. The zener bridge clamps high (at $V_Z + V_{D4} + V_{D2}$) and C2 charges via D6, D7, and D8. The input voltage forces current through R1, and C1 begins to charge positively (trace A, Figure 23-8). When C1 crosses zero volts, A1's output (trace B) goes low and the zener bridge clamps negative, discharging C2 (C2's current is trace C) via the D5–C1 path. The resultant charge removal from C1 causes it to rapidly discharge (trace A). R2–C3 provides positive feedback to A1's positive input (trace D), reinforcing this action and hanging up A1's output long enough for a complete C2 discharge. When the R2–C3 feedback decays, A1's output returns high and the cycle repeats. The frequency of this sequence is directly proportional to the input voltage derived current through R1. Drift terms include R1, C2, and the zener, as well as residual diode mismatches. In theory, all the diode drops cancel and do not contribute toward drift. The R2–C3 "one shot" time constant is not critical, as long as it allows enough time for C2 to completely discharge. Similarly, "integrator" C1's value is unimportant as long as it averages A1's negative input to zero.

Q1 and associated components form a start-up loop. Circuit start-up or input overdrive can cause the circuit's AC-coupled feedback to latch. If this occurs, A1 goes negative and wants to stay there. R3 and C4 slowly charge negative, biasing

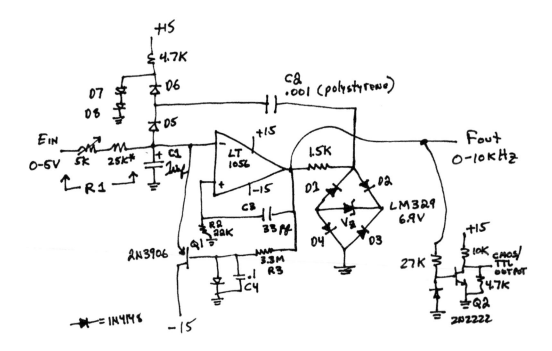

Figure 23-7.
A version of Pease's elegant voltage-to-frequency converter circuit.

Q1. Q1 turns on, pulling C1 toward the −15 V rail, initiating normal circuit action. Once the circuit starts, C4 assumes a small positive potential and Q1 goes off. Q2, a simple level shifter, furnishes a logic-compatible output.

Pease's 1972 circuit is a very elegant, practical incarnation of Figure 23-3. With care, it will meet all the customer's requirements except two. It requires a split ±15 V supply, and pulls well over 10 mA. The job now boils down to dealing with these issues.

Figure 23-9 shows my first attempt at adapting Pease's circuit to my customer's needs. Operation is similar to Pease's circuit. When the input current-derived ramp (trace A, Figure 23-10) at C1A's negative input crosses zero, C1A's output (trace B) drops low, pulling charge through C1. This forces the negative input below zero. C2 provides positive feedback (trace D is the positive input), allowing a complete discharge for C1 (C1 current is trace C). When C2 decays, C1A's output goes high, clamping at the level set by D1, D2, and V_{ref}. C1 receives charge, and recycling

Figure 23-8.
Wave forms for the Pease-type voltage-to-frequency converter.

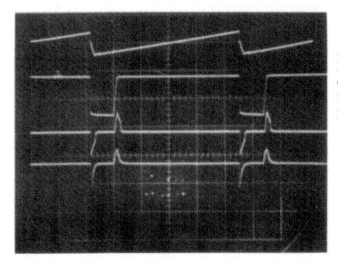

A = 0.02 V/Div.
B = 20 V/Div.
C = 20 mA/Div.
D = 20 V/Div.
Horiz. = 20 μsec/Div.

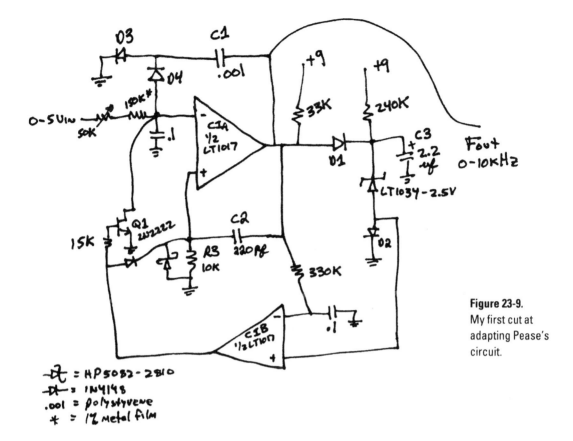

Figure 23-9.
My first cut at adapting Pease's circuit.

occurs when C1A's negative input again arrives at zero. The frequency of this action is related to the input voltage. Diodes D3 and D4 provide steering and are temperature compensated by D1 and D2. C1A's sink saturation voltage is uncompensated but small. (These temperature coefficient assumptions are first order and will require more care later.) Although the LT1017 and LT1034 have low operating currents, this circuit pulls almost 400 µA. The AC current paths include C1's charge-discharge cycle, and C2's branch. The DC path through D2 and V_{ref} is particularly costly. C1's charging must occur quickly enough for 10 kHz operation, meaning the clamp seen by C1A's output must have low impedance at this frequency. C3 helps, but significant current still must come from somewhere to keep impedance low. C1A's current-limited output ($\approx$30 µA source) cannot do the job unaided, and the resistor from the supply is required. Even if C1A could supply the necessary current, V_{ref}'s settling time would be an issue. Dropping C1's value will reduce impedance requirements proportionally and would seem to solve the problem. Unfortunately, such reduction magnifies the effects of stray capacitance at the D3–D4 junction. It also mandates increasing R_{in}'s value to keep scale factor constant. This lowers operating currents at C1A's negative input, making bias current and offset more significant error sources.

C1B, Q1, and associated components form a start-up loop which operates in similar fashion to the one in Pease's circuit (Figure 23-7).

Figure 23-11 shows an initial attempt at dealing with these issues. This scheme is similar to Figure 23-9, except that Q1 and Q2 appear. V_{ref} receives switched bias via Q1, instead of being on all the time. Q2 provides the sink path for C1. These transistors invert C1A's output, so its input pin assignments are exchanged. R1 provides a light current from the supply, improving reference settling time. This

Figure 23-10.
Wave forms for
the circuit in
Figure 23-9.

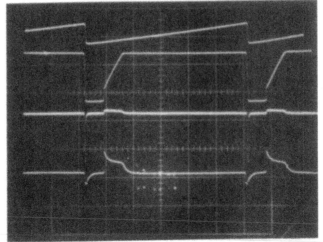

A = 50 mV/Div.
B = 2 V/Div.
C = 2 mA/DIv.
D = 1 V/Div.
Horiz. = 20 μsec/Div.

arrangement decreases supply current to about 300 μA, a significant improvement.
Several problems do exist, however. Q1's switched operation is really effective
only at higher frequencies. In the lower ranges, C1A's output is low most of the
time, biasing Q1 on and wasting power. Additionally, when C1A's output switches,
Q1 and Q2 simultaneously conduct during the transition, effectively shunting R2
across the supply. Finally, the base currents of both transistors flow to ground and

Figure 23-11. are lost. Figure 23-12 shows the wave form traces for this circuit. The basic temper-
The second try. ature compensation is as before, except that Q2's saturation term replaces the com-
Q1 and Q2 switch parator's. This temperature compensation scheme looks okay, but we're still hand
the reference, waving.
saving some   Figure 23-13 is better. Q1 is gone, Q2 remains, but Q3, Q4, and Q5 have been
power.

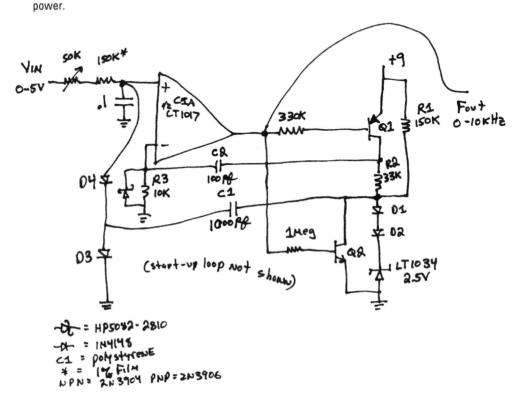

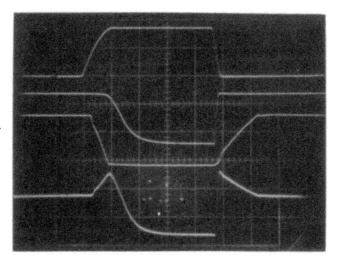

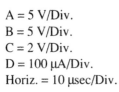

A = 5 V/Div.
B = 5 V/Div.
C = 2 V/Div.
D = 100 µA/Div.
Horiz. = 10 µsec/Div.

Figure 23-12.
Figure 11's wave forms. Traces A, B, C, and D are C1A output, Q1 collector, Q2 collector, and R2 current, respectively. Q1–Q2 simultaneous conduction problem is evident in trace D.

added. V_{ref} and its associated diodes are biased from R1. Q3, an emitter-follower, is used to source current to C1. Q4 temperature compensates Q3's V_{be}, and Q5 switches Q3.

This method has some distinct advantages. The V_{ref} string can operate at greatly reduced current because of Q3's current gain. Also, Figure 23-11's simultaneous conduction problem is largely alleviated because Q5 and Q2 are switched at the same voltage threshold out of C1A. Q3's base and emitter currents are delivered to C1. Q5's currents are wasted, although they are much smaller than Q3's. Q2's small base current is also lost. The values for C2 and R3 have been changed. The time constant is the same, but some current reduction occurs due to R3's increase.

Operating wave forms are shown in Figure 23-14, and include C1's output (trace

Figure 23-13.
A better scheme for switching the reference.

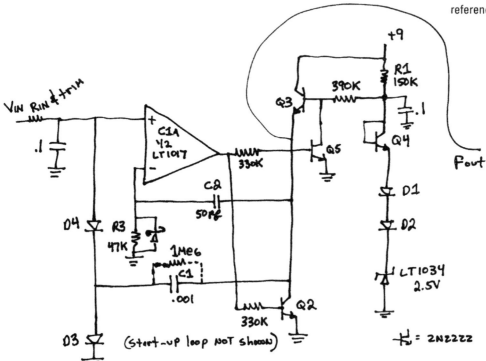

Figure 23-14.
Figure 23-13's
operation. Traces
D, E, and F reveal
no simultaneous
conduction
problems.

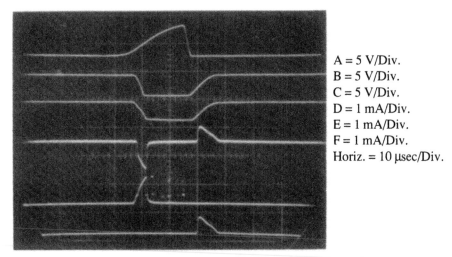

A = 5 V/Div.
B = 5 V/Div.
C = 5 V/Div.
D = 1 mA/Div.
E = 1 mA/Div.
F = 1 mA/Div.
Horiz. = 10 μsec/Div.

A), Q5's collector (trace B), Q2's collector (trace C), Q2's collector current (trace D), C1's current (trace E), and Q3's emitter current (trace F). Note that the current steering is clean, with no simultaneous conduction problems.

This circuit's 200 μA power consumption was low enough to make other specifications worth checking. Linearity came in at 0.05%, and dropped to 0.02% when I added a 1 M resistor (dashed lines) across C1. The D4–Q2 path cannot *fully* switch C1 because of junction drop limitations. The resistor squeezes the last little bit of charge out of C1, completing the discharge and improving linearity.

Power supply rejection ratio (PSRR) was not good enough. Supply shifts show up as current changes through R1. The LT1034 is relatively insensitive to this, but the Q4, D1, D2 trio shift value. As such, I measured 0.1%/V PSRR. R1 really needs to be a current source, or some compensation mechanism must be used.

Temperature compensation was next. Now it was time to stop hand waving and take a hard look. Q4 supposedly compensates Q3, with D1 and D2 opposing D3 and D4. Unfortunately, these devices operate under different dynamic and DC conditions, making precise cancellation difficult. In practice, R1's value should be established to source the current through Q4–D1–D2, which provides optimum circuit temperature coefficient. Assuming perfect cancellation, and no LT1034 or input resistor drift, we still must deal with Q2's V_{ce} saturation term. At 100 mV saturation, Q2 will drift about +0.3%/°C (see the Motorola 2N2222 data sheet), causing about a –300 μV/°C shift in the voltage C1 discharges toward. This works out to about –100 ppm/°C (C1 charges to 3 V) temperature coefficient, which will force a similar *positive* shift in output frequency. C1, a polystyrene type, drifts about –120 ppm/°C, contributing further overall positive temperature coefficient (as C1, or the voltage it charges to, gets smaller, the circuit must oscillate faster to keep the summing point at zero). So the best case is about 220 ppm/°C, and reality dictates that all the other junctions won't match precisely. Temperature testing confirmed all this. Initially, the breadboard showed about 275 ppm/°C, and, by varying R1, bottomed out at about 200 ppm/°C. This certainly wasn't production-worthy engineering but pointed the way toward a solution.

How could I reduce the temperature coefficient and fix the PSRR? Additionally, power consumption was still marginal, although linearity was close. Replacing R1 with a current source offered hope for PSRR, but reliable temperature compensation and lower power needed another approach. I pined for inspiration but got nothing. I was stuck.

Figure 23-15.
The zoo monkey
on parallel rails.

Something that *had* inspired me for a couple of months was a physician I'd been seeing. We really had a good time together—a couple of playful kids. There was much dimension to this woman, and I really enjoyed just how relaxed I felt being with her. Things were going quite nicely, and I sometimes allowed myself the luxury of wondering what would become of us.

One weekday afternoon, we played hookey and went to the San Francisco Zoo. The weather was gorgeous, no crowds, and the Alfa ran great. (On our second date it threw a fan belt.) We saw bears, elephants, tigers, birds, and ate lots of junk food. The lions got fed; they were *loud* and *hungry*. Strolling around, eating cheeseburgers, and doing just fine, we came to the monkeys.

These guys are actors; they love an audience. There was the usual array of grinning, simian catcalls, cheeping, squawking, lots of jungle bar performances, wondrous feats of balance, and other such theatrics. One character particularly caught my eye. He did a little routine between two parallel rails. First, he hung by his hands as shown in figure 23-15.

Then, very quickly, he flipped over, simultaneously rotating, so he ended up inverted (see Figure 23-16).

He did this over and over at great speed; it was his act. Standing there, watching the little fellow do his inverting routine between the rails, I saw my circuit problems simply melt. I felt very lucky. I had a good lady, and a good circuit too.

If you look inside a CMOS logic inverter, the output stage looks like Figure 23-17.

The MOS output transistors connect the output terminal to the supply or ground rail. The input circuitry is arranged so only one transistor is on at a time; simultaneous conduction cannot occur. Typically, channel-on resistance is 100–200 Ω. There are no junction effects; the transistor channels are purely ohmic. The device's input pin appears almost purely capacitive, drawing only picoamperes of bias current.

Figure 23-18 shows what happens when the CMOS inverter is dropped into the gizzard of Figure 23-13's circuit. C1 is charged and discharged via the CMOS inverter's ohmic output transistors. Q3 now drives the inverter's supply pin, and Q2 goes away. Along with Q2's departure goes its 100 ppm/°C temperature coefficient

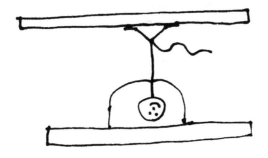

Figure 23-16.
The zoo monkey
on parallel rails,
inverted.

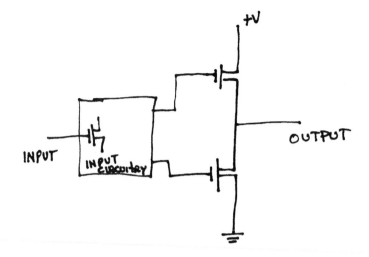

Figure 23-17.
Conceptual
CMOS inverter.

error. Also, Q2's base current is eliminated, along with Q5's base and collector current losses.

This scheme promises both lower temperature drift and lower power. Assuming ideal junction compensation, the remaining uncompensated drift terms are C1's −120 ppm temperature coefficient and the input resistor. Unfortunately, this configuration does nothing to fix the PSRR problem. The only realistic fix for that is to replace R1 with a current source. The current source doesn't have to be very stable but must run with only 2 V of headroom because the circuit has to work down to 6.5 V. The simplest alternative is the monolithic LM134. This three-terminal, resistor-programmable device will function with only 800 mV across it, although it does have a 0.33%/°C temperature coefficient. This temperature coefficient seemed small enough to avoid causing any trouble. The LT1034 shouldn't care, but what about D1, D2, and Q4? When I calculated the effect of current-source shift with temperature on these devices, I realized I had just inherited the world. It came out

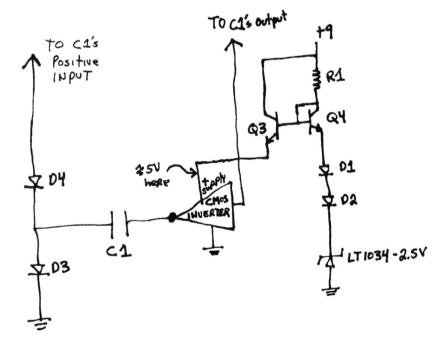

Figure 23-18.
Adding the
CMOS inverter to
the circuit in
Figure 23-13.

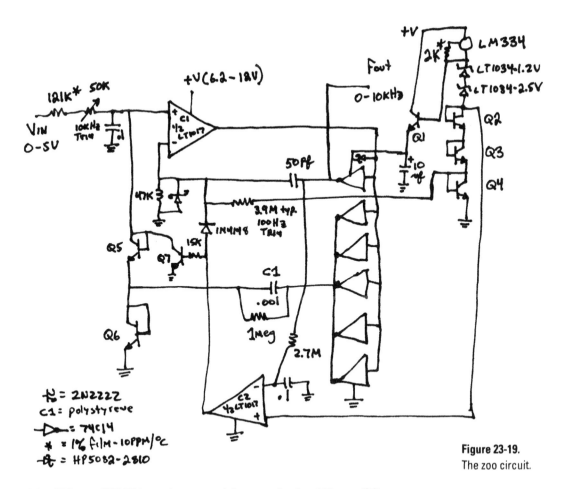

Figure 23-19.
The zoo circuit.

positive 180 ppm/°C! This tends to cancel the capacitor's –120 ppm/°C term. Additionally, increasing the LT1034's reference voltage by about 50% would pull the compensation down to +120 ppm/°C, further reducing drift. This also aids overall temperature coefficient by making the residual junction mismatches a smaller percentage of the total reference voltage. The current source's low head-room capability allows this, while maintaining operation down to $V_{supply} = 6.2$ V. The sole uncompensated term is the input resistor, which can be specified for low temperature drift.

Figure 23-19 is the final circuit. It meets or exceeds every customer specification.

A 0–5 V input produces a 0–10 kHz output, with a linearity of 0.02%. Gain drift is 40 ppm/°C, and PSRR is inside 40 ppm/V. Maximum current consumption is 145 μA, descending to 80 μA for $V_{in} = 0$. Other specifications appear in Table 2's summary. Much of this circuit should be, by now, familiar. Some changes have occurred, but nothing too drastic. The diodes have been replaced with transistors for lower leakage and more consistant matching. Also, paralleling the CMOS inverters provides lower resistance switching. The start-up loop has also been modified.

To maintain perspective, it's useful to review circuit operation. Assume C1's positive input is slightly below its negative input (C2's output is low). The input voltage causes a positive-going ramp at C1's positive input (trace A, Figure 23-20). C1's output is low, biasing the CMOS inverter outputs high. This allows current to flow from Q1's emitter, through the inverter supply pin to the 0.001 μF capacitor. The 10 μF capacitor provides high-frequency bypass, maintaining a low impedance

Figure 23-20.
Figure 23-19's
wave forms.

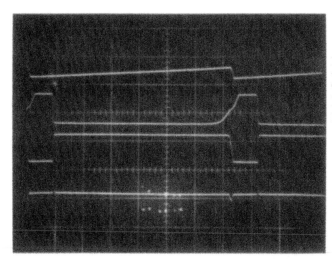

A = 50 mV/Div.
B = 5 V/Div.
C = 5 V/Div.
D = 10 mA/Div.
Horiz. = 20 μsec/Div

at Q1's emitter. Diode connected Q6 provides a path to ground. The voltage that the 0.001 μF unit charges to is a function of Q1's emitter potential and Q6's drop. When the ramp at C1's positive input goes high enough, C1's output goes high (trace B) and the inverters switch low (trace C). The Schottky clamp prevents CMOS inverter input overdrive. This action pulls current from C1's positive input capacitor via the Q5–0.001 μF route (trace D). This current removal resets C1's positive input ramp to a potential slightly below ground, forcing C1's output to go low. The 50 pF capacitor connected to the circuit output furnishes AC positive feedback, ensuring that C1's output remains positive long enough for a complete discharge of the 0.001 μF capacitor. As in Figure 23-13, the 1 MΩ resistor completes C1's discharge.

The Schottky diode prevents C1's input from being driven outside its negative common-mode limit. When the 50 pF unit's feedback decays, C1 again switches low and the entire cycle repeats. The oscillation frequency depends directly on the input voltage–derived current.

Q1's emitter voltage must be carefully controlled to get low drift. Q3 and Q4 temperature compensate Q5 and Q6 while Q2 compensates Q1's V_{be}. The two LT1034s are the actual voltage reference and the LM334 current source provides excellent supply immunity (better than 40 ppm/V PSRR) and also aids circuit temperature coefficient. It does this by utilizing the LM334's 0.3%/°C temperature coefficient to slightly temperature modulate the voltage drop in the Q2–Q4 trio. This correction's sign and magnitude directly oppose that of the –120 ppm/°C 0.001 μF polystyrene capacitor, aiding overall circuit stability.

The Q1 emitter-follower delivers charge to the 0.001 μF capacitor efficiently. Both base and collector current end up in the capacitor. The paralleled CMOS inverters provide low loss SPDT reference switching without significant drive losses. Additionally, the inverter specified is a Schmitt input type, minimizing power loss due to C1's relatively slow rising edges. The 0.001 μF capacitor, as small as accuracy permits, draws only small transient currents during its charge and discharge cycles. The 50 pF–47 K positive feedback combination draws insignificantly small switching currents. Figure 23-21, a plot of supply current versus operating frequency, reflects the low power design. At zero frequency, the LT1017's quiescent current and the 35 μA reference stack bias accounts for all current drain. There are no other paths for loss. As frequency scales up, the charge–discharge cycle of the 0.001 μF capacitor introduces the 7 μA/kHz increase shown. A smaller value

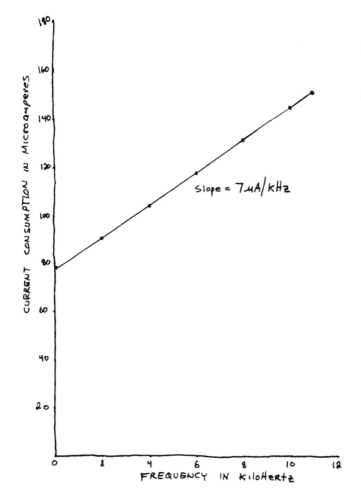

Figure 23-21.
Current con-
sumption versus
frequency for
Figure 23-19.

capacitor would cut power, but the effects of stray capacitance, charge imbalance in the 74C14, and LT1017 bias currents would introduce accuracy errors. For example, if C1 is reduced to 100 pf (along with other appropriate changes), the circuit consumes only 90 μA at 10 kHz, but linearity degrades to .05%.

Circuit start-up or overdrive can cause the circuit's AC-coupled feedback to latch. If this occurs, C1's output goes high. C2, detecting this via the inverters and the 2.7 M–0.1 μF lag, also goes high. This lifts C1's negative input and grounds the positive input with Q7, initiating normal circuit action.

Because the charge pump is directly coupled to C1's output, response is fast. Figure 23-22 shows the output (trace B) settling within one cycle for a fast input step (trace A).

To calibrate this circuit, apply 50 mV and select the value at C1's input for a 100 Hz output. Then, apply 5 V and trim the input potentiometer for a 10 kHz output.

Here's what the customer ended up getting:

Summary: Voltage-to-Frequency Converter
Output frequency	0–10 kHz
Input voltage	0–5 V
Linearity	0.02%
Drift	40 ppm/°C

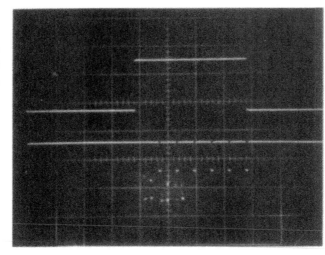

Figure 23-22.
Figure 23-18's
step response.

A = 2V/DIV
B = 5V/DIV
Horiz. = 200 μsec/DIV

PSRR	40 ppm/V
Temperature range	0–70° C
Step response	1 cycle of output frequency
Output pulse	5 V CMOS-compatible
Power supply	Single 9 V battery (6.2–12 V)
Power consumption	145 μA maximum, 80 μA quiescent
Cost	< $6.00/100,000 pieces

The zoo circuit made my customer happy, even if it is almost entirely bipolar. The inverter is the only piece of CMOS in the thing. I'm fairly certain the customer wouldn't mind if I had used 12AX7s[1] as long as it met specifications. It runs well in production, and they make lots of them, which makes my boss and the stockholders happy.

This circuit has received some amount of attention in the technical community. I am aware of some spectacularly complex mathematical descriptions of it, along with some arcane explanations of its behavior. Similarly, it has been shown that the circuit could have only been arrived at with the aid of a computer. Given this undue credit, the least I could do is come clean about the circuit's humble origins.

I hope it was as much fun to read about the circuit as it was to build it.

References

1. "Voltage to Frequency Converter," *General Electric Transistor Manual*, page 346, General Electric Company, Syracuse, New York, 1964.

2. R.A. Pease, "A New Ultra-Linear Voltage-to-Frequency Converter," 1973 *NEREM Record*, Vol. I, page 167.

3. R.A. Pease, assignee to Teledyne, "Amplitude to Frequency Converter," U.S. patent 3,746,968, filed September, 1972.

4. J. Williams, "Micropower Circuits for Signal Conditioning," 10 kHz Voltage-to-Frequency Converter, pp. 10-13, Linear Technology Corporation, Application Note 23, 1987.

5. J. Williams, "Designs for High Performance Voltage-to-Frequency Converters," Linear Technology Corporation, Application Note 14, 1986.

---

1. For those tender of years, 12AX7s are thermionically activated FETs, descended from Lee DeForest.

Part Five

Techniques, Tips, and Applications

Good analog design rests on a foundation of proven, reliable tools and methods. In this section, some analog design pros will discuss techniques and circuit models that work for them.

There's often a major gap between the theory of analog integrated circuit design and what "plays" in the real world. Derek Bowers explains what's really involved in the process, and his insights are useful whether you're an analog IC designer or user.

Current feedback amplifiers are versatile building-blocks in analog design. Dr. Sergio Franco thoroughly examines them in his chapter.

There are some special considerations when analog and digital techniques are combined in the same design. Garry Gillette explores these in his chapter on using analog techniques to extend the range of digitally-generated clocking signals.

SPICE is a controversial tool in analog design—as some of the other chapters in this book demonstrate. If it's going to be used well, its benefits and limitations must be clearly understood. Dr. E. J. Kennedy aids in this with his discussion of SPICE.

Oscillator design is frequently neglected in engineering curricula today, and as a result is often a poorly understood topic. Bob Matthys explains the basic types of crystal oscillator circuits and their design in his contribution,

One of the triumphs of analog engineering was the development of the voltage to frequency converter. Bob Pease traces the history these circuits, and describes their fundamental operating principles in the process.

The "king" of analog devices is the operational amplifier. Without an understanding of the basics of this component, successful analog design is impossible. Dan Sheingold takes care of this requirement with his clear, cogent discussion of fundamental op amp theory.

24. Reality Driven Analog Integrated Circuit Design

I first started using analog integrated circuits in hobby projects in the early 1970s, but I did not become formally involved with the design of such IC's until 1980—considerably later than most of the other authors of this book.

Despite this, the analog world in which I first started working was still mostly dominated by the basic planar bipolar process, incorporating double-diffused npn transistors, lateral pnp transistors and p-type diffused resistors. Even within such a world, the compilation of a "hitch-hikers" guide to analog integrated circuit design is a phenomenal task. However, the excellent textbooks by Gray and Meyer [1] and, later, Grebene [2] (and the second edition of Gray and Meyer) demonstrate that at least this is somewhat feasible.

It is interesting to note that the first edition of Gray and Meyer had only one reference to the MOS transistor in analog design, and fewer than eight pages were devoted to more exotic processing than the standard bipolar process mentioned above. This is not intended as a criticism, for as the authors pointed out, "The basic fabrication process consisting of six or seven mask steps and four diffusions is used for the vast majority of the analog integrated circuits produced" (page 109). This was certainly true in the year of publication (1977) and is probably true even today. In the future, however, the nature of analog IC design is going to be much more shaped and diversified by the vast promulgation of new "linear compatible" processes. This chapter thus partly concerns the adoption of new technologies into analog IC design, but this is far from my quintessential point.

Integrated circuit engineering (and most engineering, for that matter) centers around compromise. Infinite time and resources are simply not available for any given design. Many elegant and thoroughly cogitated designs have not worked as expected, or have not been commercially successful; conversely, many designs of a somewhat trivial character, from an engineering standpoint, have made companies (and occasionally the designers) a large sum of money. When I was new to this industry, one of my superiors made a remark that went something like: "If a good designer feels there is a 0.1% chance that something might go wrong in an area of his design, then he will wait until he is more confident before proceeding further." This worried me at the time, but I can no longer conceal the fact that I have rarely been close to achieving this sort of confidence before integrating my designs. I feel that the engineer who feels he, or she, has mastered an understanding of all significant variables in a particular design has, almost certainly, not considered all of the significant variables.

Experience, of course, is the key to taking shortcuts (enforced or otherwise) in an industry in which mistakes are very expensive, and I am certainly not trying to encourage lackadaisical attitudes to IC design. But this industry is also an example of one in which missed time scales are also expensive, so a good engineer must learn to compromise correctly. This chapter is therefore partly concerned with the psychology of any given project.

So, armed with a thorough knowledge of available technology, and a good feeling in his or her head for design goals and concerns, it only remains for the design engineer to perform the actual work.

Now, of course, we are at a point where design tools become very important. Some circuits have been designed purely by hand calculation; many others have been checked by means of breadboarding. The majority today, however, and certainly future designs will rely mostly on computer simulation.

In reality all three techniques are always in use. All designs start with at least some hand calculation, and the parameter extraction of simulation models is basically a simplified breadboard. And a fourth technique, instinct, is possibly the most important of all if the design is to be an innovative one.

To summarize, this chapter consists of three important sections related to the design process. How best to make use of modern technology, how to form realistic design goals, and how to use the available design tools to accomplish the task. These three sections are highly interactive, so this chapter reads a little like a badly written computer program, but I have done my best to keep it as digestible as possible.

Many years ago, upon coagulating the three laws of thermodynamics, it was realized that a more fundamental "zeroth" law was missing. Similarly, the above trilogy is missing a fundamental starting point. Put bluntly, "who wants these designs anyway?" So we need a "zeroth" section. (Jim did tell me to write this chapter in my own style!)

Section 0. What's All This Analog Stuff Anyway?

What Is an Analog Integrated Circuit?

From a physical standpoint, there is actually no such thing as a digital integrated circuit. All electronic components have complicated analog behavior, and their successful inclusion in digital systems (and conversely, attempts to model them accurately by digital means), requires much human ingenuity. The basic distinction between analog and digital circuits is that the former rely for their operation on the analog properties of circuit elements, while these properties are a nuisance in the case of the latter. This point is raised because of the increasing prevalence of IC's with both digital and analog functions on the same chip.

Integrating digital functions (of significant complexity) on a process optimized for analog circuitry is difficult, and integrating analog functions (with significant performance requirements) on a digitally optimized process is also difficult. A process engineer might add that the development of a process with some degree of optimization for both functions is at least equally difficult.

But none of these is necessarily impossible; it is a highly complex trade-off to decide whether a given "mixed-signal" function (one of the current "buzzwords" perpetuated in the description of circuits containing both analog and digital functions) is practical with existing technology, and if not, whether it is practical (or worthwhile) to develop a new technology for its realization. I am tempted to explore this in more detail, but space is limited, and I have a lot more ground to cover . . .

The Analog IC Marketplace

I am always fascinated by the number of statistics tables, pie-charts, and so on, that are generated in an attempt to gain an understanding of the analog IC marketplace, largely an offshoot of very rapidly changing customer expectations. In the early 1970s, customers were still amazed that any useful function could be totally integrated on a tiny piece of silicon. Nowadays, the same customers are often surprised that a complete system of their own definition cannot.

Analog IC's basically started with operational amplifiers and evolved into a plethora of standard circuits. Improved technology eventually led to a generation of digital-to-analog and analog-to-digital converters, and this step is being followed by more complex "mixed-signal" functions. On top of this, we have the "application specific" type of circuits, which really are just a conglomeration of building blocks that can belong to both the previous categories. The same can be said for custom circuits, except that the range of complexity in this case can be much wider—from a few transistors to many thousands (big for an analog function).

So, from a fundamental point of view, there are really only two basic partitions to consider—analog and digital (in the light of the technology available for manufacture)—when evaluating the feasibility of a given marketing proposal. For the designer, however, the market ramifications are far more severe.

The trite statement that "necessity is the mother of invention" seems somewhat obvious if one assumes that the majority of people do not waste their time trying to invent things which are clearly not needed. There are many important exceptions to this logic, however, where highly useful technology (including the bipolar transistor) was developed almost accidentally from research along a different direction. Many breakthroughs in analog IC design (such as the transconductance multiplier, invented by Barrie Gilbert) also happen in this way and end up fulfilling an important need in the market segment.

But such inspired discoveries can never be scheduled. A custom project cannot be undertaken on the presumption that some totally unheard of technique will be devised. Even if no particular section of a custom circuit presents individual difficulties, many problems can be encountered when blocks are eventually "glued" together.

The point I am trying to make here is that brilliance is not available on tap. If a particular project has tight time constraints, even a totally successful solution is not going to be optimum, and perfectionist engineers are unlikely to be satisfied in executing such projects. Again, it all comes down to the ability to compromise.

Establishing the feasibility of modern integrated circuits, from both a marketing and technical standpoint, is becoming increasingly difficult as technological advances are made. Fantastically complex linear and mixed-signal functions are now possible, given the right technology, large enough die area, and long enough development time. However, despite such managerial phrases such as "collective risk-taking" and "ambitious development schedules," it is likely to be the design engineer, or project leader, who takes the blame for a product that is too late or too expensive to be commercially successful.

But enough of my natural pessimism concerning a rather volatile marketplace. The truth is that there is a large market for linear IC's, and there are many highly successful companies (and designers) who serve it.

Regardless of how big a circuit ends up to be or how long it takes to develop, it is a certain fact that it will not be a commercial success if it doesn't work. The remainder of this chapter will concentrate on the technology and tools available to the modern circuit designer, as well as many pitfalls that can trap unwary engineers (and many wary ones as well) upon their procedure along the design obstacle course.

Section 1. Technological Peepshow
Analog IC Technology—A Brief History

The basic principle of the MOSFET was postulated in the 1930s, but the first commercially feasible devices were not demonstrated until late 1962. However, research initially along the same lines led to the development of the bipolar junction transistor by Shockley, Bardeen, and Brattain in the late 1940s. Initially, such transistors were fabricated from germanium, but the mid 1950s saw the development of silicon transistors with much lower leakages and wider operating temperature ranges. The subsequent invention of the silicon integrated circuit by Robert Noyce at Fairchild in the late 1950s sparked off the present integrated circuit industry.

The original "planar" IC process consisted basically of (quite fast) npn transistors and diffused p-type resistors. As such, the process was not really classifiable as a "digital" or "analog" process, though its initial usage was in the implementation of the resistor–transistor logic family used in mainframe computer applications.

This basic process, however, was also the same one used to fabricate op amps such as the μA702 in the early 1960s, thus beginning the era of the analog integrated circuit.

These early designs were difficult to use and restricted in their application, but by the end of the decade a second generation of op amps (and other analog functions) had emerged, including such standards as the μA709, μA741 (Fairchild), and LM101 (National).

The improvements in these later products were due to three main reasons. First, as might be expected, a considerable depth of design experience had been built up, enabling more efficient architectures to be devised. Second, as might equally be expected, the performance and control of the process had been considerably improved over the same period. But additionally, many new components were incorporated into these designs such as lateral and substrate pnp transistors, pinch resistors, epi-FETs and oxide capacitors. Such components were the combined result of design and process innovation.

This progression is not a mere slice of history; rather, it is part of the ongoing general formula for improved IC design.

The 1970s saw further developments in analog technology and a corresponding increase in the performance and variety of available functions. National Semiconductor's "Bi-Fet" process enabled precision JFETs to be added to a conventional bipolar process. Radiation Inc. (later part of Harris Semiconductor) pioneered dielectric isolation technology for very high-speed and radiation-hardened circuits. Analog Devices developed laser trimming at the wafer level, and Precision Monolithics introduced the first complete monolithic digital-to-analog converter.

Of course, digital technology had not been standing still during this time, and by 1970 very few companies were attempting to use the same bipolar process for both analog and digital circuits. The dominant reason for this was a somewhat divergent marketplace.

Digital technology was evolving down the higher speed, low voltage path of TTL and ECL logic families, where parameters such as resistor nonlinearity and transistor matching were unimportant. Conversely, analog technology was pursuing higher functionality and precision in what was (and still largely is) a 30 V world. To exacerbate matters, such techniques as gold-doping, used to improve the performance of digital circuits, could completely wreck the performance of such components as the lateral pnp transistor. Alas, it had not taken long for up-to-date digital and analog technologies to become highly incompatible.

All this, of course, before even considering the parallel development of MOS technology.

For a time, the MOSFET took a back-seat to the bipolar transistor, because the latter had been successfully produced and appeared to be adequate for immediate demands. When logic started to become more complex, the MOSFET looked increasingly attractive. The original reasons for this were straightforward:

1. A basic PMOS or NMOS process technology is simpler than a corresponding bipolar one (typically four versus six masking steps).
2. For a given set of photolithography limitations, the MOS transistor is somewhat smaller than its bipolar counterpart due to the removal of nested diffusions and reduced mask tolerance accumulations. Also the total area of a complete circuit is much further reduced because MOS devices are self isolating; and it is the isolation region which wastes much of the area of a typical bipolar IC.

Unlike bipolar technology, it cannot be said that early MOS processes were in any way well suited to supporting analog circuitry. Indeed, the threshold control on the first commercial PMOS devices was so poor that multiple supplies were required just to turn them on and off! There have been some valiant attempts to integrate analog functions in such recalcitrant technologies, particularly where a predominantly digital chip required some rudimentary analog support circuitry. But few people have seriously considered PMOS or NMOS technology as anything but a means to produce logic IC's at the lowest possible price, because of the poor transistor performance compared to bipolar devices.

The mid 1970s saw a new breed of digital circuits using CMOS (complementary metal-oxide semiconductor) technology, which offered very low power consumption and ease of use. The low power meant that large systems could now potentially be integrated, and there was thus a stronger desire than ever to include analog functions along with the digital ones. This task was made somewhat simpler by the inclusion of transistors of both polarity, but in general radically different design techniques evolved to overcome CMOS limitations.

The most important of these techniques is undoubtedly the "switched capacitor" concept, which makes use of a very fundamental property of a MOSFET: its very high input impedance. Switched capacitor circuits were originally developed for telecommunications applications, where large integrated systems required considerable analog circuitry. Nowadays, switched capacitor techniques provide a whole cornucopia of filter, digital-to-analog and autozeroing amplifier/comparator functions. These make the combination of complex digital circuitry and high-performance analog circuitry a very practical proposition.

Switched capacitor techniques are definitely not a panacea, however. Compared with a bipolar implementation, the switched capacitor functions are slower and much noisier.

So, to generalize, bipolar technology can provide extremely high performance analog circuitry, but large digital sections consume a lot of power and chip area. Conversely, CMOS produces very efficient digital circuitry but is limited in analog performance. Therefore, it does not take Sherlock Holmes to realize that a process offering both technologies would be very powerful in the integration of mixed-signal systems. Such processes are only now beginning to become prevalent in the industry, and in view of the above comments one might ask why this has taken so long. I offer the following as a partial answer.

Initially, digital requirements were modest and could be taken care of with fairly simple bipolar logic gates. As the digital requirements became more complex, a considerable amount of ingenuity was directed toward bipolar digital techniques with improved density and power efficiency. This resulted in the development of a significant number of logic families; current-hogging logic, integrated Schottky logic, integrated injection logic, linear differential logic and many more; all of which were more-or-less compatible with linear IC processing. But although most of these achieved some success, none had a good enough combination of features to be considered a general purpose logic form, and the increasing requirements eventually exhausted the ingenuity supply. Even so, BiCMOS was on the back burner for a while.

Another difficulty arose from the fact that bipolar IC's have historically been produced using silicon with a surface orientated on the <1 1 1> crystal plane. This type of silicon yields a high surface charge, which in turn helps in attaining a high field threshold. The goal is to keep the field threshold higher than the total operating voltage to eliminate parasitic MOS action between devices.

High surface charge is obviously undesirable for MOS devices, because of the difficulty of controlling the device threshold voltage, and thus MOS (including CMOS) has traditionally been built on <1 0 0> silicon. With modern ion-implant technology, it is easy to increase field threshold with a light surface implant, and thus bipolar transistors can now be made without compromise on <1 0 0> material. It has actually been claimed that such bipolar transistors have lower flicker noise than those fabricated on conventional <1 1 1> silicon, but I have not seen any convincing evidence of this.

A final factor involves the logistics of BiCMOS. Even within companies producing both analog and digital products, it is rare for individual design or process groups to have experience (or interest) in both technologies. There has been some recent alleviation of this problem due to the increasing use of high speed BiCMOS in purely digital applications (notably fast static RAMs) but these processes are definitely not optimized for analog work and lack the capacitors and resistors desirable in a good all-round process.

The bottom line is that much organized effort is needed within a company to make BiCMOS worthwhile; but now that some industry-wide inertia has been established, the benefits of BiCMOS technology seem certain to ensure its widespread adoption.

So Where Are We Today?

At this point it should be clear that the "good old days" of six or seven mask steps of analog processing are rapidly disappearing. Not that all current design is "state-of-the-art," or even close to it, but if advanced technology is available, then sooner or later it will become dominant. As I have pointed out, the extreme diversity of current analog processes makes this an awkward section to write, but I feel that at least a half-decent approach is to review the types of components available to the thoroughly modern analog IC designer.

Bipolar Transistors

The conventional linear bipolar process, still very much in use, offers npn transistors with roughly 500 MHz F_t, lateral pnp transistors with an F_t of 3–7 MHz, and substrate pnp's (often confusingly referred to as vertical pnp's) with an F_t around 15 MHz. Early IC texts quoted the lateral pnp transistor as being a very low gain device, but this is not true for modern processes with well-annealed surface oxides. Such a pnp built on an op amp type process with 5 Ω-cm epi material is now capable of a peak beta in the range of 50–200, and this peak normally occurs at a collector current of about 10 μA for a small device. At lower currents, the beta reduces but is usually still quite usable even in the picoamp range. At high currents, high level injection drastically reduces the current gain, making the lateral pnp extremely unsuitable as a power device unless made unreasonably large. Also, the low F_t (and high excess phase) limit this transistor to low frequency applications. The substrate pnp, of course, has its collector unalterably connected to the negative supply (or more correctly, the p-type substrate, which almost always has to be the negative supply) but has found uses as an output emitter follower due to its improved F_t and power handling characteristics. These improvements are not dramatic, though, and such an output stage quickly falls apart at currents much in excess of a few tens of milliamps, or at frequencies above a few megahertz.

Several processes have evolved from this standard type linear bipolar process which include a multiply diffused pnp transistor with characteristics more closely compatible with the npn transistor.

The usual method of achieving this is to diffuse in a p-well to act as the pnp collector, and an extra n-type diffusion for the pnp base. The pnp emitter can usually double as the npn base, and a p-type buried layer (if included) has also been used as part of the isolation diffusion. Dielectric isolation simplifies this procedure, and this is why the first commercially successful examples of such a process used this type of isolation. A further method of creating such a "complementary" process is to turn the bipolar process upside down to optimize for the pnp transistor. While this does complicate the fabrication of the npn device, more liberties can be taken here because of an inherent performance advantage.

Such processes take from 10 to 18 mask steps and provide pnp transistors with F_t's in the range of 150–600 MHz, a vast improvement over lateral types.

Another trend is toward lower supply voltages for linear integrated circuits. If the wide dynamic signal range is not needed, this approach enables considerably smaller, and therefore faster, bipolar transistors to be fabricated. Additionally, such processes enable "analog LSI" circuits to be integrated and also enable a fair amount of bipolar logic to be combined on the same chip.

Typical low-voltage analog bipolar processes (usually a 12 V breakdown is provided to facilitate operation from ± 5 V supplies) feature very fast npn transistors with F_t's from 1.5 to 8 GHz or so. Due to the small geometries involved, lateral pnp transistors can often be surprisingly fast, F_t's of 80 MHz being not uncommon.

Again, there is a trend toward the integration of a true complementary pnp transistor, and F_t's of 1–4 GHz are now available on some of these processes.

Conventional dielectric isolation (almost an oxymoron) cannot provide the epitaxial thickness control required for these processes, so all are basically junction isolated, although the capacitance of the sidewalls is often reduced using a plasma or reactive-ion etched trench. Recent advances in oxygen implantation and wafer-bonding indicate that dielectrically isolated versions of these processes are soon to become a reality, but I am quickly venturing beyond the scope of this chapter.

Junction FETs

JFETs (with the exception of GaAs MESFETs—I'll cover those later) are almost the "forgotten legion" of the semiconductor industry.

The JFET was proposed by Shockley in 1952 and later demonstrated by Dacey and Ross, but despite its advantage of offering a very high input impedance, it was not easy to integrate on the first linear bipolar IC processes. The silicon JFET has never found any volume application in digital circuits, which is another contributing factor to its low profile.

Early integrated JFETs had poor I_{dss} control and V_{gs} matching, but in 1972 National Semiconductor developed an ion-implanted p-channel JFET process which was completely compatible with linear bipolar processing. This has been subsequently trade-marked by National as the "Bi-FET" process. The biggest current use of JFETs in analog design is to provide high input impedances with reasonably good matching and noise characteristics. Almost all current JFET processes are variations on the original National idea.

The p-channel device is the most popular, because it can be conveniently fabricated by two extra mask steps in the same isolation pocket used by a standard npn transistor. Typical of such FETs are a pinch-off voltage of 1–2 V, a transconductance parameter of 5-12 μA/V^2, a breakdown of 40 V and a maximum usable frequency of around 60 MHz. Low voltage processes have also been developed (notably by Tektronix) and offer considerably faster results.

The n-channel JFET is more difficult to integrate but, because of the higher mobility of electrons compared with holes, has a transconductance parameter about three times higher than the p-channel device. At least two companies (Analog Devices and Texas Instruments) have developed linear compatible n-channel processes, and it will be interesting to see if this sets a trend.

MOSFETs

As previously discussed, modern linear MOS technology is essentially the same set as linear CMOS technology, with a subset consisting of BiCMOS technology and a further emerging subset consisting of complementary BiCMOS (CBiCMOS?) technology. Furthermore, the question of operating voltage is somewhat diverse here since although the vast majority of current CMOS processes are 5 V ones, those that have been optimized for linear applications frequently can operate at up to 44 V.

To add further confusion, the "sex" of the process is also not as easy to deal with as with the case of bipolar transistors. p-well CMOS (the first to become available) is basically a PMOS process with the addition of N-channel MOSFET in a p-well. n-well CMOS, as might be expected, is the opposite. It is easier to achieve higher breakdown with n-well CMOS, so as VLSI circuits shrink, and even meeting 5 V becomes a problem, n-well is starting to predominate in this area. However, for a given set of design rules, p-well CMOS logic is theoretically faster and also has more production history than n-well. The net result is that neither process can be ignored (one might expect, with the importance of breakdown for linear circuits, that n-well CMOS would predominate here. It doesn't, mostly because of backgating problems pertaining to the logic sections).

While state-of-the-art gate lengths for digital CMOS circuits are currently 0.8 μm or so (at the time of writing; if you've inherited a first edition, please don't laugh), processes intended for analog use vary from 10 μm or so for a 44 V process down to about 1.3 μm. Some analog companies do have their sights on submicron geometries but mostly for switched-capacitor type circuits where a lot of digital circuitry can be included on chip for error correction and calibration.

There are three basic uses for MOSFETs in analog design: First, as a linear component in much the same way that bipolar transistors are used in conventional bipolar circuits; second, to produce the switches and op amps used in switched capacitor circuits; and finally, the MOSFET is often used as an adjunctive component (for cascoding, switching, biasing, etc.) in a BiCMOS process where the majority of analog processing is performed by the bipolar transistors. By far the most strenuous task is as a purely linear component.

One of the biggest surprises to befall engineers (particularly those with a bipolar background) when attempting a linear CMOS design is the severity of the body, or backgating effect. Fortunately, it is usually the "welled" device which suffers the worse for this and in many applications the backgate can be connected to the source to overcome this effect. Even so, for wide body-to-source excursions, the unwelled device can suffer a threshold shift equivalent to 10-20% of the full supply voltage, rendering it useless, for example, as an output source follower. A more publicized effect is the much lower effective transconductance (particularly for the p-channel), compared with a bipolar transistor. Oddly, many textbooks then use this as a reason why CMOS op amps always have pitiful gain. In fact, of course, poor gain is caused by poor loading for both CMOS and bipolar op amps; in CMOS you just have to work a little harder, that's all.

One consolation is that both the transconductance and body effect can be modeled, and therefore simulated, quite accurately.

The actual transconductance, of course, depends on a number of things including W/L, I_d, carrier mobility, and oxide thickness. These are all fairly fundamental except for the latter, which varies widely depending on the supply voltage requirement. Thicker oxides, up to 1500 Å or so, are used for 36 V devices while only a few hundred angstroms are commonplace for 5 V ones. Coupled with the shorter channels available on low voltage processes, this enables considerably higher values of transconductance to be obtained (in a practical area) compared with higher voltage processes. This effect is completely absent in bipolar design. Additional speed-limiting effects (which are not absent in bipolar design) are reduced stray capacitances and reduced dissipation for low voltage processes. In linear CMOS then, higher voltage processes incur a much more severe speed penalty than in the corresponding bipolar situation.

MOSFETs also have all sorts of other quirky behavior which dyed-in-the-wool bipolar loyalists love to point out: noise, subthreshold operation, short channel effects, hot electron breakdown, threshold instability. The list goes on. But a glance at any *IEEE Journal of Solid-State Circuits* shows that high performance analog MOS design is very possible nevertheless. It all boils down to a good theoretical background, a good understanding of the available process, and a little carefully applied analog stubbornness.

Resistors

Resistors of the diffused, junction-isolated type have always been available on integrated circuits. Perhaps it is the ease of manufacturing them that has led most companies to eschew the development of a better component.

The basic drawbacks of diffused resistors include limited value range, high temperature coefficient, and value sensitivity to operating voltage and current.

p-Type diffused resistors typically have temperature coefficients around +1300 ppm/K and voltage coefficients around –350 ppm/V.

Usually enough diffusions are available to make low and moderate value resistors, but high values consume a drastic amount of die area. For example, the 100–200 Ω/square of a bipolar npn base diffusion starts to become impractical for resistors above 20 kΩ or so, particularly if there are many of them. The corresponding P-channel source/drain diffusion in CMOS is usually even lower in sheet resistance, making matters even worse.

Some tricks can be played, of course. Pinch resistors are commonly used in bipolar circuits to provide sheet resistances around 50 times higher, but severe breakdown problems and temperature/voltage coefficients result. Epi-FETs (bipolar) and p-well resistors (CMOS) can remove the breakdown problems but are extremely crude resistors in all other respects.

A more honest approach to achieving a high resistor value is to add an extra ion-implant step. Ion-implanted resistors can achieve wide value ranges, but above 1–2 kΩ/square tend to have very poor temperature and voltage coefficients.

To remove the voltage coefficient, necessary for accuracies of 12 bits and above, resistors deposited on the silicon rather than fabricated within it must be used.

The major requirement of the type of material used to form deposited resistors is that it be compatible with normal silicon processing, particularly the metallization system. Common resistive materials include polysilicon, silicon-chrome (cermet), and nichrome, and a large part of the reason for their popularity is that considerable experience has already been gained in their usage.

Polysilicon deposition is already part of many IC processes, notably silicon gate CMOS and polysilicon emitter bipolar. Unfortunately, the 20 Ω/square available with standard heavy phosphorus doping makes these resistors too low in value for most applications. The temperature coefficient is also rather high (around 1000 ppm/K). Doping 5000 Å polysilicon to around 600 Ω/square with boron can yield a resistor with less than 100 ppm/K of temperature coefficient, and such resistors are better suited to most designs.

Nichrome resistors are easily capable of temperature coefficients of less than 100 ppm/K, but they also tend to have low sheet resistances, generally in the range of 50–200 Ω/square. Sichrome resistors are about an order of magnitude higher than this but are trickier to deposit if low temperature coefficient is needed.

One big advantage of the last two (and most thin film) materials is that they can be trimmed by means of a laser. This and other trimming techniques will be considered in due course.

Capacitors

The p-n junction forms a natural depletion capacitor, but voltage coefficient, breakdown voltage, and leakage problems greatly limit its application. Virtually all analog IC processes, therefore, include a quality capacitor of some kind. The dielectrics used are almost always silicon dioxide or silicon nitride (or more correctly trisilicon tetranitride).

Silicon dioxide can be grown over a diffusion or polysilicon, which then forms the bottom plate. The top plate is then deposited above this and is normally another layer of metal or polysilicon. For a thickness of 1000 Å, such capacitors feature a capacitance of roughly 0.32 fF/μm^2. This makes capacitors above a few tens of picofarads costly in terms of die area. Such capacitors feature extremely low temperature coefficients and dielectric absorption.

Deposited silicon dioxide or silicon nitride (the latter boasts approximately twice the capacitance per unit area) can also be used, but such capacitors have poor dielectric absorption, making them unsuitable for switched capacitor or sample/hold circuits. Deposited dielectrics also allow metal/metal capacitors on processes having two layers of metallization.

One drawback of IC capacitors is that a stray capacitance exists to substrate on the lower plate. This can be a significant fraction of the total capacitance when a diffusion is used for this purpose. The parasitic is much less if polysilicon is employed as the bottom plate and even less for most metal/metal capacitors.

Inductors

There is still no easy way to fabricate a reasonably valued inductor on an integrated circuit. Microwave circuits use twin level metallization and airbridge techniques to fabricate spiral inductors with inductances of a few tens of nanohenries, but these are not much use for mainstream analog design.

Four-Layer Structures

Four-layer structures were (intentionally) invented by Shockley and have been (unintentionally) rediscovered by about a thousand IC engineers.

Four-layer devices are fabricated as discrete components to handle extremely large switching currents and are called thyristors, controlled silicon rectifiers (CSRs) and (wrongly) silicon controlled rectifiers (SCRs). The triac (a four-quadrant thyristor) and diac (a two-terminal breakdown activated triac) are also members of this group. It is the ability to handle such large currents combined with an extreme reluctance to turn off, that make them such a nuisance in IC design.

In CMOS, four-layer structures exist naturally as parasitics across the supply rails. To turn them on, generally a junction has to forward bias to the substrate, and if this happens it is usually due to input transients beyond the supplies, poor supply sequencing, or natural switching transients in the absence of adequate substrate pick-ups. Cures to such problems are buried in the design rules and folklore of many semiconductor companies but always involve intricacies of device spacing and good substrate connections. Failure to pay attention to this can result in severe, often destructive, latch-up, which may even prevent evaluation of a design.

Bipolar engineers have always felt somewhat pleased to be left out of such fun and games, but these days are also changing. Junction-isolated complementary bipolar processes also feature potentially pernicious four-layer devices, and similar care must be exercised as in CMOS. As for BiCMOS, well, what would you expect…?

Mixed-Signal Technologies

It is now technically possible to integrated silicon anything alongside silicon just about anything else. There are even GaAs-upon-silicon processes being developed that will enable mixed process chips to be integrated. Having every device available that I have discussed would be a powerful process indeed; but this chapter is about reality. The two main thrusts within analog process enhancement are currently increased speed (which is largely evolutionary), and the inclusion of digital functions on an otherwise analog process. The latter has become known as "mixed-signal" technology.

As mentioned earlier, there were many attempts (not all unsuccessful) to merge analog and digital functions by means of purely design techniques. Indeed, integrated injection logic was initially touted as a design technique, until it was realized that few analog bipolar processes possessed a high enough inverse beta to provide reliable gate operation. This logic form also had single input gates with multiple outputs (as did integrated Schottky logic), necessitating a rethinking in digital design methodology in addition to process modification. It is difficult to find engineers who are enthusiastic enough to attempt the integration of a significant amount of logic using some weird and wonderful logic form. It is even more difficult to persuade them to do it more than once. The ease of use of CMOS logic, coupled with the availability of documentation and design tools, I feel, will make it win out in the end (If you doubt this, try finding a digital simulator that will handle multiple-level logic, or an auto-router that can route the differential signals needed for ECL). I therefore see a bright future for BiCMOS technologies.

Early examples of BiCMOS were adaptations of an existing digital process to include a pnp or npn transistor, usually with mediocre characteristics, with as few possible extra steps. Such processes are useful but, in a way, buy digital simplicity at the expense of an analog pain-in-the-neck.

More recent BiCMOS processes have improved their bipolar transistors considerably, though it is still unusual to have both polarities available. Additional analog luxuries (actually necessities for many types of circuit) such as thin-film resistors and quality capacitors are starting to appear on BiCMOS processes. Also, two distinct classes are becoming apparent: those that will run from standard 12 or 15 V analog supplies and those intended only for low-voltage applications. I believe there is ample room for the coexistence of both types.

Recently, several BiCMOS processes have emerged that integrate both high-quality npn and pnp transistors (some have christened this CBiCMOS). It can be argued that this is an unnecessary step, because, for example, a PMOS device can be used as a level-shift for an npn transistor. But it turns out that some of the CBiCMOS processes are not as complicated as might be thought, since considerable sharing of process steps with the MOS devices can be accomplished. It will be interesting to see how things develop over the next decade, but I feel that the extra versatility of CBiCMOS will give it a very firm place in the future.

Trimming Integrated Circuits

On a good IC process, transistor V_{be}'s can be matched to around 250 μV, but for good yield 1 mV is probably a more realistic design value. Resistors, if made wide enough, can yield to a matching accuracy of about 0.04%, but again, designing for a figure of 10 times as great is likely to result in improved yield and smaller die size. Such tolerances are insufficient for many of today's chips, where 12-bit accuracy or better is routinely required. Additionally, many circuits (such as bandgap references and anything involving time constants) require absolute tolerances on one or more components. This presents a problem in an industry in which process engineers are usually reluctant to promise anything better than 20%. To ameliorate these situations, some form of on-chip trimming must be utilized.

One interesting method that is rapidly gaining importance is to include an on-chip EPROM or EAROM. Such a device normally would control some type of correction DAC and thus is not trivial to implement. One advantage of such a technique is that a considerable amount of trimming can be done after packaging, thus nullifying the effects of assembly shift.

Currently, however, the three most used trim methods in the industry are fusible link, zener-zap, and laser trim. All these techniques can also be performed to some extent after packaging, but the trend is to perform trimming at wafer sort for economic reasons.

A fusible link is a conductive link that can be blown open by means of a pulse of current. Such links normally short out sections of a resistor placed in such a manner as to predictably affect the parameter to be trimmed. Thus, at least one test pad is required for each link, making large amounts of trimming expensive in terms of die area and cumbersome in terms of the probe card. The usual material for fabricating the links is aluminum or polysilicon, since these materials are essentially free, though regular thickness CMOS gate material does not make an especially good fuse. 1 μm thick aluminum metallization fuses at about 100 mA/μm width, so a fair amount of current is required to open a link. Because of this, other materials such as tungsten are sometimes used instead and can be "programmed" at much lower currents. A drawback of the fusible link technique is that it is not possible to know exactly what effect opening a link will have until you blow it, somewhat like testing photographic flashbulbs. This can result in some yield loss.

Zener-zap trimming is a similar concept, except that the idea is to close the links rather than open them. The obvious advantage over fusible links is that they can be externally shorted to test their effect before actually blowing them closed.

In the late 1960s it was discovered (at Fairchild) that if the base–emitter junction of a bipolar transistor was avalanched at a high current, then a permanent short would result. The technique was first applied to integrated circuit trimming by Intersil and Precision Monolithics in the mid 1970s. A typical small geometry bipolar transistor will typically "zap" at about 150 mA of current, and it takes about 25 V to generate this for a period of about 1 msec. The resulting "short" has a resistance of a few ohms and consists of a thin strand of aluminum buried under the passivation. Like fusible links, normally a test pad is required for each zener, but some clever techniques involving diode steering have been used to reduce this.

Although this technique has become known as "zener-zap," the shorting process occurs purely as a result of heat and electric field. The semiconductor junction is convenient to prevent conduction of an unshorted element but plays no part in the shorting process. It is therefore possible to short (low valued) diffused resistors as well as zeners. It is even possible to gradually reduce the value of such resistors by controlled current pulsing, resulting in a continuous trim capability. Such techniques are currently in use by Motorola to reduce offset on some of their JFET op amps.

Laser trimming involves the use of a laser to burn away part, or all, of a thin-film resistor. This technique can therefore be used to trim links or as a continuous trim. Suitable choice of resistor geometry can produce trims that are as coarse or fine as necessary, and the resulting area penalty is not large. Laser trimming thus has great advantages over the aforementioned methods of trimming.

The main disadvantage of laser trim is cost. Production laser equipment can cost half a million dollars or so and requires considerable maintenance. Also, of course, thin-film resistors must be included as part of the process, which rules out many standard "foundry" type processes. Test time is also lengthened with laser trim, especially as complicated trim algorithms are often necessary. However, the ability to perform a vast number of trims often enables a considerable price premium to be obtained for the finished product.

One area of concern with laser trimming is the long-term stability of a partially trimmed resistor. Early attempts at trimming often cracked the surface passivation, causing extreme reliability problems. This has been solved by adjusting the oxide thickness beneath the resistors so that constructive interference occurs, enabling trimming to be performed at low power. A secondary effect is the annealing of resistor material around the laser cut, which can cause slightly different drift rates between resistors that have been trimmed by different amounts. This effect, as might be expected, is a strong function of the resistor material used. Many companies have developed proprietary film compositions and stabilizing agents to reduce this effect to negligible proportions.

Integrated Exotica

Silicon is not the only semiconductor material suitable for integrated circuit realization. The main reason that it overtook germanium as the preferred material for transistor fabrication was its ease of manufacture—largely due to the availability of a high-quality (water insoluble!) thermal oxide.

Compound semiconductors (such as gallium arsenide) have been much touted for their potential advantages over silicon, but their processing difficulties, and vast number of possible varieties, to my mind put them in a class of "integrated exotica."

A compound semiconductor is a compound of two or more elements, usually (but not always) symmetrically straddled about group 4 of the periodic table. Most explored are the 3–5 compounds (such as gallium arsenide and indium phosphide), but there are useful 2–6 compounds (such as cadmium sulphide) and even 4–4 compounds (such as silicon carbide). Compound semiconductors are not new to electronics: the cat's whisker detector (patented in 1906) used galena (lead sulphide) as its semiconductor. Copper oxide rectifiers were in use in the first half of this century, as were cadmium sulphide photocells (early work on MOSFETs was based on such materials until it was decided that silicon was a better way to go). The major reason to turn to such materials today is to improve the speed of both digital and analog circuits.

Because silicon is an indirect-gap semiconductor with electron mobility limited by phonon scattering, it is not difficult to find alternative semiconductors with much higher electron mobilities. Gallium arsenide, for example, has roughly a six times improvement over silicon in this regard (the hole mobility is actually worse than that of silicon, which is why we don't see any complementary GaAs devices).

Circuits designed using suitable compound semiconductors therefore have potentially higher operating speed than their silicon counterparts. But here comes the twist. Difficult process techniques coupled with the late start of the compound IC industry result in almost no standardization of such circuits. At least silicon has a defined mainstream technology as a starting point. The most "mature" sector of the compound industry at the moment is GaAs MESFET technology, which I feel compelled to at least cover in some detail. Heterojunction bipolar transistors are also becoming a highly practical proposition, and it would be foolish to dismiss this as a purely research technology in the future.

As with silicon, today's compound active devices potentially consist of bipolar transistors, JFETs and MOSFETs. The lack of a stable thermal oxide on most usual compounds results in surface states that are "pinned," making MOSFETs extremely difficult (but not necessarily impossible) to manufacture. The two major compound devices in use are therefore BJTs and JFETs. The techniques for producing both of these are very different from those of silicon, because compound semiconductors have an unfortunate habit of dissociating at diffusion temperatures, and the conventional methods of creating junctions are rendered impractical.

In the heterojunction BJT, techniques such as molecular beam epitaxy are used to create junctions not merely of different doping but also of different compounds. This enables the bandgap energies of the emitter, base, and collector regions to be different, allowing optimization of parameters such as emitter efficiency and breakdown voltage. For example, a high emitter efficiency allows a heavily doped base region without sacrificing current gain, which in turn means that the base region can be made very thin. Such a transistor would naturally be expected to exhibit a high cutoff frequency, and figures of 80 GHz have already been achieved. Many different materials and fabrication methods are being investigated for such devices, and there is no real attempt at standardization. I hope that one day such processes will be commonplace in the IC industry.

The GaAs MESFET IC process is an altogether different story, with many manufacturers able to deliver foundry capability on an almost routine basis. As such, it is' the closest thing to a "mature" compound semiconductor technology around.

The MESFET (metal-Schottky FET) is basically a JFET with the gate junction replaced by a Schottky barrier diode. This is a much easier device to fabricate than one requiring a true p-n junction. The GaAs MESFET behaves similarly to an n-channel silicon JFET (with much increased transconductance), except for anomalies caused by interaction with the substrate. To nullify the effects of stray capacitance, GaAs circuits are built on a nearly intrinsic "semi-insulating" substrate. This creates many peculiar effects, such as side-gating, hysteresis, and even low frequency oscillation. Such effects are not too important for the digital designer but cause havoc with analog designs. Furthermore, these effects are unpredictable and layout dependent, making them difficult to design around.

Processes specifically aimed at microwave applications get around such problems by using a very high pinch-off voltage (3–5 V), which also provides an increased transconductance.

While more general purpose analog functions can be designed on microwave processes, the pinch-off voltages used for digital processes (0.6 V or so) are easier to handle. Also, designing with a purely depletion mode device is difficult for both analog and digital functions, and most modern processes also provide an enhancement mode MESFET. Clearly, attempting to enhance such a device beyond the Schottky forward voltage drop (around 0.8 V for GaAs) generates a lot of gate current, so the enhancement device is not such a godsend as might first appear. Typically, the "threshold" of such a device varies from 0 to 300 mV, which further complicates the design procedure.

Most GaAs processes can accommodate 1 μm gate lengths with 1/2 μm becoming more common. This feature size is limited by the same constraints as with silicon technology and will diminish as improved equipment becomes available.

Besides MESFETs, many processes include capacitors (usually metal–silicon nitride–metal) and resistors (usually nichrome but sometimes formulated from a refractory metal such as titanium).

In general, the exotica field has not taken off as many (particularly the exotica proponents) have predicted. This is still very much a space to watch, however.

Technological Quandary

The notion of digital circuitry being highly technology driven, with analog being the poor brother subsisting on fifteen-year-old processing, is obviously changing somewhat. Companies producing state-of-the-art analog products, like their corresponding digital counterparts, will require similarly advanced technology in the future. But the diversity of analog and mixed-signal processes is far greater than those used for purely digital purposes, and every engineer will have access, or lack of it, to different aspects of this. Furthermore, by no means all products require advanced processing. Many successful analog products are still being produced using basic bipolar processing or metal gate CMOS. The digital world is somewhat more cutthroat, but similar examples can still be found (who needs a 50 MHz watch chip?).

There is a strong temptation to use technology just because it is there, what I refer to as the "Mount Everest syndrome" after the rather ridiculous remark of the first person who climbed it. Mask steps are definitely not free, and though it is true that economies can be made by standardizing many products on a given process, I have yet to see this happen effectively for a wide range of analog designs.

Trimming, particularly laser trimming, is another example of a technology that can easily run wild. Again, trimming costs money, and using it to "patch up" a shoddy piece of work is rarely excusable, unless time pressure is extreme or the test department is short of things to do. Trimming is a powerful tool for fine tuning circuits where there is no practical alternative, and this is how it should be used.

Analog companies are now in a technological quandary. It takes a really large company to even think of investing in all the areas I have mentioned, and even given the necessary resources, this is not likely to be a profitable strategy. The better way to look at things is to decide which product areas suffer most from not having whatever technologies available, and to make a pragmatic effort to limit both to manageable sizes. It may also be wiser to buy technology than to develop it, at least as long as sufficient outside capacity is likely to be available.

With the increasing costs of staying in the technology game, expect a corresponding increase in the occurrence of "strategic alliances" between companies. Usually this involves a company possessing a technology which is then licensed or made available to other noncompeting companies. Given the present rate of progress, joint development efforts, even between companies ostensibly in competition, are likely to become commonplace.

And a final sobering thought: the design engineer must be expected to stay abreast of all this, if his designs are to remain competitive.

Section 2. Philosophical Considerations

Getting Started on the Project

So a project has been defined. It has been determined that there is a market for it and that suitable technology is available to make it feasible. The next step, obviously, is to get things underway. This can be a surprisingly difficult task.

All too often I have seen formal definitions of projects followed by formal team meetings, full of equally formal reasons why things were going to take so long and many formal suggestions for wasting time and money. This is the point at which everyone who is involved (or who feels he should be involved) sees an opportunity to have his say. It is amazing how many such "contributors" are conspicuous in their absence, should the project run into difficulties.

My own designs don't seem to have any such formal beginning, since most of them fall out of a conglomeration of ideas that have been stewing in my brain for some time. Eventually, I will get around to submitting a schematic to mask design, or performing the layout myself, at which point product, process, and test engineers get involved, and there is no hiding the fact that something definite is happening. This is a great way of beginning standard products that have no critical time pressure, but I have to admit that it is not much use for a rigidly scheduled custom design. Even within my own little world, I have had 90% complete designs waiting months for some inspiration to finish them. Sometimes a missing piece of the jigsaw will drop into place while I'm relaxing or eating or taking a bath (remember Archimedes—I think "Eureka" is a Greek word meaning the water's too hot). This usually motivates me to formally launch the project as quickly as possible. Otherwise, I just have to be content that nothing is ever as perfect as I would like it, or 90% of my designs will stay unrealized forever.

But whatever motivates you to finally get things moving, the important thing is to be seen to be getting on with it. Draw a schematic, do some simulations, ask some embarrassing questions. Concrete progress has a psychological impact on oneself and others. If it is not obvious that the design engineer is making progress, it is unlikely that people with support functions to perform will be motivated to do anything obvious either.

Getting the project started should be like crossing the road. By all means look carefully in both directions first, but once you have decided to go, for heaven's sake do it.

Knowing When to Stop

As a direct result of the large number of "things" that must be made to go right to successfully complete an IC project, the point at which any given task can be determined to have been completed is extremely nebulous. I have designed many integrated circuits, some (supposedly) straightforward and some quite risky, but there are still to me two very scary times. The first is the final sign-off before sending the tape for mask tooling, and the second is the time that the first wafers are powered up. The latter is just pure nervousness, of course, but the former is a sort of heavy commitment that should not be taken lightly. There always seems to be something one would like to check further, but there has to come a point at which the design is finalized.

Most companies have a checklist of things that have bitten designers from time to time, but completing one of these is likely to be of more psychological than practical use.

Perhaps the most aggravating concerns are areas that could not be completely checked out, or where breadboard and/or simulation results were not quite conclusive.

This is also the time for pondering over such concerns as voltage drops across metal lines, possible electromigration problems, stray capacitance of cross-unders, thermal effects, and the list goes on. It is important here to distinguish caution from paranoia, something with which all good designers that I know still have some trouble.

But try to be realistic. Any last area of concern which might stop the chip working must obviously be given some priority. A partially sick IC is bad enough to have, but a totally useless one is far worse. Reliability hazards such as excessive current density or potential latch-up should also be scrupulously avoided. Certainly, if time permits, recheck all the little things that are on your mind (and probably a few kind colleagues have provided a dozen other suggestions, should you happen to run out of things to worry about), but you have to stop somewhere. Most of my serious errors have been things that never occurred to me before the silicon came out (what I call the "oh, we should have thought of that" syndrome), and all the finer details that had been bothering me disappeared at that point.

I remember once dreaming up a possible latch-up scenario on a circuit that had been tooled but not yet processed. I feverishly worked out a three mask fix for the problem and hurriedly tooled new plates. Some two months later I received the silicon with the wrong masks on, which made the whole last minute panic seem somewhat pointless. To rub the salt in further, there were additional problems unrelated to this issue which necessitated an all layer change. If there is a lesson here I am not sure I have learned it, for I think I would probably do the same thing again given a similar situation. Maybe I'd better add my ten cents worth to that checklist . . .

Sometimes Things Go Wrong

The vast number of variables involved in the completion of an IC project leave plenty of room for mistakes. In the semiconductor industry, very small errors can take several months and cost many tens of thousands of dollars to correct, even without considering any consequently lost business.

Some IC designers become very good at troubleshooting, which usually means rescuing some ill-fated project from almost certain disaster. Some unkind observations could be made concerning the reasons why such skills are actually acquired, but nevertheless the art of fixing a circuit is certainly quite different from its initial design.

Of course, a whole book could be written about IC troubleshooting, and one day probably will, but in lieu of this I would refer the reader to the excellent series of articles by Bob Pease [3]. There is a rare feeling of satisfaction in determining the cause of an obscure problem, and an even rarer feeling in correcting it. One consolation for the IC designer confronted with a problematical circuit lies in the fact that there is usually no shortage of chips upon which to experiment!

I personally have a reputation (of which I am not particularly proud or ashamed) of taking a "shoot-from-the-hip" approach to design. When it comes to troubleshooting, however, I have learned to be very methodical. The other lesson I have learned is to use only the best available equipment, maintained in the best possible condition. There are enough problems involved in analyzing a circuit some fraction of an inch square without having to worry about triggering the oscilloscope or tracking down the short circuit on the probe card.

It sounds like one of those "constructive management" kind of statements to say that a royally screwed-up circuit represents a learning opportunity, but it is true nevertheless. Careers have been made and broken due to problem solving skills or the lack of them.

But having completed one's detective work to the best of one's ability, it is unlikely that all doubt will have been removed concerning the proposed fixes.

The same judgment discussed in the previous section will have to be used once more; and of course one is now certain to be under considerably more time pressure compared with the first mask release. But please, stay methodical; and above all make sure that everyone concerned understands the risk factors involved in a possible redesign. Also, try to resist the temptation to take further unnecessary gambles. There are probably many possible improvements that have occurred to you since the original design, but unless time is plentiful, save them for the next chip. I think there is an unwritten Newton law stating that "for every design there is an equal and opposite redesign"; believe me, I've been guilty of this sin on several occasions myself.

Of course, rather than become the "Red Adair" of integrated circuits, it would be better to avoid mistakes in the first place. More realistically, an attempt to predict the kind of things that might prove disastrous can pay off heavily in the long run. While this may seem like a somewhat ridiculous statement, my experience is that many mistakes (and most serious ones) arise from risks that were unnecessarily taken or that were not sufficiently evaluated. I am sure that all designers sometimes have that uncomfortable feeling about a particular design section, and though I am convinced there is nothing wrong or unusual about this, it should indicate an area for special attention. It is often tempting to say "what the hell," but if you adopt this attitude too often, keep your probe station (and your résumé) in good shape.

But as I pointed out in the introduction, true design creativity mandates some calculated risk-taking, and I offer the following guidelines to help with the calculated part.

1. If an area of circuitry is new or obviously risky, try to determine how necessary it is to take such a risk. If saving 10% die area or so over a more conventional approach, for example, is its only benefit, then why bother? Often, a designer will come up with some cute little trick to maybe save a few transistors, or do something a little more elegantly. Examples of this are the Wilson current source and the Baker clamp. These examples work, and the reason we know that is because they have been integrated so many times. There are other examples that, to me, look a lot riskier than these two. When I first saw the biasing loop around the input stage of a 741 I was surprised that it didn't oscillate. It doesn't and so is a good example of a risk that paid off. Ideas that I have along these lines don't usually get incorporated on custom chips or circuits with rigid deadlines, however.

   The overall risk of a project is the sum total of risk of all its building blocks, so keep extremely dubious ideas for test chips or less critical designs. The cute little widget that you've just dreamed up may not be in general use because nobody has previously thought of it, but maybe it has been tried before with poor results. For example, how many designs do you see using the composite npn/pnp device? This is described in many textbooks but is virtually impossible to keep from oscillating.

2. Pay particular attention to design areas that are absolutely critical to the success of a design. For example, an A/D converter with which I was involved had an error in the output buffer cell. This meant that to get at the output information, many probes were necessary. There were some other errors, but this one in particular made evaluation extremely laborious. Even worse is a mistake which prevents proper operation of the whole circuit. Shorts across power supplies and failure of regulator circuits fall into this category. And the number of bias line circuits I've seen with oscillation problems is quite

amazing. I'm all in favor of taking gambles to improve circuit performance, but how marvellous does a bias line need to be?

This in general has been referred to as the "onion" syndrome. Trouble-shooting circuitry like this is indeed like peeling an onion: fixing one set of mistakes reveals the next layer, and so on.

3. I once read an article somewhere advocating that blank areas of silicon should be left wherever problems were most likely to arise. I remember thinking at the time that this would have meant an awful lot of blank silicon on most of the designs I had seen. But there are things that can be done up front to make life easier in the future without really sacrificing anything.

Make critical resistors easy to change. Diffused resistors can be changed with just a contact mask if wide enough and metal overlaps the anticipated contact area. Thin film resistors can usually be changed with at most two masks, so represent a good choice for an adjustable component. Compensation capacitors can be made oversize in case they are not quite adequate, but if they are metal capacitors, be careful not to route signals directly through the top plate. This will allow them to be cut back by scraping away metallization.

Metallization in general needs some attention if trouble-shooting is not to be a total nightmare. If the design can be broken into sections (and most large ones can), arrange the interconnect to allow each section to be powered up separately with the minimum of surgery. Also, if dual level metal is used, remember that probing the lower level is difficult at best and impossible if covered over by the second layer. Small vias can be inserted at known critical points in the circuit, and these will greatly facilitate voltage mapping if it proves necessary.

Knowing When to Give Up

I would love to have some statistics on the percentage of integrations that succeed, compared to the failures. Of, course, everybody has his or her own idea of what constitutes failure. Even if the design works correctly, a marketing person might regard it as a failure if it sold miserably in the market. Similarly, a process engineer would not vote it "chip of the year" either if it was impossible to manufacture. For the purposes of this discussion, I am limiting my thoughts to technical failures; and even without the statistics, I know there are a lot of them.

Run-of-the-mill digital design is to the point at which the first-time success rate is very high, and there is some straightforward analog work in the same category. Most analog projects do, however, require some rework before going into production. Some require a lot of rework, or even complete redesign, and these are the type of projects that can have severe impact on the fortunes of companies and the careers of engineers. I must confess that I have been trapped in the middle of several of these, and the big problem is knowing if and where to give up. It is a severe ego blow to admit that something is just too difficult, especially after spending considerable time and money on it. It always seems that just one more revision is needed, and it is very hard to make rational judgments at this stage.

All sorts of factors interplay at this point, and most are nearly impossible to quantify. Is a market still available? If so how big is it now? What is the design engineer's confidence of correcting the problems? What is everybody else's?

My experience is that it is extremely difficult to know when to terminate such a project. I have had some really messy designs that eventually became huge sellers. I have had some that never worked correctly after swallowing up considerable effort and capital. I have had some in-between projects which I have managed to get to work but which never recovered their expenditure. Hindsight works incredibly well here, and there is always an obvious point at which the design should have been abandoned. But at the time nobody ever seems to be able to identify it.

Optimistically, experience is designed to keep engineers out of such situations. Pessimistically, they will always occur; particularly in areas in which the state-of-the-art is being advanced. It is possible to keep one's head (and dignity) in such a position, as I am sure all successful engineers have discovered at some time. This is just part of the game, that's all.

Section 3. The CAD Revolution
Limitations of Traditional Design Methodology

At this point I will make the potentially controversial statement that available analog design tools have not even closely advanced at the same rate as the technology (or at the same rate as digital design tools).

The traditional way of prototyping an analog IC is to build a breadboard, and this technique is still in use at many analog design companies.

The breadboard relies, of course, on the availability of packaged kit parts, consisting of representative transistors available on the final integrated circuit. This is not too unrealistic for something like a conventional bipolar process, but when more types of component (JFETs, superbeta transistors, and the like) are added, the library of kit parts quickly becomes excessive. Breadboarding becomes a true nightmare when MOS devices are included. One advantage of the MOSFET is that a wide range of transconductance control is possible by varying both the length and width, and to successfully duplicate this on a breadboard requires a virtually infinite array of kit parts. Additionally, because of threshold control problems, circuitry such as ratioed current mirrors or scaled bias lines will require different geometries to be available in the same package, further confounding the kit part issue. There are somewhat cumbersome ways around this (unlike bipolar transistors, MOSFETs can be connected in series/parallel arrangements like resistors), but this makes for a complicated breadboard.

Also, I am of the opinion that if it is worth constructing a breadboard then it should be trustworthy.

I am always somewhat perturbed by such statements as, "I'm sure that will go away when we integrate it," or "well, what do you expect from a breadboard?" After all, one is trying to figure out what to expect from the final silicon, and if one does not believe that the breadboard is going to be representative, then why build it?

Attempting to breadboard high frequency circuits is also very dubious because of the inevitable stray capacitances introduced at almost every node. This tended not to matter so much when the response limitation came mostly from lateral pnp transistors but with today's complementary processes this is no longer the case. Building breadboards with unsocketed kit parts alleviates this situation but makes changes to the board very difficult.

Breadboards, however, do have some very useful features. First, they can be measured in real time in exactly the same way that the final IC will be evaluated. This is much faster than attempting to run multiple simulations to emulate the same evaluation. Second, the final breadboard can be loaned to a customer or patched into an overall system for further confirmation of the suitability of the design. This unfortunately highlights another problem of breadboarding, which is the delicacy of the resulting breadboard. All the advantages of a quick evaluation are gradually lost if the breadboard has to be repaired at frequent intervals. With the current availability of PC board routing equipment, it might be worth spending some effort in producing a cleaned up version (or two) if any protracted evaluation is predicted (I have never had the patience to do this myself).

Computer simulation has also, perhaps surprisingly, failed to keep pace with the technology. Only now are the first "mixed-mode" simulation packages becoming available in a rather slow response to the needs of the already established mixed-signal IC industry. Most simulators still use the basic "enhanced" Gummel-Poon bipolar transistor model used by SPICE in the 1970s. Further, almost all the progress on MOS modelling has been highly biased toward digital circuitry, not surprisingly perhaps, but this is of little consolation to an analog engineer struggling with discontinuous models.

Nevertheless, progress is being made in the simulation area, and in view of its inevitable importance, I will devote some space to this subject.

Circuit Simulation

There are many simulation techniques used in the integrated circuit industry. For example, at the process and device level there are SUPREM and SEDAN from Stanford University, and for distrubuted parasitic analysis there is GRID from Grid Software Inc. For the purposes of this chapter, though, I shall restrict the discussion to electrical circuit analysis.

Over the years, many programs intended to perform various aspects of circuit simulation have been created. For analog integrated circuit engineering, however, the *de facto* standard is definitely SPICE from the University of California, Berkeley.

SPICE (Simulation Program with Integrated Circuit Emphasis) dates from 1972 but evolved from an earlier simulation program called CANCER (Circuit Analysis for Nonlinear Circuits Excluding Radiation). In 1975 SPICE2, a much more powerful version, was released. This can be considered a real breakthrough in circuit simulation.

SPICE2 included some features that made it extremely easy to use compared with other simulators. Such features included a "free format" input listing, ability to handle inductors and voltage sources, dynamic memory management, and a highly intelligent transient analysis algorithm. SPICE2 also included advanced models for bipolar junction transistors, diodes, junction FETs, and MOSFETs.

All the versions of Berkeley SPICE were public-domain programs available at purely nominal charge, since public funds were used to support the development. However, many software vendors have recognized the need for a fully supported, adapted, and improved commercial circuit simulator. The first mainframe-based versions of such programs included HSPICE from Meta-Software, I-SPICE from NCSS Timesharing, and PRECISE from Electronic Engineering Software. Nowadays, most mainframe versions have been adapted for use on workstations, and some are also available for the personal computer.

The first PC-based version of SPICE was PSPICE from MicroSim Corporation. It has been followed by several others, such as IS-SPICE from Intusoft.

The commercial versions of SPICE generally include extra features over Berkeley SPICE2. Schematic capture interfaces and graphics postprocessors are common additions. Other areas of improvement include better convergence and additional active models. In general, though, care has been taken to maintain as much compatibility as possible with the Berkeley SPICE format.

It should be noted that though dominant, the SPICE-type simulator has not been the only one to achieve some success. For example, the IBM ASTAP simulator uses a different circuit representation, known as Sparse Tableau, which allows access to all desired state variables at the expense of run time and memory usage. This program was only available to IBM users, however, which is probably the main reason why its success was limited.

Although SPICE-type simulators can be used for the simulation of digital circuits, they are far too slow for the efficient analysis of LSI complexity designs. This area has been served by a whole class of digital simulators which take advantage of the limited nature of digital systems. In particular, such a simulator does not need to calculate voltages to tiny fractions of a percent, does not need to handle feedback, and only needs to model specific gate topologies. Such simulators easily achieve an order of magnitude speed improvement over SPICE.

So we have two types of simulator to keep everybody happy. Until, of course, we need to simulate the up-and-coming mixed-signal designs.

An incredible amount of technical persiflage has been generated with respect to "mixed-mode" simulation, and now would be a good place to define some terminology. I have used the term *mixed-signal* to describe circuitry consisting of both analog and digital functions. In essence, a digital function is quantized in the amplitude domain, whereas an analog function is not. I use the term *mixed-mode* to describe a simulator that is capable of performing both digital and analog simulations simultaneously. Here, however, the most important difference between the two functions is quantization with respect to time. To interact with the digital section, the analog simulator must be "event driven," and the digital section also has to be able to act upon continuous time information extracted from the analog simulation. Thus the two simulations proceed simultaneously, periodically holding each other up.

Despite early efforts in this direction, such as SPLICE from the University of California, Berkeley, DIANA from Leuven University, and SAMSON from Carnegie-Mellon University, there is currently no universal approach to this problem. Most commercial suppliers of simulation software are working on mixed-mode lines, the usual approach being to interface an existing digital simulator to an existing analog one, which has resulted in several cooperative deals between software companies.

Another simulation trend is to allow behavioral modeling, where descriptive rather than physically defined models can be entered into the simulator. This allows for behavioral modeling of digital sections within an analog simulator (as well as behavioral modeling of complete analog blocks). This approach tends to be intermediate in speed between a full analog simulation and one involving a digital simulator, though the ability to model large blocks of analog circuitry in a behavioral fashion can often reduce run times to reasonable proportions. An example of such a simulator is SABER from Analogy Incorporated.

Limitations of Simulation

I have come to the conclusion that most engineers love to complain about simulators. It is the best example of a love–hate relationship that I know. Most of the complaints I would classify as annoyances rather than true limitations, but nevertheless this section deals broadly with things that get in the way of a successful simulation.

It is impossible to have too thorough a knowledge about simulators. I am not suggesting that anyone necessarily analyze the SPICE source code (though some engineers have), but certainly an understanding of the algorithms and models involved is essential if maximum benefit is to be obtained from simulation.

As soon as one uses the words *simulation* or *model*, one is talking about an approximation to real life behavior. Furthermore, such an approximation is not necessarily intended to mimic all aspects of real-life behavior simultaneously. This is one advantage of simulation—being able to single out certain aspects of circuit behavior in a manner independent of the others. Actually, the only computation that even attempts a universal circuit simulation is the transient analysis, and even then only within some fairly severe numerical constraints.

In the real world, when one powers up a breadboard, one is effectively performing a transient analysis. Everything bounces around a little and maybe comes to rest at some final DC condition. But perhaps it oscillates, possibly at several different frequencies; both the real world and a computer transient analysis allow such things to happen.

But in simulation, as every good schoolchild knows, the first thing that is examined is the so-called DC operating point.

The operating point ignores all AC effects such as inductance and capacitance and also uses fixed values for all power supplies. The operating point will also cheerfully report hundreds of amperes of supply current without destroying the circuit—a big advantage of simulation over breadboarding. And it won't oscillate, so an op amp for example can be successfully debugged for static errors before any attempt is made at frequency compensation. The first problems occur when the simulation cannot find an operating point or finds an apparently wrong one. An "apparently wrong" result I define as one which is not as the designer expected but which cannot be traced to an obvious circuit problem.

Failure to find an operating point is a failure of the iteration algorithm (Newton–Raphson being used by essentially all current simulators) to converge to a final solution. Actually, no iterative technique can ever converge completely, so some more realistic criteria must be established to decide when to terminate the iteration process. Such criteria are established by ABSTOL, RELTOL, and VNTOL or their equivalents, and the respective simulator manual should adequately explain the exact mechanism it uses to establish convergence.

The first treatment that is usually tried on a nonconvergent circuit is to relax these limits. Occasionally this will result in convergence, and the designer usually consoles himself by concluding that the default limits were too tight in the first place. Some circuits are like this, where numerical round-off in one section of the circuit can produce drastic changes in another (remember, all the active models contain exponential functions). But my experience is that the SPICE default limits (but not necessarily those of commercial simulators) are set at about the right value and that it should be possible to obtain good results without relaxing or tightening them.

The vast majority of convergence problems occur because one or more of the nonlinear models used has been evaluated in a region that yields spurious results. Ideally, of course, the models should be valid for all regions of operation. Most of them though, particularly high-level MOS models, have become so complicated that discontinuities can occur even under normal operating conditions, let alone with some of the ridiculous voltages and currents that exist temporarily during an operating point calculation. We are dealing with an implicit process: node voltages are examined, a set of internal operating currents are derived from the models, and these are used to calculate a set of branch conductances. The matrix is solved, new node voltages result, and the process repeats itself. If any of the operating currents of any of the nonlinear devices at any point of the analysis is calculated wrongly, then nonconvergence can result.

Other causes of nonconvergence, not necessarily related to model deficiencies, include numerical overflow and incomplete equations. Numerical overflow (or underflow) occurs when the simulator attempts to generate numbers outside the computer's floating point range. Exponential functions are notorious for this, and

SPICE bipolar transistor models have limiting functions to reduce the overflow possibility, and a minimum conductance (GMIN) is added to prevent underflow. Attempting to use very small resistor values can also generate overflow since extremely large conductance values will be generated, which can produce even larger current perturbations during iteration.

The SPICE topology check partly checks for incomplete equations, typically floating nodes or inputs incapable of accepting current. This topology check does not catch everything, however. Take, for example, two or more MOS gates tied together but to nothing else. Such an occurrence has happened to me on many occasions when I have forgotten to tie down a digital input. This will pass the topology check but clearly has no DC solution and will result in nonconvergence. Such errors are usually fairly obvious from looking at the "last node voltages tried." A node like this will usually shoot off to some silly voltage and can be spotted fairly easily. In fact, many convergence problems in general can be traced to a wrong connection that resulted in a circuit that was impractical to solve. Sometimes though, nodes will head off in strange directions without any circuit misconnections. These can again be easily spotted, and the only real remedy here is to try using NODESET on the offending nodes. Setting a stubborn node close to where it should end up may not help as much as might be expected, for one correct voltages amidst a sea of wrong ones is not necessarily going to help convergence. But setting the node to a voltage in the opposite direction to where it last ended up might help considerably. Setting all node voltage to their approximate correct value can also ensure convergence, and if this sounds a bit ridiculous, please read on.

Anyway, if you have searched for all the above and still get nonconvergence, the chances are that you have stumbled across some discontinuity buried in one of the models or possibly a rough edge in the simulator's algorithms. Most simulators have optional additional algorithms to circumvent such problems, which usually involve steady ramping of the power supplies. These can result in a solution, but may take a long time.

Another technique I have used successfully is to try different temperatures. Often there is some oddball temperature at which the models will behave themselves. The resulting node table can then be used as a NODESET table for convergence at the correct operating temperature. I know of no commercial simulator that includes a program to do this, but it is easy to write one, and makes the initial setting of all node voltages less ridiculous than it might at first seem.

The next case to consider is where the operating point has converged, but the results don't make sense. The first thing to check is that convergence really has occurred. As mentioned earlier, the simulator will stop when certain criteria involving maximum voltage change and current change and percentage voltage change fall within the limits of VNTOL, ABSTOL, and RELTOL, respectively. These may not be realistic for your particular circuit, especially if the defaults have been relaxed to obtain the convergence. Try tightening them to see if the result is more sensible. A result with relaxed tolerances can be used as the starting point for one with tighter limits (now you see why it's worth writing that NODESET program) if convergence is a problem with the tighter tolerances.

Wrong component values will also cause equally wrong results, and it is easy to miss the odd decimal point here and there. Wrong model parameters are also easy to miss. I once had a BJT model with the saturation current set to 1E16 instead of 1E-16, which caused overflow errors. Another one that had me tearing my hair out was a value for an emission coefficient set to a very small value; nonconvergence was the result. My advice is to leave your hair where it is and check the input deck very carefully.

Sometimes a circuit will have a valid operating point that just was not predicted but which the simulator finds. The classic example is a circuit that has bistable behavior. NODESET is the best way to persuade the simulator to look for another state.

Having found an operating point, analyses such as DC transfer and AC are usually straightforward. But remember that neither of these analyses is a complete simulation.

The DC transfer, like the operating point, ignores all AC effects. Bistable circuits cause real problems here because they don't actually have a DC transfer characteristic. It is better to analyze these with a slow transient ramp.

The AC analysis uses the DC operating point as a basis for extracting small-signal linearized models of all the active elements. An AC sweep is then performed using the linearized models. "Small signal," however, means "small" as far as the models are concerned. As far as the simulator is concerned, an input signal of 1 MV will produce the same frequency response as that of 1 μV, the only difference being the magnitude of the output response. Thus the signal magnitude is useful for scaling the output amplitude, or the relative amplitudes of several different inputs, but has no other value. Also it should be noted that the AC analysis will only run after an operating point has been successfully found, but if the operating point is not correct, neither will the AC analysis be.

As mentioned, transient analysis is the closest thing to a complete simulation that can be run. To the extent that the algorithms and models are complete, the transient analysis mimics a real life simulation of what happens when a circuit is powered up or hit with a succession of pulse, ramped, or sinusoidal inputs. The analysis does start from the DC operating point, but after that all DC and AC effects are taken into account. As might be expected, therefore, this is also the most computationally intensive analysis.

Unfortunately, transient analysis algorithms that were even close to complete would require ridiculous amounts of computer time. So compromise has to be made, and the major areas of compromise are in the timestep control and numerical integration method.

Essentially, wherever the circuit exhibits complicated behavior, a very fine timestep must be used by the simulator, but to save computation time a coarser step can be used in areas of relative inactivity. In the original version of SPICE, timesteps were provided by the user based on the user's expectations for the behavior of the circuit. In practice, these expectations became modified as experience was gained, and so even using the simulator became an iterative exercise, unless one was willing to accept wrong results. SPICE2, and all commercial simulators I know of, use a timestep that is automatically adjusted based on the number of iterations taken at the last timestep and the resulting local truncation error. This works very well, but sometimes gets a little too enthusiastic in its quest for accuracy. Failure of the analysis can result, with an error message something like "timestep too small in transient analysis." This means that the simulator has given up trying to converge to its specified accuracy by reducing the timestep. Often such a situation occurs because of unrealistically fast edges in the input waveforms, and these should given a sanity check. Modeling the power supplies with a little inductance and resistance can also keep transient currents to reasonable limits. There are also some numerical cures; again try relaxing RELTOL or increasing the allowed number of iterations per timestep (in SPICE this is the ITL4 option and defaults to ten).

The default numerical integration used by virtually all simulators is the TRAPEZOIDAL method and in general will give good results. Sometimes, however, this method causes some numerical "ringing" which looks like circuit oscillation. A classic sign of this is a wave form that settles nicely and then breaks into a bizarre oscillation. Other methods in use are called GEAR methods (after C. W.

Gear of the University of Illinois), the simplest of which is backward-Euler integration. GEAR methods can have the opposite problem to TRAPEZOIDAL integration: they can cover up real circuit oscillations. In general if both methods are stable or both show oscillation, then it is fairly safe to assume that this represents true circuit behavior. If TRAPEZOIDAL and GEAR disagree, then it is definitely worth trying to determine why.

By far the most fundamental limitation of all simulators is the accuracy of the models used. No matter how painlessly an analysis is performed, it can never be more lifelike than the models that were used to generate it. Some simulators allow for user-defined models, the "you don't like it, you fix it" approach. This is a great feature, but semiconductor device modeling is far from easy or straightforward.

As a generalization, the original Berkeley models are physically based models combined with some empirical patch equations. This is undoubtedly better than nothing, but as technology advances the empirical patching starts to dominate the physical model aspects and one loses one's intuitive feel for what goes on. Also, the underlying physical assumptions are often not realistic for modern IC processes. The SPICE BJT model, for example, is based on a one-dimensional physical model which can never be correct for a planar transistor. This model also has no means of generating substrate current, one of the biggest nuisances in a real circuit.

Some of the more recent models (notably the BSIM MOS model) are so empirical that the only realistic way of obtaining the parameters is by means of an extraction and optimization program. Such programs will happily play around with all manner of fundamental parameters until they obtain what they think is a best fit. At this point, it is difficult to assess the model validity merely by examining the parameters. I have seen all sorts of strange curves resulting from such blind modeling. The BSIM model, for example, is brilliant at generating parabolic output curves which result in negative output impedances at certain critical points. It is no fun to have the computer tell you that your "simple" feedback circuit has three stable states.

Model generation in this fashion as yet has no way of generating models with continuous temperature functions. This is not too important where only a few discrete temperatures are needed (as in digital work) but is hopelessly inadequate for analog circuits.

What is needed in the future is a thorough reevaluation of simulation modeling in the light of new technology. Hopefully, this will happen before this technology, too, becomes obsolescent.

The "Building Block" Approach

The early days of analog design saw almost every component "handcrafted "to satisfy the designer of its suitability for the particular application. As circuitry became more complicated, this approach migrated to one where a library of component cells was specified (with the option of customized components where needed), and a mask designer was left to hook them together. Even so, a fairly detailed placement sketch was usually provided by the designer, and every crossunder included in the final layout was scrupulously analyzed for any possible impact on performance. Such manual checking is admirable for small circuits but rapidly becomes impractical for today's systems-on-a-chip. Even if the time and effort were available for a full manual check, it would be unlikely to catch all the problems. Present design verification software needs considerable help to adequately check an analog layout, and by far the most helpful is the modular or building block approach.

The idea of the building block approach is to have access to sections of circuitry that are already laid out, proven, and characterized. These can vary from single transistors to large circuits such as complete amplifiers and A/D converters. These circuits can be entered onto schematics in their block format and subsequently recognized by verification software purely by the physical location of their external connections. Such an approach is a very powerful method of producing large custom circuits very efficiently. A side benefit is that customers can be given the performance of the block to include in their custom design without having to divulge the actual circuit details.

Such an approach has been used in custom circuit design for some time but is only recently becoming prevalent in the design of standard products. There are many reasons for this.

Most small circuits (up to 200 transistors or so) require a very high degree of performance optimization to be successful in the marketplace. Even commodity circuits, such as general purpose op amps, require considerable effort to integrate them in a small enough die size to ensure profitability. I really do not see this changing drastically in the future. What is changing is the average complexity of standard circuits, and this will force many standard product designers to adopt a less labor intensive approach to the design of them.

Overall die area is always a major concern for complex circuits. As pointed out, products in which die area is one of the most important considerations will always require extensive hand optimization; however the overall overhead of using predefined blocks may not be as large as might be expected. As the average circuit gets larger, the obvious geometric inflexibility of a particular building block becomes less important. Interconnect overhead is being reduced by the increasing use of multiple interconnect systems (notably dual layer metallization). Also, there is no fundamental reason why circuit blocks cannot be computer massaged for a better overall fit with little potential impact on performance.

Another point concerns the type of optimization carried out by experienced designers. This often includes such techniques as the design of unusual components, deliberate use of parasitics and common-pocketing to reduce space and capacitance. Such techniques require intimate familiarity with the process being used, something which has traditionally been acquired, and shared, by designers over a significant period of time. Today's designer is likely to be faced with many different processes, each considerably more complicated than those only a few years ago. Acquiring the depth of experience necessary to confidently plunge into the unknown is therefore becoming much more difficult. Also, the vastly different processes being used by different companies (even now there is nothing approaching a standard complementary bipolar or Bi-CMOS process) make sharing of knowledge less useful than in previous times.

Just as digital designers have become used to designing at the macro level, the same thing is happening to analog design. As well as the potential time savings, this is another step toward making digital and analog design tools more compatible. Maybe one day there will be little distinction between digital and analog IC engineers, or even systems and IC engineers in general (there have been several start-up companies founded to make a reality out of such concepts). But I sort of doubt this somehow, and I am not considering a career change quite yet.

Conclusion

I have often heard new designers express wishes that they had been in the analog design field in the early days of integrated circuits. With the extreme benefit of hindsight, designing a successful product seemed so much easier back then. Hindsight, however, is a very dangerous thing, because it conveniently ignores a vast trail of failures.

The field of electronic engineering in general, in my opinion, is still one of the most rapidly progressing industries in the world. While it is true that there is much competitive pressure to keep abreast of such progress, the horizons are widening in a seemingly multi-dimensional fashion. It is now the routine duty of the silicon wizard to produce circuitry that would have been the realm of a team of systems engineers only a decade ago. The opportunities, and means to take advantage of them, are definitely here.

If the semiconductor industry is maturing, as some would have it, then it is certainly not at the expense of a stagnation of technology.

So be (cautiously) optimistic and (recklessly) enthusiastic about the future of analog integrated electronics. Despite my sometimes cynical and skeptical nature, I assure you that I share such optimism and enthusiasm.

References

1. P. R. Gray and R. G. Meyer, "Analysis and Design of Analog Integrated Circuits," Wiley, New York, 1977.

2. A. B. Grebene, "Bipolar and MOS Analog Integrated Circuit Design," Wiley, New York, 1984.

3. R. A. Pease, "Troubleshooting Analog Circuits," Parts 1–11, *EDN Magazine,* January–October 1989.

25. Current–Feedback Amplifiers

In their effort to approximate the ideal op amp, manufacturers strive not only to maximize the open-loop gain and minimize input-referred errors such as offset voltage, bias current, and noise, but also to ensure adequate bandwidth and settling-time characteristics. Amplifier dynamics are particularly important in high speed applications like bipolar DAC buffers, subranging ADCs, S/H circuits, ATE pin drivers, and video and IF drivers. Being voltage-processing devices, conventional op amps are subject to the speed limitations inherent to *voltage-mode* operation, stemming primarily from the stray capacitances of nodes and the cutoff frequencies of transistors. Particularly severe is the effect of the stray capacitance between the input and output nodes of high-gain inverting stages because of the Miller effect, which multiplies this capacitance by the voltage gain of the stage. By contrast, *current-mode* operation has long been recognized as inherently faster than voltage-mode operation. The effect of stray inductances in an integrated circuit is usually less severe than that of its stray capacitances, and BJTs switch currents more rapidly than voltages. These technological reasons are at the basis of *emitter coupled logic, bipolar DACs, current conveyors,* and the high speed amplifier topology known as *current-feedback* [1].

For true current-mode operation, all nodes in the circuit should ideally be kept at fixed potentials to avoid the slow-down effect by their stray capacitances. However, since the input and output of the amplifier must be voltages, some form of high speed voltage-mode operation must be provided at some point. This is achieved by driving the nodes with push–pull emitter follower stages to rapidly charge or discharge their stray capacitances and by employing gain configurations inherently immune to the Miller effect, such as the cascode configuration.

The above concepts are illustrated using the simplified AC equivalent of Figure 25-1 as a vehicle. The circuit consists of the emitter follower input stage Q_1, the current mirror Q_2 and Q_3, the cascode gain stage Q_3 and Q_4, and the emitter follower output stage Q_5. The feedback signal is the current fed from the emitter of Q_5 back to the emitter of Q_1 via R_2, indicating series-shunt feedback. A qualitative analysis reveals that the open-loop characteristics are set primarily by the equivalent impedance z between the collector of Q_4 and ground. The resistive component of z sets the open loop DC gain, and the capacitive component controls the open loop dynamics.

Variants of the basic topology of Figure 25-1 have long been used in high speed applications such as active probes. Its adaptation to op-amp–like operation requires an input stage of the differential type. Moreover, to ensure symmetric rise and fall times, each stage must be capable of complementary push–pull action, and the npn and pnp transistors must have comparable characteristics in terms of the cutoff

Based on an article which appeared in EDN Magazine (January 5, 1989) © Cahners Publishing Company 1990, A Division of Reed Publishing USA

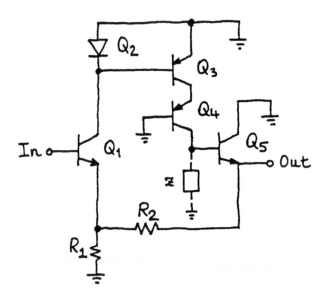

Figure 25-1.
The current-feedback concept.

frequency f_t. Traditionally, monolithic pnp transistors have been plagued by much poorer performance characteristics than their npn counterparts. However, the development of truly complementary high speed processes has made it possible to achieve monolithic speeds that were previously available only in hybrid form. The unique features and operation of the current-feedback (CF) amp are best appreciated by comparing them against those of its better known counterpart, the conventional op amp.

The Conventional Op Amp

The conventional op amp consists of a high input-impedance differential stage followed by additional gain stages, the last of which is a low output-impedance stage. As shown in the circuit model of Figure 25-2a, the op amp transfer characteristic is

$$V_o = a(jf)V_d \qquad (1)$$

where V_o is the output voltage; $V_d = V_p - V_n$ is the differential input voltage; and $a(jf)$, a complex function of frequency f, is the *open-loop gain*.

Connecting an external network as in Figure 25-2b creates a feedback path along

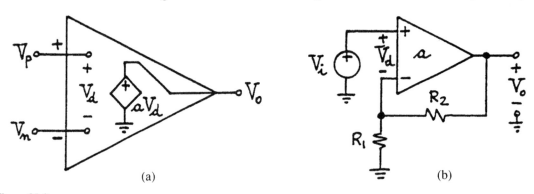

(a)   (b)

Figure 25-2.
Circuit model of the conventional op amp, and connection as a noninverting amplifier.

which a signal in the form of a *voltage* is derived from the output and applied to the noninverting input. By inspection,

$$V_d = V_i - \frac{R_1}{R_1 + R_2} V_o \qquad (2)$$

Substituting into Eq. (1), collecting, and solving for the ratio V_o/V_i yields the *noninverting amplifier* transfer characteristic

$$A(jf) = \frac{V_o}{V_i} = \left(1 + \frac{R_2}{R_1}\right)\frac{1}{1 + 1/T(jf)} \qquad (3)$$

$$T(jf) = \frac{a(jf)}{1 + R_2/R_1} \qquad (4)$$

where $A(jf)$ is the *closed-loop gain*, and $T(jf)$ is the *loop gain*. The designation *loop gain* stems from the fact that if we break the loop as in Figure 25-3a and inject a test signal V_x with V_i suppressed, the circuit will first attenuate V_x to produce $V_n = V_x/(1 + R_2/R_1)$, and then amplify V_n to produce $V_o = -aV_n$. The gain experienced by a signal in going around the loop is thus $V_o/V_x = -a/(1 + R_2/R_1)$. The *negative* of this ratio is the loop gain, $T = -(V_o/V_x)$. Hence, Eq. (4).

The loop gain gives a measure of how close A is to the ideal value $1 + R_2/R_1$, also called the *noise gain* of the circuit. By Eq. (3), the larger T, the better. To ensure a substantial loop gain over a wide range of closed-loop gains, the manufacturer strives to make a as large as possible. Consequently, since $V_d = V_o/a$, V_d will assume extremely small values. In the limit $a \to \infty$ we obtain $V_d \to 0$, that is, $V_n \to V_p$. This forms the basis of the familiar op amp rule: *When operated with negative feedback, an op amp will provide whatever output is needed to force V_n to follow V_p.*

Gain–Bandwidth Trade-off

Large open-loop gains can physically be realized only over a limited frequency range. Past this range, gain rolls off with frequency. Most op amps are designed for

Figure 25-3.
Test circuit to find the loop gain, and graphical method to determine the closed-loop bandwidth f_A.

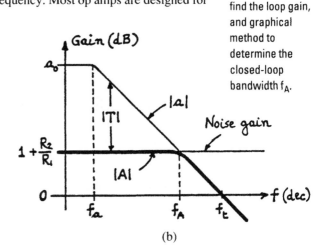

(a)   (b)

a constant rolloff of –20 dB/dec, so that the open-loop response can be expressed as

$$a(jf) = \frac{a_o}{1 + j(f/f_a)} \tag{5}$$

where a_o represents the *DC gain*, and f_a is the *–3 dB frequency* of the open-loop response. For instance, the popular 741 op amp has $a_o \approx 2 \times 10^5$ and $f_a \approx 5$ Hz.

Substituting Eq. (5) into Eq. (4) and then into Eq. (3), and exploiting the fact that $(1 + R_2/R_1)/a_o \ll 1$, we obtain

$$A(jf) = \frac{1 + R_2/R_1}{1 + j(f/f_A)} \tag{6}$$

$$f_A = \frac{f_t}{1 + R_2/R_1} \tag{7}$$

where f_A is the *closed-loop bandwidth*, and $f_t = a_o f_a$ is the open-loop unity-gain frequency, that is, the frequency at which $|a| = 1$. For instance, the 741 op amp has $f_t = 2 \times 10^5 \times 5 = 1$ MHz.

Equation (7) reveals a *gain-bandwidth trade-off*. As we raise the R_2/R_1 ratio to increase the closed-loop gain, we also decrease its bandwidth in the process. Moreover, by Eq. (4), the loop-gain is also decreased, leading to a greater closed-loop gain error.

These concepts can also be visualized graphically. By Eq. (4) we have $|T|_{dB} = 20 \log |T| = 20 \log |a| - 20 \log |1 + R_2/R_1|$, or

$$|T|_{dB} = |a|_{dB} - |1 + R_2/R_1|_{dB} \tag{8}$$

indicating that the loop gain can be found graphically as the *difference* between the open-loop gain and the noise gain. This is shown in Figure 25-3b. The frequency at which the two curves meet is called the *crossover frequency*. It is readily seen that at this frequency we have $T = 1/\underline{-90°} = -j$, so that Eq. (3) yields $|A| = (1 + R_2/R_1)/|1 + j| = (1 + R_2/R_1)/\sqrt{2}$. Consequently, the crossover frequency is also the –3 dB frequency of the closed-loop response, that is, the closed-loop bandwidth f_A.

We now see that increasing the closed-loop gain shifts the noise-gain curve upward, thus reducing the loop gain, and causes the crosspoint to move up the $|a|$ curve, thus decreasing the closed-loop bandwidth as well as the loop gain. Clearly, the circuit with the widest bandwidth and the highest loop gain is also the one with the lowest closed-loop gain. This is the voltage follower, for which $R_2/R_1 = 0$ so that $A = 1/[(1 + j(f/f_t)]$.

Slew-Rate Limiting

To fully characterize the dynamic behavior of an op amp, we also need to know its *transient response*. If an op amp with the response of Eq. (5) is operated as a unity-gain voltage follower and is subjected to a suitably small voltage step, its dynamic behavior will be similar to that of an RC network. Applying an input step of magnitude ΔV_i as in Figure 25-4a will cause the output to undergo an exponential transition with magnitude $\Delta V_o = \Delta V_i$, and with the time constant $\tau = 1/(2\pi f_t)$.

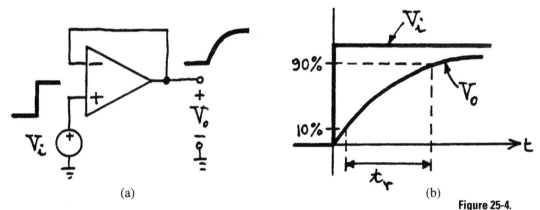

(a)　　　　　　　　　　　　　　(b)

Figure 25-4.
The voltage
follower and its
small-signal step
response.

The *rise time* is defined as the amount of time t_r it takes for the output to swing from 10% to 90% of the step size. For an exponential transition we have $t_r = \tau \times \ln(0.9/0.1) = 2.2\tau$. For the 741 op amp we have $\tau = 1/(2\pi \times 10^6) \approx 160$ nsec, and $t_r \approx 350$ nsec.

The rate at which the output changes with time is highest at the beginning of the transition, when its value is $\Delta V_o / \tau$. Increasing the step magnitude increases this initial rate of change, until this rate saturates at a value called the *slew-rate* (SR). This effect stems from the limited ability of the internal circuitry to charge or discharge capacitive loads, especially the internal frequency compensation capacitor.

To illustrate, refer to the circuit model of Figure 25-5, which is typical of many op amps. The input stage is a transconductance amplifier consisting of the differential pair Q_1–Q_2 and the current mirror load Q_3–Q_4. The remaining stages are lumped together as an integrator block consisting of an inverting amplifier and the compensation capacitor C. Slew-rate limiting occurs when the transconductance stage is driven into saturation, so that all the current available to charge or discharge C is the bias current I of this stage.

For example, the 741 op amp has $I = 20\ \mu A$ and $C = 30$ pF, so that SR $= I/C = 0.67$ V/μsec. The step magnitude corresponding to the onset of slew-rate limiting is such that $\Delta V_i / \tau =$ SR, or $\Delta V_i =$ SR $\times \tau = (0.67\ V/\mu sec) \times (160\ nsec) = 106$ mV. As long as the input step is less than 106 mV, a 741 voltage follower will respond with an exponential transition governed by $\tau \approx 160$ nsec, whereas for greater input steps the output will slew at a constant rate of 0.67 V/μsec.

Figure 25-5.
Simplified slew
rate model of the
conventional op
amp.

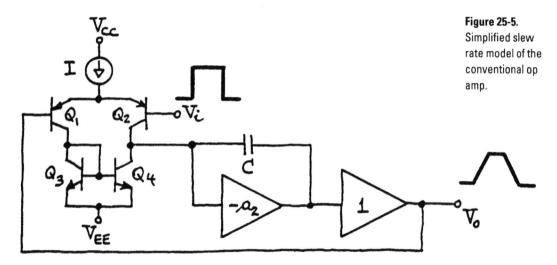

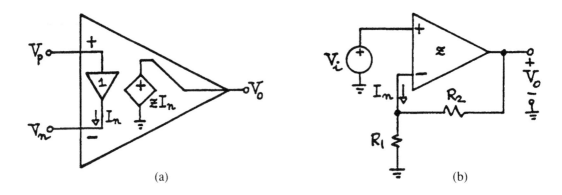

(a)   (b)

Figure 25-6.
Circuit model of
the current feed-
back amplifer,
and connection
as a noninverting
amplifer.

Figure 25-6. An important dynamic parameter in high speed applications is the *settling time*, that is, the time it takes for the output to settle and remain within a specified band around its final value, usually for a full-scale output transition. Clearly, slew-rate limiting plays an important role in the settling-time characteristic of a conventional op amp.

The Current–Feedback Amplifier

As shown in the circuit model of Figure 25-6a, the architecture of the current-feedback amplifier differs from the conventional op amp in two respects [1]:

1. The input stage is a *unity-gain voltage buffer* connected across the inputs. Its function is to force V_n to follow V_p, very much like a conventional op amp does via negative feedback. However, because of the low output impedance of this buffer, current can easily flow in or out of the inverting input, though we shall see that in the steady state (nonslewing) condition this current is designed to approach zero.

2. Amplification is provided by a transimpedance stage which senses the current delivered by the buffer to the external feedback network and produces an output voltage V_o such that

$$V_o = z(jf)I_n \qquad (9)$$

where $z(jf)$ is the transimpedance gain of the amplifier, in volts per amp or ohms, and I_n is the current out of the inverting input.

To appreciate the inner workings of the CF amp, it is instructive to examine the simplified circuit diagram of Figure 25-7. The input buffer consists of transistors Q_1 through Q_4. While Q_1 and Q_2 form a low output-impedance push-pull stage, Q_3 and Q_4 provide V_{be} compensation for the push–pull pair, as well as a Darlington function to raise the input impedance.

Summing currents at the inverting node yields $I_1 - I_2 = I_n$, where I_1 and I_2 are the push–pull transistor currents. A pair of Wilson current mirrors, consisting of transistors Q_9-Q_{10}-Q_{11} and Q_{13}-Q_{14}-Q_{15}, reflect these currents and recombine them at a common node, whose equivalent capacitance to ground is denoted as C. By mirror action, the current through this capacitance is $I_c = I_1 - I_2$, or

$$I_c = I_n \qquad (10)$$

The voltage developed by C in response to this current is then conveyed to the output via a second buffer, made up of Q_5 through Q_8. The block diagram of Figure 25-8 summarizes the salient features of the CF amp.

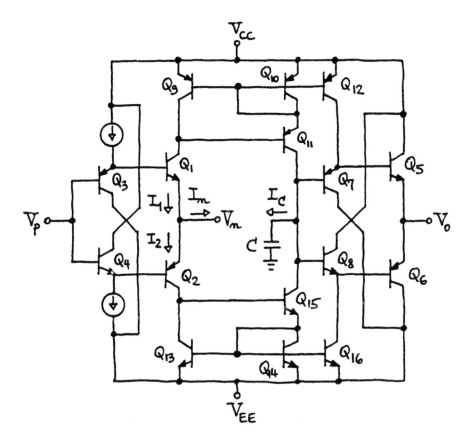

Figure 25-7.
Simplified circuit
diagram of a
current feedback
amplifier.
(Courtesy of
Comlinear
Corporation.)

When the feedback loop is closed as in Figure 25-6b, and whenever an external signal tries to imbalance the two inputs, the input buffer will begin sourcing (or sinking) an imbalance current I_n to the external resistances. This imbalance is then conveyed by the Wilson mirrors to the capacitor C, causing V_o to swing in the positive (or negative) direction until the original imbalance I_n is neutralized via the negative feedback loop. Clearly, I_n plays the role of the error signal in the system.

To obtain the closed-loop transfer characteristic, we exploit the fact that the input buffer keeps $V_n = V_p = V_i$. Applying the superposition principle, we have

$$I_n = \frac{V_i}{R_1 \| R_2} - \frac{V_o}{R_2} \tag{11}$$

This confirms that the feedback signal, V_o/R_2, is now in the form of a *current*. Substituting into Eq. (9), collecting, and solving for the ratio V_o/V_i yields

$$A(jf) = \frac{V_o}{V_i} = \left(1 + \frac{R_2}{R_1}\right) \frac{1}{1 + 1/T(jf)} \tag{12}$$

$$T(jf) = \frac{z(jf)}{R_2} \tag{13}$$

where $A(jf)$ is the *closed-loop gain* of the circuit, and $T(jf)$ is the *loop gain*. This designation stems again from the fact that if we break the loop as in Figure 25-8a, and inject a test voltage V_x with the input V_i suppressed, the circuit will first convert

Figure 25-8.
Current feedback
amplifier block
diagram.

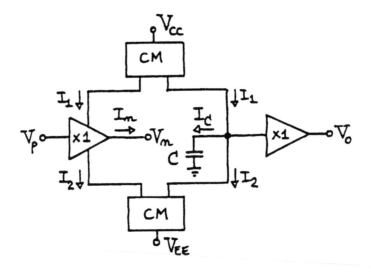

Figure 25-8.
Current feedback
amplifier block
diagram.

V_x to the current $I_n = -V_x/R_2$, and then convert I_n to the voltage $V_o = zI_n$ so that $T = -(V_o/V_x) = z/R_2$, as expected.

In an effort to ensure substantial loop gain and thus reduce the closed-loop gain error, the manufacturer strives to make z as large as possible relative to the expected range of values of R_2.

Consequently, since $I_n = V_o/z$, the inverting-input current will be very small, though this input is a low-impedance node because of the buffer. In the limit $z \to \infty$ we obtain $I_n \to 0$, indicating that a CF amp *will provide whatever output is needed to ideally drive I_n to zero*. Thus, the familiar op amp conditions $V_n \to V_p$, $I_n \to 0$, and $I_p \to 0$ hold also for CF amps, though for different reasons.

No Gain–Bandwidth Trade-off

Figure 25-9.
Test circuit to
find the loop gain,
and graphical
method to
determine the
closed loop
bandwidth f_A.

The transimpedance gain of a practical CF amp rolls off with frequency according to

$$Z(jf) = \frac{z_o}{1 + j(f/f_a)} \qquad (14)$$

where z_o is the *DC* value of the transimpedance gain, and f_a is the frequency at which rolloff begins. For instance, the CLC401 CF amp (Comlinear Co.) has $z_o \approx 710 \ \text{k}\Omega$

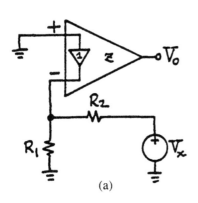

(a)

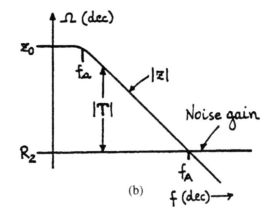

(b)

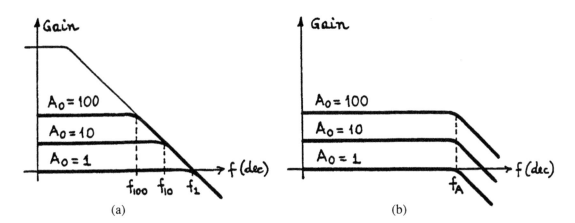

(a)

(b)

Figure 25-10.
Comparing the
gain-bandwidth
relationship of
conventional op
amps and current
feedback
amplifiers.

and $f_a \approx 350$ kHZ. Moreover, since $f_a = 1/(2\pi z_o C)$, it follows that $C = 1/(2\pi z_o f_a) \approx 0.64$ pF.

Substituting Eq. (14) into Eq. (13) and then into Eq. (12), and exploiting the fact that $R_2/z_o \ll 1$, we obtain

$$A(jf) = \frac{1 + R_2/R_1}{1 + j(f/f_A)} \qquad (15)$$

$$f_A = \frac{z_o f_a}{R_2} \qquad (16)$$

where f_A represents the *closed-loop bandwidth*. With R_2 in the kilohm range, f_A is typically in the 100 MHz range. Retracing previous reasoning, we see that the noise-gain curve is now R_2, and that f_A can be found graphically as the frequency at which this curve meets the $|z|$ curve, see Figure 25-9b.

Comparing with Eqs. (6) and (7), we note that the expressions for $A(jf)$ are formally identical; however, the bandwidth f_A now depends only on R_2, indicating that we can use R_2 to select the bandwidth, and R_1 to select the gain. The ability to control gain independently of bandwidth constitutes a major advantage of CF amps over conventional op amps, especially in automatic gain control applications. This important difference is highlighted in Figure 25-10, where $A_o = 1 + R_2/R_1$ denotes the DC value of the closed-loop gain.

Absence of Slew-Rate Limiting

The other major advantage of CF amps is the inherent absence of slew-rate limiting. This stems from the fact that the current available to charge the internal capacitance C at the onset of a step is proportional to the step *regardless* of its size. Indeed, applying a step of magnitude ΔV_i induces, by Eq. (11), an initial current imbalance $\Delta I_n = \Delta V_i/(R_1//R_2)$, which the Wilson mirrors then convey to the capacitor. The initial rate of charge is thus $\Delta I_c/C = \Delta I_n/C = \Delta V_i/[(R_1//R_2)C] = [\Delta V_i(1+R_2/R_1)]/(R_2 C) = \Delta V_o/(R_2 C)$, indicating an exponential output transition with time-constant $\tau = R_2 C$. Like the frequency response, the transient response is governed by R_2 alone, regardless of the closed-loop gain. With R_2 in the kilohm range and C in the picofarad range, τ will be in the nanosecond range.

The time it takes for an exponential transient to settle within 0.1% of its final

Figure 25-11.
Test circuit to
investigate the
effect of R_o.

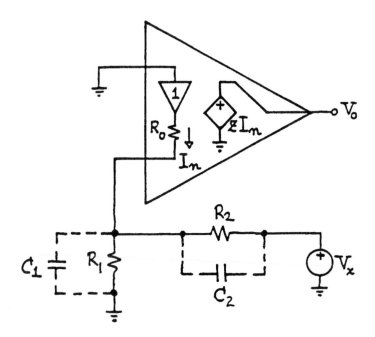

value is $t_s = \tau \ln 1000 \approx 7\tau$. For instance, in the case of a CLC401 CF amp with R_2 = 1.5 kΩ we have $\tau = R_2 C = 1.5 \times 10^3 \times 0.64 \times 10^{-12} \approx 1$ nsec, so that $t_r = 2.2\tau \approx$ 2.2 nsec, and $t_s = 7\tau \approx 7$ nsec. These values are in reasonable agreement with the data sheets values $t_r = 2.5$ nsec and $t_s = 10$ nsec.

The absence of slew-rate limiting not only allows for faster settling times but also avoids slew-rate related nonlinearities such as intermodulation distortion. This makes CF amps attractive in high-quality audio amplifier applications.

Second-Order Effects

The above analysis indicates that once R_2 has been set, the dynamics of the amplifier are unaffected by the closed-loop gain setting. In practice it is found that bandwidth and rise time do vary with gain somewhat, though not as drastically as with conventional op amps. The main cause is the nonzero *output impedance* of the input buffer, whose effect is to alter the loop gain and, hence, the closed-loop dynamics. Denoting this impedance as R_o, we shall refer to Figure 25-11 to investigate the effect of R_o as well as the effect of external capacitances, either at the input or in the feedback path.

Consider first the case in which the external network is purely resistive so that $C_1 = C_2 = 0$. The circuit first converts V_x to the current $I_x = V_x/(R_2 + R_1//R_o)$, then it divides I_x to produce $I_n = -I_x \times R_1/(R_1 + R_o)$, and finally it converts I_n to the voltage $V_o = z I_n$. Eliminating I_x and I_n and letting $T = -V_o/V_x$ yields

$$T(jf) = \frac{z(jf)}{Z_2} \tag{17}$$

$$Z_2 = R_2 \left(1 + \frac{R_o}{R_1 // R_2} \right) \tag{18}$$

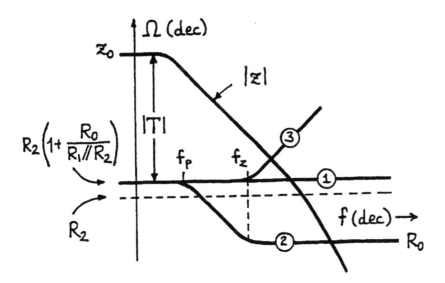

Figure 25-12.
Noise-gain
curves for the
case of (1) purely
resistive feed-
back, (2) a capac-
itance in parallel
with R_2, and (3) a
capacitance in
parallel with R_1.

Clearly, the effect of R_o is to *increase* the noise gain from R_2 to $Z_2 = R_2[1 + R_o/(R_1//R_2)]$. This is shown in Figure 25-12, curve 1. Consequently, both the band-width and the rise time will be reduced by a proportional amount. Replacing R_2 in Eq. (16) with Z_2 as given in Eq. (18) we obtain, after simple manipulation,

$$f_A = \frac{f_t}{1 + (R_o/R_2)A_o} \qquad (19)$$

where $f_t = z_o f_a/R_2$ is the extrapolated value of f_A in the limit $R_o \to 0$, and $A_o = 1 + R_2/R_1$ is the closed-loop DC gain. This equation indicates that bandwidth reduction due to R_o will be more pronounced at high closed-loop gains. This is shown in Figure 25-13.

Example 1. A certain CF amp has $R_o = 50 \, \Omega$, $R_2 = 1.5 \, k\Omega$, and $f_t = 100 \, MHz$. Find the bandwidths corresponding to $A_o = 1, 10,$ and 100.

Solution. By Eq. (18) we have $f_A = 10^8/[1 + (50/1500)A_o] = 10^8/(1 + A_o/30)$. The bandwidths corresponding to $A_o = 1, 10,$ and 100 are, respectively, $f_1 = 96.8 \, MHz$,

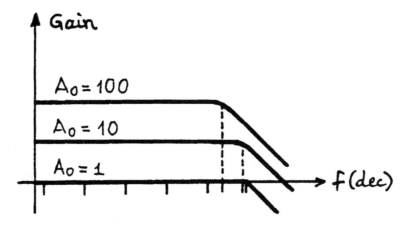

Figure 25-13.
Effect of R_o upon
f_A as a function
of A_o.

$f_{10} = 75.0$ MHz, and $f_{100} = 23.1$ MHz. We observe that these values still compare favorably with a conventional op amp, whose bandwidth would be reduced, respectivly, by 1, 10, and 100.

If desired, the external resistance values can be predistorted to compensate for the bandwidth reduction at high gains. Turning Eq. (19) around yields the required value of R_2 for a given bandwidth f_A and DC gain A_o,

$$R_2 = \frac{z_o f_a}{f_A} - R_o A_o \qquad (20)$$

and the required value of R_o for the given DC gain A_o,

$$R_1 = \frac{R_2}{A_o - 1} \qquad (21)$$

Example 2. Redesign the amplifier of Example 1 so that $f_{10} = 100$ MHz.

Solution. Since with $R_2 = 1.5$ kΩ this device has $z_o f_a / R_2 = 100$ MHz, it follows that $z_o f_a = 10^8 \times 1500 = 1.5 \times 10^{11}$ ΩHz. Then, for $A_o = 10$ and $f_{10} = 100$ MHz, we need $R_2 = 1.5 \times 10^{11}/10^8$ - $50 \times 10 = 1$ kΩ, and $R_1 = 1000/(10 - 1) = 111$ Ω.

Besides a dominant pole at f_a, the open-loop response of a practical amplifier presents additional poles above the crossover frequency. As shown in Figure 25-12, the effect of these poles is to cause a steeper gain rolloff at this frequency, further reducing the closed-loop bandwidth. Moreover, the additional phase-shift due to these poles decreases the phase margin somewhat, and this may cause some peaking in the frequency response and ringing in the step response.

Finally, it must be said that the rise time of a practical CF amp does increase with the step size somewhat, due primarily to transistor current gain degradation at high current levels. For instance, the rise time of the CLC401 changes from 2.5 nsec to 5 nsec as the step size is changed from 2 V to 5 V. In spite of second-order limitations, CF amps still provide superior dynamics.

CF Application Considerations

Although the above treatment has focused on the noninverting configuration, the CF amp will work as well in most other resistive feedback configurations like inverting amplifiers, summing and differencing amplifiers, I-V and V-I converters, and KRC active filters [2]. In fact, the derivation of the transfer characteristic of any of these circuits proceeds along the same lines as conventional op amps. Special consideration, however, merit the cases in which the external network includes reactive elements, either intentional or parasitic.

Consider first the effect of the *feedback capacitance* C_2 in parallel with R_2 in the basic circuit of Figure 25-11. Replacing R_2 with $R_2//(1/sC_2)$ in Eq. (18) and expanding, it is readily seen that Z_2 now has a pole at $f_p = 1/(2\pi R_2 C_2)$ and a zero at $f_z = 1/[2\pi(R_o//R_1//R_2)C_2]$. The corresponding noise-gain curve is shown in Figure 25-12, curve 2, indicating that the crossover frequency is now pushed into the region of substantial phase shift due to the higher-order poles of z. If the overall shift reaches $-180°$ at this frequency, the loop gain will become $T = 1/\underline{-180°} = -1$, making A

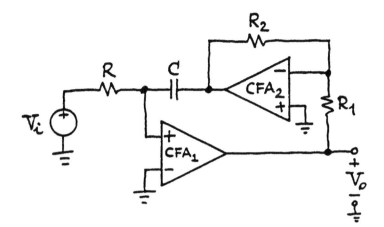

Figure 25-14.
Actively
compensated CF
integrator.

infinite, by Eq. (12). When this condition is met, the circuit will oscillate. Even if the phase shift fails to reach $-180°$, the closed-loop response may still exhibit intolerable peaking and ringing. Hence, *capacitive feedback must be avoided with CF amps*. To minimize the effect of stray feedback capacitances, manufacturers often provide R_2 internally.

CF Amp Integrators

To synthesize the integrator function in CF form, which provides the basis for dual-integrator-loop filters and oscillators as well as other popular circuits, we must use configurations that avoid a direct capacitance between the output and the inverting input. One possibility is offered by the Deboo integrator, which belongs to the class of KRC filters and is therefore amenable to CF amp realization. Its drawback is the need for tightly matched resistances, if lossless integration is desired. The alternative shown in Figure 25-13 not only meets the given constraint but also provides *active frequency compensation*, a highly desirable feature to cope with Q-enhancement problems in dual-integrator-loop filters [2]. Using standard op amp analysis techniques, it is readily seen that the unity-gain frequency of this integrator is $f_o = (R_2/R_1)/(2\pi RC)$. This circuit can be realized in a cost effective manner using a dual CF amp, such as the OP-260 (Precision Monolithics).

Stray Input–Capacitance Compensation

Next, let us investigate the effect of the *input capacitance C_1* in parallel with R_1 in the basic circuit of Figure 25-11. Replacing R_1 with $R_1//(1/sC_1)$ in Eq. (18) and expanding, it is readily seen that Z_2 now has a zero at $f_z = 1/[2\pi(R_o//R_1//R_2)C_1]$. The corresponding noise-gain curve is shown in Figure 25-12, curve 3. If C_1 is sufficiently large, the phase of T at the crossover frequency will again approach $-180°$, bringing the circuit to the verge of instability.

As in the case of a conventional op amp, the CF amp can be stabilized by using a feedback capacitance C_2 to introduce sufficient phase lead around the loop to compensate for the phase lag due to the input capacitance C_1. Though it was said earlier that capacitive feedback should be avoided with CF amps, this no longer holds when we want to combat the effect of an input capacitance.

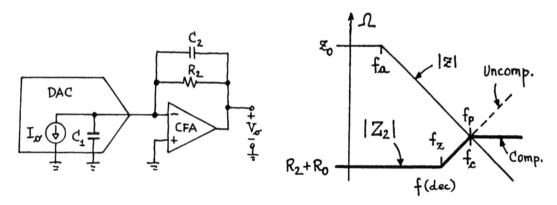

Figure 25-15.
DAC output
capacitance
compensation.

The destabilizing effect of the stray input capacitance is of particular concern in current-mode DAC output amplification, where C_1 is the output capacitance of the DAC, typically in the range of a few tens to a few hundreds of picofarads. The situation is depicted in Figure 25-15a. The use of C_2 creates a noise-gain pole at $f_p = 1/(2\pi R_2 C_2)$. For a phase margin of 45°, C_2 is chosen to make this pole coincide with the crossover frequency f_c. Referring to Figure 25-15b, one can show that if f_z is sufficiently lower than f_c then $f_c \approx \sqrt{z_o f_a f_z}/(R_o + R_2)$. Letting $f_c = 1/2\pi(R_o//R_2)C_1$ and imposing $f_p = f_c$ yields

$$C_2 = \left(\frac{R_o}{2\pi R_2 z_o f_a} C_1 \right)^{1/2} \tag{22}$$

Example 3. A DAC having $C_1 = 100$ pF feeds a CF amp having $R_2 = 1.5$ kΩ, $f_t = 150$ MHz, and $R_o = 50$ Ω. Find C_2 for a phase margin of 45°, and estimate the bandwidth of the amplifier.

Solution. Since $f_t = z_o f_a/R_2$, it follows that $z_o f_a = R_2 f_t = 1.5 \times 10^3 \times 150 \times 10^6 = 2.25 \times 10^{11}$ ΩHz. Then $C_2 = [50 \times 100 \times 10^{-12}/(2\pi \times 1.5 \times 10^3 \times 2.25 \times 10^{11})]^{1/2} = 1.54$ pF. The bandwidth is $f_A \approx 1/(2\pi R_2 C_2) = 1/(2\pi \times 1500 \times 1.54 \times 10^{-12}) \approx 69$ MHz. The value of C_2 may be increased for a greater phase margin, but this will also reduce the bandwidth of the amplifier.

Noise in CF Amp Circuits

Since CF amps are wideband amplifiers, they generally tend to be noisier than conventional op amps. The noise characteristics are specified in terms of three input noise densities: the *voltage density e_n*, the *inverting-input current density i_{nn}*, and the *noninverting-input current density i_{np}*. Since the BJTs of CF amps are generally biased at much higher current levels than conventional op amps, CF amps tend to exhibit lower voltage noise but higher current noise. Moreover, since the inputs are dissimilar because of the input voltage buffer, so are the current densities. Consequently, the data sheets report i_{nn} and i_{np} separately.

Figure 25-16 shows the noise model of a CF amp with resistive feedback. To find the overall *input noise density e_{ni}*, we use the superposition principle to find the

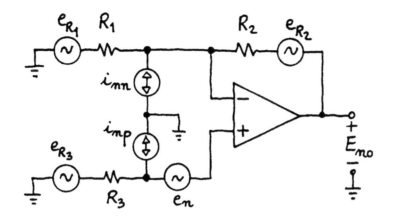

Figure 25-16.
Noise model of a resistive CF amp circuit.

contributions from the individual noise sources, and then we add up these contributions in rms fashion. The result is [2]

$$e_{ni}^2 = e_n^2 + R_3^2 i_{np}^2 + (R_1 \| R_2)^2 i_{nn}^2 + 4kT[R_3 + (R_1 \| R_2)] \tag{23}$$

where each component has been put in a form that lends itself to be amplified by the same noise gain $A(jf)$. The total *rms output noise* E_{no} above a given frequency f_L is

$$E_{no} = \left(\int_{f_L}^{\infty} |A(jf)|^2 e_{ni}^2 \, df \right)^{1/2} \tag{24}$$

where $A(jf)$ is given in Eq. (15). Expressing the noise densities as

$$e_n^2 = e_{nw}^2 (f_{ce}/f + 1), \quad i_{nn}^2 = i_{nnw}^2 (f_{cin}/f + 1) \text{ , and } \quad i_{np}^2 = i_{np}^2 \times (f_{cip}/f + 1),$$

and substituting into Eq. (23) and then into Eq. (24) yields [2]

$$E_{no} = \left(1 + \frac{R_2}{R_1} \right) \times \left[\begin{array}{l} e_{nw}^2 \left(f_{ce} \ln \dfrac{f_A}{f_L} + 1.57 f_A - f_L \right) + R_3^2 i_{npw}^2 \left(f_{cip} \ln \dfrac{f_A}{f_L} + 1.57 f_A - f_L \right) \\[2mm] + (R_1 \| R_2)^2 i_{nnw}^2 \left(f_{cin} \ln \dfrac{f_A}{f_L} + 1.57 f_A - f_L \right) + 4kT[R_3 + (R_1 \| R_2)](1.57 f_A - f_L) \end{array} \right]^{1/2} \tag{25}$$

Example 4. Let the circuit of Figure 25-16 be a CLC401 CF amp configured for a 20 dB noninverting gain with $R_1 = 166.7 \ \Omega$ and $R_2 = 1.5 \ \text{k}\Omega$. Moreover, let $R_3 = 100 \ \Omega$. Find E_{no} for the case in which noise is observed over a 10 sec period.

Solution. We have $f_L = 1/10 = 0.1$ Hz. Using the data sheets values $z_o = 710 \ \text{k}\Omega$ and $f_a = 350$ kHz, we obtain $f_t = z_o f_a/R_2 = 165.7$ MHz. Substituting into Eq. (19), along with the data sheet value $R_o = 50 \ \Omega$ yields $f_A = 124$ MHz. Substituting the data sheet values $e_{nw} \approx 2.4 \ \text{nV}/\sqrt{\text{Hz}}$, $f_{ce} \approx 30$ kHz, $i_{npw} \approx 2.6 \ \text{pA}/\sqrt{\text{Hz}}$, $f_{cip} \approx 30$ kHz, $i_{nnw} \approx 17 \ \text{pA}/\sqrt{\text{Hz}}$, and $f_{cin} \approx 40$ kHz into Eq. (25) yields $E_{no} \approx 0.57$ mV rms, or $E_{no} = 0.57 \times 6 = 3.4$ mV peak-to-peak.

References

1. *A New Approach to Op Amp Design*, Comlinear Corporation Application Note 300–1, March 1985.

2. Sergio Franco, *Design with Operational Amplifiers and Analog ICs*, McGraw–Hill Book Company, 1988.

3. *Current-Feedback Op Amp Applications Circuit Guide*, Comlinear Corporation Application Note OA–07, 1988.

Garry Gillette

26. Analog Extensions of Digital Time and Frequency Generation

In many cases it is possible to obtain desired analog accuracy and resolution in the frequency or time domain simply by counting the period of a fixed high frequency clock or by phase-locking a voltage controlled oscillator to a selectable harmonic of a fixed low frequency reference. Familiar examples are the digital watch, integrating voltmeter, and the tuner found in most radios and TVs. Accuracy of the programmed or measured value is referenced to that of a crystal oscillator, which can in turn be phase-locked to an even more accurate standard if desired. Since these designs typically use counters and control logic, they are excellent candidates for a CMOS digital integrated circuit, and they form the basis of many high volume consumer products.

For higher frequency and higher performance applications when increased resolution or improved jitter is required, analog extensions to these basic techniques are used to extend the range provided by the crystal-controlled clock. Examples requiring this type of extension are the computer-controlled frequency source in a space communications network, signal generators and analyzers, and the timing control in a modern 100 MHz VLSI test system.

Wide dynamic range analog signal performance in the presence of complex digital processing creates the opportunity for the two domains to interfere. For example, in precision low frequency analog design it is mandatory to maintain a good star ground discipline, yet when digital logic is present there are high edge currents flowing in device power supply and output pins which can readily induce noise into an otherwise clean analog ground by means of induction or a misplaced bypass capacitor. In these designs the overall performance may be limited by the extent to which these wide dynamic-range low frequency and digital frequency signals can simultaneously coexist and not interfere in undesired ways.

An example of such a system is the Digiphase synthesizer. The Digiphase synthesizer was the first indirect synthesizer which provided frequency resolution less than the reference frequency while maintaining very low output spurious and low random phase noise. Prior to its development, frequency resolution of 1 kHz was typically obtained from a "divide-by-N" loop with a relatively slow settling time and essentially no cancellation of voltage-controlled oscillator (VCO) noise and power line frequency spurious. While indirect synthesis was acceptable for voice communication, for high performance applications a direct synthesis technique was the only other alternative, with a very high cost and complexity. For phase sensitive applications such as space vehicle communications and over-the-horizon radar, the large number of high-Q tuned circuits used in the direct synthesis technique made it subject to unacceptable phase "roll" from temperature drift.

The Digiphase technique requires no tuned circuits other than the one used in the VCO. The first design covered the range of 40–51 MHz in 1 Hz increments with

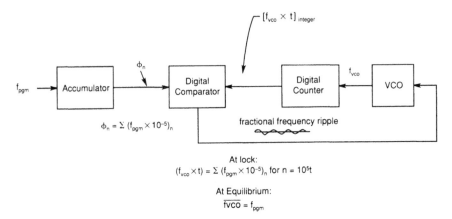

Figure 26-1.
Digiphase integer
cycle phase
locking.

sideband phase spurious of less than -80 dBc, and random phase noise of -110 dB/Hz. Its high sampling rate of 100 kHz provided the loop-gain bandwidth needed to control the static and dynamic output phase of the VCO with a stability of a few picoseconds over a temperature range of several degrees. Because the output was derived from a single VCO, phase continuity was guaranteed while following a computer-generated phase trajectory, such as that required for Range, Doppler, and Doppler rate in the Deep Space Network to guarantee acquisition of a remote signal buried in noise.

The Digiphase technique was an invention of Noel Braymer, who was a founder and resident inventor at Dana Labs from its inception in 1960 through the early 1970s. The technique is based on very accurately controlling the phase of a VCO to be exactly that of a computed value which is periodically updated in a digital phase register. (See Figure 26-1.) The computed phase can of course have very good resolution, since it is generated by digital computation. It also can be computed to have a constant rate of change, which corresponds to a constant output frequency, or it can be programmed to have a quadratic increase in phase, which corresponds to a linear frequency ramp. The digital value in the phase register is updated every 10 μsec, and contains the number of whole and fractional cycles desired from the output phase of the VCO from some initial time reference. Since the VCO frequency range was 40–51 MHz, every 10 μsec the computed phase increased between 400 and 510 integer cycles of phase and increased a fractional cycle amount determined by the fractional 100 kHz portion of the programmed frequency below the 100 kHz digit. For a fixed frequency this increment value will remain constant, and the phase register will simply be increased each 10 μsec by this constant increment. It should be noted, however, that it is the long-term phase of the VCO that is under control, even though the phase register is always increasing at a constant rate equal to the desired frequency times 10^{-5} cycles every 10 μsec. For example, if the programmed frequency is 47.654321 MHz, the number of whole cycles added to the phase register each 10 μsec is 476, and the fractional increase would be 0.54321 cycles.

Obviously the value accumulated in the phase register becomes very large. In applications involving a constant output frequency, the slower moving higher order portion can be truncated, since only the difference between phase updates is being dynamically controlled. In deep space applications it is necessary to keep track of every cycle of phase, since phase corresponds to range. Because the resolution must be sufficient for the best atomic standards, the phase register for that application was extended to a resolution of 10 picocycles (1 μHz frequency resolution), although

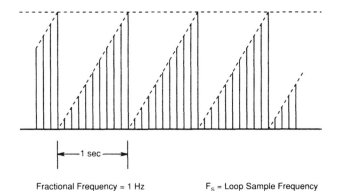

Figure 26-2.
Digiphase phase
detector output.

1 sec

Fractional Frequency = 1 Hz

F_S = Loop Sample Frequency
= 100 kHz

the higher order bits of phase are carried in a computer and are not utilized for local tracking of the accumulated VCO phase.

In the Digiphase technique, control of integer cycles of VCO phase is accomplished simply by accumulating them in a counter whose value is periodically compared to an updated integer cycle portion of the phase register. If the compared value is below the calculated value, a phase error is generated which speeds up the VCO, and if its compared value is above the calculated value, a phase error is generated which slows down the VCO. A locked phase-lock loop will not allow the output phase of the VCO to vary more than a fraction of one cycle from the programmed value. If a frequency counter is used to check the output frequency of the VCO, its accuracy will be seen to be quite good, since in 1 sec it will be within 1 Hz.

If a fixed frequency is all that is desired, the integer cycle portion of the counter, equality, and phase register can be replaced today by a programmable-modulus counter. The carry from the fractional cycle computation in the phase register triggers the programmable-modulus carry-in to increase the count by one during the sample period of the overflow. This technique is referred to as "fractional-N" in references. In the earlier Digiphase it was necessary to have a decade-ranged symmetric analog FM "search" voltage control with the capability to add or subtract more than 1 cycle per 10 μsec sample interval, so the integer cycle counter, equality, and phase register were necessary. Also, this design preceeded the availability of integrated 50 MHz "count-by-N" or programmable-modulus MSI ECL by many years. Both Digiphase and fractional-N implementations have identical analog loop requirements for low spurious and phase noise.

The second aspect of the Digiphase technique is that of using an ultralinear (in time delay) phase detector in the phase-lock loop and then subtracting from its analog output signal the exact amount of loop phase ripple signal which would occur if the VCO were exactly at its programmed frequency. To understand the phase compensation technique, consider the case in which the portion below 100 kHz of the programmed frequency is programmed to 1 Hz. (See Figure 26-2.) The calculated phase in the fractional cycle portion of the phase register will increase by 10^{-5} cycles at each 10 μsec sample update, until in 1 sec it overflows and creates a carry of one cycle into the integer cycle portion of the phase register, at which time the fractional cycle portion overflows to zero and resumes its linear 1 Hz ramp.

If the VCO were exactly at its programmed frequency, the digital value in the fractional portion of the phase register would be exactly proportional to the output ripple signal from the loop phase detector. With suitable scaling of the frequency-dependent gain of the phase detector, the value in the fractional cycle portion of the

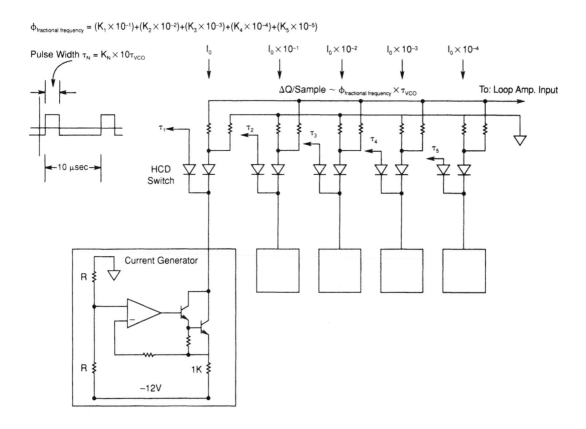

$\phi_{\text{fractional frequency}} = (K_1 \times 10^{-1}) + (K_2 \times 10^{-2}) + (K_3 \times 10^{-3}) + (K_4 \times 10^{-4}) + (K_5 \times 10^{-5})$

Pulse Width $\tau_N = K_N \times 10\tau_{VCO}$

$I_0 \quad\quad I_0 \times 10^{-1} \quad\quad I_0 \times 10^{-2} \quad\quad I_0 \times 10^{-3} \quad\quad I_0 \times 10^{-4}$

$\Delta Q/\text{Sample} \sim \phi_{\text{fractional frequency}} \times \tau_{VCO}$

To: Loop Amp. Input

$\tau_1 \quad\quad \tau_2 \quad\quad \tau_3 \quad\quad \tau_4 \quad\quad \tau_5$

—10 μsec—

HCD Switch

Current Generator

R

R   1K

−12V

Figure 26-3. The time-modulated DAC used by Digiphase.

phase register is converted into an analog signal using a time-modulated DAC, which is called a TAC. (See Figure 26-3.) The TAC output is subtracted from the output of the loop phase detector to exactly cancel its time varying ripple output at the fractional frequency overflow rate. If the ripple cancellation is exact, the phase sidebands at the fractional frequency will be suppressed in the VCO control signal. The single-frequency phase spurious level achieved in production Digiphase synthesizers was −80 dBc (relative to one radian[2] of the output VCO phase). To achieve this level, extreme care in analog signal-to-noise ratio was necessary.

While a high performance analog design problem, the actual amount of hardware required by Digiphase is comparatively small, resulting in a much lower cost than alternative approaches. In order to guarantee no phase continuities from switching, a single VCO was used to cover the whole range of 40–51 MHz. The result was a large VCO gain constant and a narrow percentage bandwidth, which allowed the design of the phase-lock loop to always be in a linear region and remain locked over the whole 11 MHz output range. The 100 kHz sample rate allowed a reasonable 150 μsec loop settling time for a 1 MHz step in programmed frequency. The actual latency time in frequency switching is only that of programming the phase register, which is less than 10 μsec when synchronized to the phase register update time. For a programmed linear swept-frequency program, the VCO output frequency is totally linear and void of phase discontinuities. When the output of two synthesizers was mixed together while both were synchronized to a frequency sweep over the range 40.0 MHz to 40.1 MHz, at a 1 MHz/sec sweep rate, the measured phase difference tracked within ± 1 degree. This capability is useful in phase-coherent "chirp" radar or coherent swept-frequency signal generators.

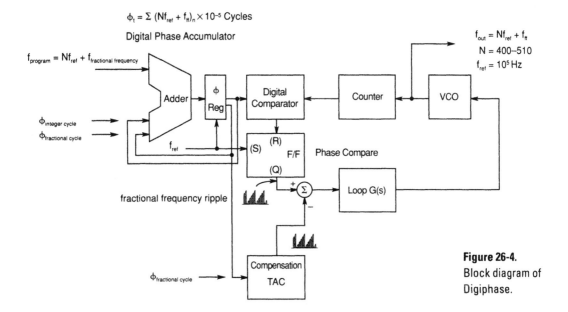

$$\phi_t = \Sigma \, (Nf_{ref} + f_{ff})_n \times 10^{-5} \text{ Cycles}$$

Digital Phase Accumulator

$f_{out} = Nf_{ref} + f_{ff}$
$N = 400\text{--}510$
$f_{ref} = 10^5\,Hz$

$f_{program} = Nf_{ref} + f_{fractional\ frequency}$

Adder

ϕ Reg

Digital Comparator

Counter

VCO

$\phi_{integer\ cycle}$

$\phi_{fractional\ cycle}$

f_{ref}

(R)

(S)   F/F

(Q)

Phase Compare

fractional frequency ripple

Σ + −

Loop G(s)

$\phi_{fractional\ cycle}$

Compensation TAC

Figure 26-4.
Block diagram of
Digiphase.

To provide a frequency range of 0–11 MHz, the 40–51 MHz VCO output was mixed with a fixed 40 MHz signal derived from a crystal oscillator frequency reference. This provided a ninth-order worst case mixing ratio and less than –90 dBc sideband spurious in the balanced hot carrier diode mixer. The ninth-order mixer spur was limited to the range in output frequency of 10–11 MHz, due to the narrow range of frequencies used with the VCO. The output of the mixer was low-pass filtered with a four-pole Butterworth filter and amplified by a wide-band DC amplifier. Since the output phase was very stable and the temperature coefficient of input voltage on the amplifier was designed to be very low, the 1 Hz frequency control front panel knob could be switched in and out from zero programmed frequency to establish a precise DC zero output voltage, which could then be used to monitor output sideband phase noise directly using low frequency instrumentation, even though the VCO was running at 40 MHz.

A divided-down output was provided as an option, which covered the range of DC to 110 kHz. The VCO output was divided by 100 to cover the range of 400–510 kHz, and a fixed reference frequency of 400 kHz was mixed with it to obtain the output frequency. The output mixer was designed as a well-balanced full-wave phase detector using transistors as switches and a transformer wound with trifilar wire. The output spurious radiations were 40 dB below that of the main loop (–120 dBc), and the phase noise within the loop bandwidth was also improved by the same amount (–150 dB/Hz). The frequency resolution was 10^{-2} Hz.

To achieve high performance, the loop phase detector was constructed from a constant-current pulse established by a flip-flop whose first edge was created from the 100 kHz reference and whose second edge was clocked at the VCO axis crossing coincident with the equality of calculated and measured integer phase. (See Figure 26-4.) The correspondence between output charge-per-sample at the output of the phase detector and input edge delay was required to be linear and accurate within 10 ppm (0.2 psec in time or 1 fC in charge) at the loop sample rate of 100 kHz. In addition an exact conjugate of the charge output per sample of the phase detector was necessary from the compensation TAC. The value of this charge is digitally computed and updated for each 100 kHz sample. Each decade-weighted current pulse had a width proportional to its digital value in quanta of VCO cycles. The charge transferred per sample was inversely proportional to VCO frequency and directly proportional to the digital value of the fractional portion of the phase register. It was adjusted to cancel the output charge from the loop phase detector at each harmonic of the fractional cycle rate. Accuracy and linearity were less than 10 ppm (1 fC/sample).

Both the phase detector output and the TAC pulses were summed into an integrator, so that during the time when the phase detector current pulses were off, a stable voltage was present on the output of the integrator. The output of the integrator was connected to a track-and-hold follower amplifier which held the output during the short fraction of the 10 µsec sample interval that the current pulses occurred. This assured that the ripple associated with the inexact time of arrival of the conjugate TAC pulses would not be transferred to the VCO input. In addition, to further assure no transfer of the 100 kHz track-and-hold ripple, a 100 kHz notch filter followed the sample circuit at the input of a voltage follower, and the output of the follower was filtered with a seven-pole passive equal-ripple filter with zeroes in the filter transfer at 100 kHz and 200 kHz. This guaranteed that the loop contribution to noise would be greatly attenuated at the input to the VCO for noise offset frequencies beyond the loop bandwidth of 10 kHz. These passive filters were designed to provide greater than 100 dB of attenuation at 100 kHz while adding minimal phase shift for frequencies within the loop bandwidth.

VCO Design

In a high performance indirect synthesizer, the VCO phase noise characteristics are very important, since these will determine the output phase noise for offset frequencies greater than the loop bandwidth. In signal generator applications, the phase noise in a 1 Hz bandwidth measured several hundred kilohertz away from the output frequency is an important performance specification when making adjacent channel receiver tests, and the VCO will determine the noise "floor" in making this measurement. In addition, the gain constant in megahertz-per-volt for a varactor-tuned VCO is naturally quite nonlinear, requiring loop-gain compensation dependent upon programmed frequency, or selection between multiple overlapping-range narrow-band VCOs to cover a wide bandwith of output frequencies. The generalized phase noise plot shows that for low offset frequencies the noise in a 1 Hz bandwidth falls at –9 dB per octave, and for large offset frequencies is flat (0 dB/octave) with frequency. In between these regions there may or may not exist a region of –6 dB per octave rolloff. This –6 dB/octave segment can either be a result of the thermal noise level in the resonant circuit of the oscillator $[(NF_{osc}kT)/(2\,P_{osc})]$ increased by the gain of the resonant circuit $[(f_o/2Q)^2]$, or the equivalent thermal noise voltage at the input to the VCO integrated by the gain transfer of the VCO $[f_{vco}/f_{offset}]$. (See Figure 26-5.)

The additional $1/f$ "flicker" phase noise contribution was first explained by Dr. D. Halford at NBS in Boulder, Colorado, in 1967. A chemist by training, he was stymied in obtaining a low open-loop $1/f$ phase noise contribution in crystal oscillators and multipliers for use in his planned hydrogen maser project, and he experimented with all types of devices and circuits to improve the $1/f$ (variance) phase noise close to the carrier frequency. From this empirical data he postulated that all transistors had about the same intrinsic $1/f$ phase noise performance (–110 dB to –120 dB) when extrapolated to 1 Hz offset from the signal frequency, for which he invented the notation L(1). If this was the case, then only local degenerative feed-

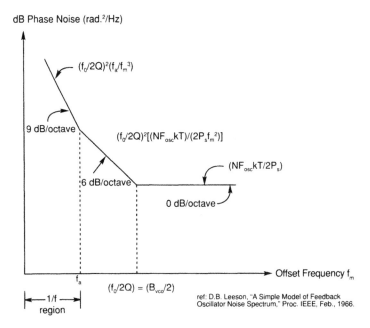

Figure 26-5.
Model of
oscillator noise.

dB Phase Noise (rad.²/Hz)

$(f_0/2Q)^2(f_a/f_m^3)$

9 dB/octave

$(f_0/2Q)^2[(NF_{osc}kT)/(2P_sf_m^2)]$

$(NF_{osc}kT/2P_s)$

6 dB/octave

0 dB/octave

Offset Frequency f_m

f_a

$(f_0/2Q) = (B_{vco}/2)$

1/f
region

ref: D.B. Leeson, "A Simple Model of Feedback
Oscillator Noise Spectrum," Proc. IEEE, Feb., 1966.

back could suppress it. By adding emitter degeneration he was able to demonstrate
improvements of more than 30 dB and consequently discovered a method to improve the short-term stability of all future atomic standards and crystal oscillators.

In the VCO circuit this $1/f$ phase noise component is increased in amplitude by
the factor VCO gain = $(f_{vco}/2Q)^2$. For $f_{vco} = 5 \times 10^7$, $2Q = 100$, the gain is 114 dB,
and the extrapolated data at a 1 Hz offset frequency from the carrier is +3 dB. This
calculates to an open loop $1/f$ flicker noise value of $L(1) = -111$ dB, which is in
agreement with the results above. (See Figure 26-6.) The data also indicate that for
VCOs the phase noise for small offset frequencies from the output frequency can
only be reduced by local degenerative feedback or a high-Q in the resonant circuit.
It is also very desirable to maintain a low VCO gain-constant, since sensitivity to
thermal and power line noise at the VCO input is reduced. This can be accomplished
by using a varactor to tune over a narrow bandwidth of a mechanically tuned
high-Q cavity, or by switching between many narrow-band VCOs to cover a wider
range. Because of an imposed requirement for phase-continuous agile frequency
sweeping, both of these techniques were ruled out in the Digiphase design.

The use of a single wide-bandwidth VCO presents many conflicting requirements. Mechanically rigid inductors with total magnetic shielding tend to have less
than optimum Q, and at the time very high-Q varactors did not exist. In addition,
mechanical shock and radiation susceptibility were also mandatory considerations.
With a gain constant of 3 MHz-per-volt at the input to the VCO, it only takes a few
nanovolts of induced ground loop voltage at 60 Hz to induce a −80 dBc sideband.
Mechanical vibration from fans and audio frequency noise can easily produce much
larger effects if the electrostatic and magnetic fields surrounding the resonant circuit
are not mechanically rigid in space. In an early experiment using a VCO inductor
shielded only by a large aluminum can, vibrations from aircraft taking off at a local
airport were clearly observed, and microphonic noise from voices in the laboratory
were clearly detected by an FM receiver. To minimize these effects, a coil form
which totally enclosed the coil in ferrite was used, and this in turn was mounted in a
close-fitting metal can. The coil itself was sealed in place to reduce the possibility
of mechanical motion with respect to the form. (After a bad experience in produc-

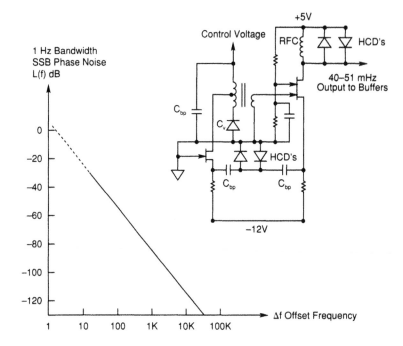

Figure 26-6.
Plot of the VCO
phase noise.

tion, tests were added to assure the absence of microcracks in the ferrite as a result of assembly, since cracks such as these were traced to units with rogue phase noise performance.)

Ground-loop-induced voltages from magnetic flux splatter caused by power tranformer core saturation were minimized by specifying the maximum field to be 8 kG and the core to be made from a good grade of silicon steel. A mu-metal shield was tooled (mu-metal cannot be formed after it has once been annealed) to trap as much as possible of the remaining residual flux. Three electrostatic shields were used between the primary and secondaries to return current from line common mode voltage and winding voltage imbalances.

Loop Design

Many references are available on the subject of phase-lock loop anaylsis, but the requirement for low DC error at the output of the phase detector leaves few options. The loop dynamics were kept quite straightforward to avoid nonlinear out-of-lock behavior. The VCO provides output phase-rate proportional to input voltage, and as such is equivalent to an integrator with transfer function K_1/s, where K_1 is the gain in (radians/volt-second). To more readily calculate gain at loop frequencies, the gain-constant is also described in units of hertz per volt, which provides the transfer function gain by dividing K_1 directly by the loop frequency in units of hertz.

Just as in an op-amp rolloff analysis, with one (VCO) pole already at the origin, and with an additional pole near the origin required for high DC gain, the solution is to add a zero in the transfer safely in-band from the loop zero-dB frequency, taking into account nonlinear VCO gain, phase shift within the loop bandwidth added by loop filters, and low noise. (See Figure 26-7.) This was realized by adding the zero in the feedback path of the loop transimpedance amplifier, ultimately made up of the loop integrator and track-and-hold buffer amplifier.

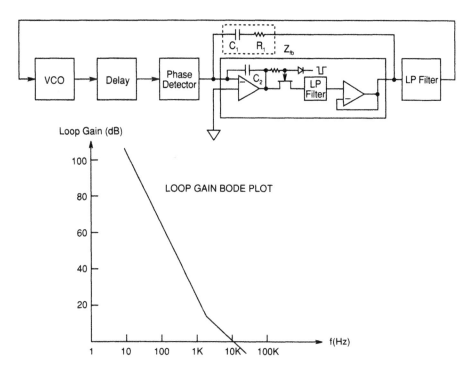

Figure 26-7.
Block diagram of
the PLL and a
bode plot of its
gain.

The summing-node at the loop amplifier input is a very critical point in the loop, since it must not transmit AC ripple or excess noise at the VCO input. At the summing node there must be high loop-gain bandwidth which contains no low-frequency zeroes in the loop-gain transfer that might contribute sample-to-sample memory storage problems. A low $1/f$ noise voltage high-beta pnp bipolar differential input stage with optimum noise resistance approximately matched to that of the feedback resistor was selected. This allowed a first-stage emitter current to be high enough to avoid a Darlington input and maintain a high transistor f_t. Since power supply rejection and low DC temperature drift performance were also required, it was designed as a differential amplifier with high common-mode rejection. Early op amps were ruled out because of noise, settling time, bandwidth, and output resistance. The dynamic summing-node and amplifier output impedance were required to be low enough to absorb the 1 nsec rise-time 6 mA pulses of current from the phase detector output without pulse-to-pulse memory or loop in-band fractional frequency ripple in the voltage presented to the VCO. A loop-gain 0 dB frequency of 10 MHz allowed margin for stable operation with existing printed circuit layout dimensions. Care was taken to create a star ground system to reduce injected noise in the output voltage. An attempt was made to connect (reference) the "cold" end of the varactor in the VCO to the loop summing-node to reduce the effects of summing-node voltage noise. The requirement of bypassing the VCO tuned circuit current from the summing-node and the question of where to reference capacitors at the input of the track-and-hold amplifier made this impractical.

When these low-level noise contributions were finally visible much later in the project, a lower limit on the TAC spurious compensation was observed which was perplexing to say the least. Ultimately it was discovered that the slight difference in time-of-arrival of the compensating current pulses between the phase detector and the TAC created a fractional frequency ripple voltage on the output of the loop amplifier, which resulted in a minimum of –54 dBc output spurious. The track-and-hold buffer amplifier was then added in series with the output of the loop amplifier to assure transfer of only the resultant output difference after nonsimultaneous cancellation. The loop amplifier was converted into a high gain integrator to minimize sample-to-sample memory, and the track-and-hold was designed to hold during the time interval of the current pulses. An "m-derived bridged-T" filter with the zero frequency set for 100 kHz was used in place of the hold capacitor, and the "on" resistance of the sample FET was used as the filter input resistance to minimize thermal noise. A low input noise resistance voltage-follower provided low sample ripple and a low output resistance to drive the main loop transimpedance feedback and VCO input filter. The $(kT/C)^{0.5}$ noise voltage of the hold capacitance was still significant, but it was reduced somewhat by the amount of minor-loop gain afforded with the loop amplifier connected as an integrator. Even so, the peak step voltage on the output of the loop amplifier for a full cycle (6.28 radians) of fractional frequency phase compensation was 50 mV. A –80 dBc sideband (relative to one radian^2) would only require 1.6 µV of fractional frequency bounce at the amplifier output due to nonlinear or memory effects at loop frequencies below 10 kHz.

The seven-pole equal-ripple low-pass filter was inserted in series with the VCO input to remove out-of-band thermal noise and any 100 kHz ripple. The filter was designed with zeroes set at 100 kHz and 200 kHz, and contributed a minimal amount of phase shift within the 10 kHz loop bandwidth. The filter was designed with a 50 Ω impedance to minimize thermal noise. Filter coil forms with high relative permiability ($\mu_r = 5000$) and physically similar to those used on the VCO provided good mechanical integrity.

Loop dynamics were established by the transconductance (Gm) of the phase detector, the transimpedance zero at a time constant of $R_1 C_1$, and the gain of the VCO of 3.5 MHz/V. C_2 was adjusted for minor-loop stability to be 20% of C_1. After many iterations of repackaging all the components in the loop for improved grounding and shielding, a second fractional frequency spurious floor of approximately –66 dB was reached, which resisted all attempts to improve it. After many months of rechecking every element in the loop for its contribution to this floor, no improvement was observed. All experiments were done without the direct benefit of instrumentation, since oscilloscopes were useless in observing the small voltage nonlinearities which might cause such an error, and any intrusions in the loop were unacceptable. Amplifier loop dynamics were modified many times without improvement in the fractional frequency spurious. Thermal effects in the final clocking of the phase detector flip-flops were postulated, and if present would have been impossible to change.

Microwave oscillations were suspected but were not observed. These had been discovered earlier in the discrete bipolar transistor switch drives used in the many hot carrier diode current switches and were traced to using 39¢ consumer type transistors which were unsafe at any speed. A careful search with a small probe and a gigahertz-bandwidth spectrum analyzer yielded nothing. A "calibrated finger" test was applied as well to the tops of all transistor cases with no change in results. This had proven to be as effective as the spectrum analyzer in the earlier investigations.

Enclosures were tightened for RF shielding, and the ground inductance on all interconnect coaxial cables was reduced. At one time ground loops between modules had been a serious cause of interaction, but these were removed with use of a thick aluminum baseplate. Power supply isolation between modules was improved by installing feedthrough capacitors and RF chokes at the exit point of all power connections inside modules.

All of the loads driven by the VCO, which included the digital logic, the two output mixer-amplifier-filter-attenuator paths, and two independent clockings of the final phase detector edge, were all independently buffered by four stages of high-level dual-gate MOSFET amplifiers in each path. These were ultimately packaged in independently shielded high aspect ratio boxes to maximize physical isolation from output to input. The purpose of these buffers was to assure that the reverse transfer from any of the VCO load circuit currents could not in any way corrupt the axis crossings of the VCO output. If the buffers were less than perfect it would be possible for the fixed frequency load circuits to interact with the VCO (output) frequency circuits, and mix together to create a minute (0.2 psec) pulling at the output of the phase detector. In early primitive versions of the packaging, a large amount of pulling was observed, and it was not until all exposed antennas were eliminated and the buffers installed that low spurious phase sidebands were possible.

The output signal at each stage of these buffers was transformer coupled (using trifilar #30 wire on small "3E2A" high-u toroids) and voltage limited with back-to-back hot carrier diodes to reduce the possibility of AM to PM conversion of power supply noise. The dual-gate FETs were especially good at minimizing reverse transfer effects. Regular ECL gates and flip-flops were checked and found to provide no more than 15 dB reverse isolation from load voltage effects at 50 MHz, which made them poor buffers. By various substitution experiments it was demonstrated that ultimately there were no discernable loading effects when using the dual-gate FET buffers and waveguide-beyond-cutoff packaging.

After what seemed to be an interminable project delay, an unplanned discovery was made while modifying the loop amplifier rolloff values. Normally the material for capacitor C2 was carefully selected to be a "low" dielectric absorption material such as polystyrene, and in this one experiment it was changed to "high" dielectric absorption dipped silver mica for convenience. The observed spurious from non-linearity in the fractional frequency compensation basically disappeared, and after many substitution experiments with glass, film, and mica dielectrics it was determined that the normally "high" dielectric absorbtion silver mica gave the best results. In retrospect it was determined that when integrating 6 mA current pulses with 1 nsec risetimes, silver mica could indeed be a better approximation to a perfect capacitor. The other materials, which are documented to have much better dielectric absorption at lower frequencies, had "soak" time constants in the region of the fractional frequencies with just sufficient memory to cause a small (5 μV) voltage memory between samples, and empirically the dipped silver mica did not. Finally, after several complete mechanical and electical packaging iterations, and many circuit revisions which spanned several years in time, the fractional frequency component of phase-error was finally observed to disappear into random noise while being adjusted.

Even though it had taken much longer than anticipated, it was tremendously exciting to finally observe some proverbial light at the end of the tunnel. Before that moment indirect experimentation had basically not given any indication of ultimate yielding or progress, and all suspected areas had been unsuccessfully checked many times over for some clue to the problem. The necessity of having the loop locked and all elements in final packaged form for each indirect experiment compounded the problem of diagnosis and the time required to achieve a solution.

Systematic Noise

The first requirement for low systematic noise is a bullet-proof ground system. A (Kelvin) star ground is the only one that is acceptable, and all others present varying compromises. The simultaneous requirements of a low-inductance RF ground and a star ground were best met by partitioning a large aluminum baseplate into sections which supported the ground current return to a common central point in the baseplate for each power supply ground and voltage output load path. The general-purpose logic, loop analog circuitry, VCO and buffers, and fixed-frequency sections were separated physically in modules tooled from aluminum extrusions. The bottom of each extrusion formed a local ground reference, and the baseplate then connected those to a central point in the baseplate from which all power supply voltages were referenced. While there was some possibility of ground current flowing between modules, the connections between them were carefully isolated at line frequencies to minimize this component. In addition, the much lower resistance of the baseplate limited differential voltage drops between modules. This technique allowed a very good RF ground to exist for module circuits and still maintained a good low frequency analog ground for front-panel inputs and outputs, and power supplies. The 50 Ω attenuator in series with the DC to 11 MHz output amplifier was isolated from the front panel to minimize voltage drops in series with the potentially low level RF output ground.

In addition to VCO intrinsic noise, noise can be injected into the output phase of the VCO from virtually any circuit in the phase-lock loop. From experiments which added controlled amounts of ripple to logic power supplies in the final clocking of the phase detector, it was determined early in the project that 1 μV of ripple was all that could be tolerated at line harmonic frequencies. Just to be on the safe side, it was decided to use the same electrical design for all supplies to minimize ripple. Magnetic pickup from the power transformer was minimized with twisted pair sense and force lines on grounds and output voltages, and physical partitioning of the baseplate ground system was used to isolate ground current returns. The power supplies were designed as high performance DC reference buffer amplifiers to provide a very high signal-to-noise ratio and low output impedance.

Shielding

Tooled extrusions provided a convenient and low cost means of electrostatically shielding RF circuits. Only four modules were required in the design, and these were mounted on the baseplate with a tremendous amount of room remaining for a planned future decade extension of frequency to 510 MHz. All RF signals were connected with shielded cables via holes in the bottom of the modules, and power was connected though a D-Series connector with appropriate filtering on the inside of the module. A simple miniature in-line coax connector provided easy disconnect of coax cables without great expense. If runs of over 6 in. were necessary, a snap-on ground clip was provided on the baseplate to provide a low inductance ground on the outside ferrule of the in-line connection. The digital logic section was shielded with sheet metal, since its susceptibility to small amounts of stray EMI was low. This shielding technique seemed to work quite well and provided modular flexibility and ease of repair.

Thermal Noise

All capacitors in the loop can contribute $(kT/C)^{0.5}$ voltage noise, resistors $(4kTBR)^{0.5}$ thermal voltage noise, current generators $(2qIB)^{0.5}$ shot noise current, and the loop amplifier also has an equivalent noise resistance referred to its input. From the previous discussion, the VCO noise will be dominant for offset frequencies greater than the phase-lock loop bandwidth.

The feedback resistor and capacitor become the noise source impedance at the input of the VCO, and the amplifier increases this due to its equivalent noise resistance. The 1 Hz bandwidth single-sideband voltage noise in the 2.2 kΩ feedback resistor is

$$en/(Hz)^{0.5} = (2kTR)^{0.5}$$
$$en/(Hz)^{0.5} = 4.27 \text{ nV}/\sqrt{Hz}$$

The sensitivity of the phase detector output at a 50 MHz VCO frequency is:

$$I_{pd}/\text{radian} = (Q/\text{radian})f_{sample} = [(I_{det} \times t_{vco})/2\pi] \times f_{sample}$$
$$I_{pd}/\text{radian} = 1.91 \text{ }\mu\text{A/radian}$$

The noise current per Hz generated by the feedback resistor is:

$$= 4.27 \text{ nV}/(Hz)^{0.5} / 2200 \text{ }\Omega$$
$$= 1.94 \times 10^{-12} \text{ A}/(Hz)^{0.5}$$

The current phase noise in a 1 Hz bandwidth referred to 1 radian at the phase detector is:

$$\text{phase noise/Hz} = 1.016 \times 10^{-6}$$
$$= -120 \text{ dB/Hz}$$

A noise figure on the amplifier of 2 dB will establish a thermal noise floor of $-120 + 2 \text{ dB} = -118 \text{ dB/Hz}$ for the loop amplifier and phase detector gain.

Shot Noise

In addition, there is a shot noise contribution from the current generators in the phase detector of

$$(2qIB)^{0.5} = 4.4 \times 10^{-11} \text{ A/Hz}$$

Since this is large compared to the thermal noise current of the feedback resistor, it was necessary to minimize the duty cycle of this current duration at the sample rate. By gating the current for only a small fraction of the total sample period, we reduced the amount of the phase detector shot noise at the summing-node by an order of magnitude to 4.3×10^{-12} A/Hz, which added another noise floor at approximately -113 dB/Hz in the loop.

VCO Noise Contribution

At 10 kHz offset frequency the measured VCO phase noise is -115 dB/Hz. For offset frequencies greater than 10 kHz the noise decreases at 9 dB per octave until bottoming out at a level determined by the signal power level:

$$\text{Single sideband phase noise to signal} = (NF_{vco})(kT/2)P_{vco} \text{ where}$$
$$NF_{vco} = (\text{VCO} + \text{buffer Noise Figure}) = 6 \text{ dB}$$
$$P_{vco} = \text{VCO signal power level} = 0 \text{ dBm}$$
$$kT/2 = -177 \text{ dBm/Hz}$$
$$\text{SSB Phase noise to signal} = 6 - 177 - 0$$
$$= -171 \text{ dBm/Hz (relative to 1 rad}^2 \text{ per hertz bandwidth)}$$

For offset frequencies below 10 kHz, phase noise increases at 9 dB per octave. Between 10 kHz and 3.3 kHz, the loop gain increases at 6 dB per octave and the VCO phase noise increases at 9 dB per octave, producing a net increase of 3 dB per

Figure 26-8.
Closed loop
phase noise
performance.

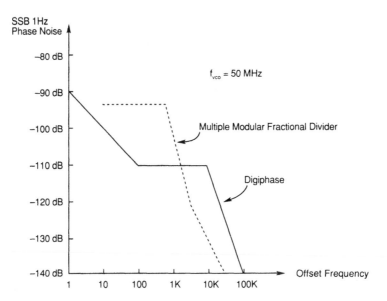

octave in phase noise contribution from the VCO. At 3.3 kHz the resultant loop phase noise increases by 5 dB to −110 dB, where for lower offset frequencies the loop gain increases faster than the VCO noise, resulting in an improved cancellation of the VCO noise at lower frequencies by 3 dB per octave until the loop amplifier 1/f noise voltage corner frequency is reached below 100 Hz.

The resultant system phase noise is less than −110 dB/Hz throughout the phase-lock loop bandwidth, typically −113 dB/Hz. (See Figure 26-8.)

For high performance indirect synthesizers, it is necessary to pay attention to noise contributions from all possible sources, since at any one offset frequency there may be several components. Testing the very best receivers requires a large signal-to-noise ratio. Much of the complexity in direct synthesis techniques results from the budgeting of noise contributions from the many elements adding to output phase noise. Even though still complex, the 100 kHz sample-rate single loop indirect technique has considerably fewer contributing factors to manage.

From the above analysis, the best optimization can be obtained by maintaining the loop sample-rate as high as possible and the VCO gain-constant as low and linear as possible to allow maximum Q and power in the resonant circuit. For this reason early mechanically tuned signal generators used high-power oscillators tuned by a high-Q cavity. A superconducting cavity with a Q greater than one million has been utilized for the ultimate in low phase noise.

Analog Search Control

A tough requirement for digitally controlled signal generators is that of providing continuous analog FM control around a digital center frequency from a front panel control, and also providing a linear voltage input for frequency modulation with a programmable full-scale frequency deviation gain. An appendage from the nondigital past, a typical front-panel "search" capability assumed use of a potentiometer to allow tuning through the center frequency of a narrow-band filter while observing the transfer on a wide dynamic range voltmeter. The full-scale change in programmed frequency was decade-ranged to minimize added noise and nonlinearities from the

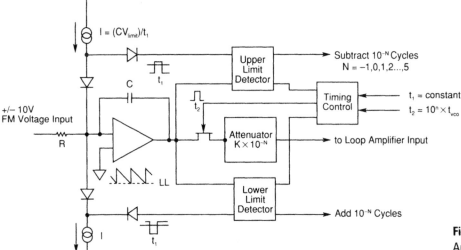

Figure 26-9.
Analog search
FM control.

analog FM voltage control. On direct synthesizers this was typically accomplished by selectively replacing the output of one of the decade units with a linear VCO that covered the output frequency range of the decade. Indirect phase-lock synthesizers had no comparable way to achieve this result.

Using the flexibility of the Digiphase technique, a linear integrating multivibrator was designed which allowed bipolar voltage input control of the output frequency, which was adjusted to be 100 kHz at full scale input voltage. (See Figure 26-9.) The analog FM input voltage was integrated until a full-scale limit was detected in either polarity on the output of the integrator, and, at a time synchronized to that of the 100 kHz sample interval of the synthesizer, the integrator output was reset by a very accurate charge dump at its input summing-node with a polarity opposite that of the input signal and with a gain equal to that required for one cycle of phase modulation to the loop. The DC drift and random noise of the integrator amplifier were minimized by careful wideband balanced differential design.

Using the appropriate integrator reset signal to digitally add/subtract decade-weighted (10^{-N}, where $N = -1,0,1,2,....,5$) increments of cycles in the phase register, the digital value in the register was augmented at a rate proportional to the analog FM control input. This was sufficient to cause the average output frequency to be under control of the FM input, but it also caused phase jumps of exactly 10^{-N} cycles in the synthesizer output each time a reset occurred. To exactly compensate for this, a decade-ranged current derived from the voltage on the output of the integrator was gated for a constant number of VCO cycles and then subtracted from the output of the loop phase detector. This provides an exact analog of the function performed by the TAC. The analog FM input controlled the rate at which phase was increased in the phase register. Phase detector ripple compensation was obtained by integrating the FM input, which then was an exact analog signal representative of phase. As with the TAC, the timing on the pulses was synchronized to occur at the same time as the compensation in the main loop during the hold time of the track-and-hold amplifier.

This resulted in a very linear analog FM control which could symmetrically and continuously change the VCO output frequency around its 1 Hz resolution programmed value. The analog FM bandwidth was that of the loop itself, and the full

scale ranges covered ± 1 MHz to ± 1 Hz. With the ability to add or subtract increments of $\pm 10^{-N}$ ($N = -1,0,1,2,...,5$) cycles per transfer to the phase register, the 1
MHz full scale range covered a frequency range 10 times that of the loop sample
rate. With such a large gain from the analog FM input on the ± 1 MHz range, the
random noise on the output of the integrator raised the loop noise floor considerably. In addition, a slight gain error in the attenuator generated spurious at the integrator reset rate, since the gain on the high (±1 MHz) FM range was 10 times
greater than that of the TAC. For each lower decade-range, this noise was reduced
by 20 dB, until on the ± 10 kHz range and below, negligible added noise was contributed to the main loop performance.

Removing Fractional Spurs

Using recent advances in oversampling A/D conversion it, is now possible to shape
the spectrum of error energy so that fractional frequency ripple energy is pushed out
in offset frequency from the carrier. (See Figure 26-10.) Based on this technology, a
CMOS integrated fractional-N divider has been successfully developed. Reported
by Brian Miller and Robert Conley of Hewlett-Packard at the 1990 Frequency
Control Symposium ("A Multiple Modulator Fractional Divider"), no fractional
spurs were observed using only a simple loop filter and VCO without a TAC.

The technique is based on the sigma-delta modulator technique used in interpola-
tive A/D converters, in which the input is greatly oversampled with a coarse (1 bit)
converter and the result digitally filtered to eliminate out-of-band quantization noise.
The S/N is enhanced by use of recursive filtering to shape the quantization noise
present at the converter output so that most of the noise energy lies outside the band
of interest and is removed during filtering.

A three-stage sigma-delta modulator was used to process the digital phase value
added per sample $[N(k)+f(k)]$ into a new value $[N(k) + N_{div}(k)]$. The resulting
single-sideband phase spectral density is

$N(k)$ = Whole cycle value of phase, where $f(k)$ = Fractional cycle of phase
$$L(f) \approx [(2\pi)^2/12F_{ref}][2\pi f/F_{ref}]^{2(n-1)} \text{ radian}^2/\text{Hz}$$

F_{ref} = loop sample rate, f = offset frequency, m = number of modulator sections.

This phase spectral density appears as colored noise to the loop amplifier and must
be filtered before presentation to the VCO. The topology of the digital processing
reduces to a forward path of accumulators and a reverse path of differentiators. In
this example the sample rate was 200 kHz, and the loop bandwidth was approximately 750 Hz. The results published when compared to the Digiphase noise normalized to 50 MHz show that, above 2 kHz offset frequency, the single sideband
synthesis noise is much higher, although there are no single frequency spurious
components. The normalized VCO noise for the published results is improved 23 dB
over that of the Digiphase, which could be a result of improved Q (190 vs. 50). The
practical implications of eliminating the TAC are significant, since all of the digital
processing, low frequency division, and a microprocessor interface were designed
into one CMOS IC. A 0.5–1 GHz phase-lock loop, including the VCO, was packaged on a single 5 × 6 in. PC board. The processor can be programmed for other
functions, such as specialized sweep and FM or PM. The potential for even higher
integration is possible in the future if required. In this example a frequency of 1 GHz
is achieved with a resolution of one part in 2.38×10^{-11} (0.0238 Hz) with low
noise. These results represent the current optimum for use of digital extensions to
analog frequency generation.

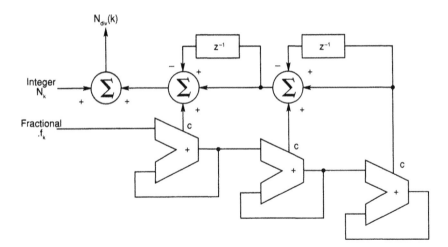

Figure 26-10.
Multiple modulator fractional divider.

Time Synthesizers

In many systems applications it is necessary to program the time of arrival of data presentation. Particularly in high data-rate systems, the time spent in cabling and system delays can easily be multiples of the system clock period. In digital integrated circuit test systems, it is now common to expect the time of arrival of a clock or data edge to be precise within a small fraction of the risetime of a pin driver This may only be the time it takes for a signal to travel 0.3 in. (for $e_r = 4, 0.3$ in. $= 50$ psec) on its way to the device-under-test (DUT). Mechanical cable length and digital/analog delay matching cannot achieve such a low timing skew. Even if this were acceptable, the electrical connection lengths to the DUT cannot be controlled when handling equipment is used.

In digital VLSI test systems it is necessary to control the time of arrival of pin transitions with up to six events-per-pin on 512 pins. In addition, the time of arrival of each edge must be programmable from the device test program on a test cycle basis at cycle rates of 100 MHz. Typical worst case edge placement inaccuracy is ± 100 psec, with a resolution of less than 25 psec. While in principle it would be possible to store program values for every possible edge transition, the calibration time for this is prohibitive. Typically calibration consists of accurately establishing channel timing for each edge at a single point; then using the timing system linearity to achieve accuracy at all other program values. This adds a requirement to the timing system that it must also be quite linear, typically within ± 25 psec for all values, with jitter and noise (no one mentions to within how many sigmas) sufficiently low that the accuracy specification is maintained.

The accurate placement of some 3000 events updated at a 100 MHz rate is a good application for a digitally extended time synthesizer. Time synthesis is direct synthesis, since time delay can be generated by a fixed frequency source (typically a crystal oscillator) or by using analog delay lines. The time synthesis technique is quite similar to that for Digiphase frequency synthesis: A digital calculation is continuously updated to generate a very accurate digital value for each of the desired sequential edge transition times (referenced to the start of the test). Using a counter for integer cycles of delay of a reference clock, and adding a variable delay for interpolating between cycles, each output analog edge is created at a time delay consisting of a sum of integer clock cycles from the beginning of the test, followed by the appropriate analog interpolation delay necessary to cause the edge to occur at its calculated time. (See Figure 26-11.) The updated values used to calculate time are obtained from the device program. Typically delay program values are specified from the beginning of the current test cycle, which is a time

$$t_N = \sum_{1}^{N-1} P_n + t_0$$

where P_n = programmed cycle time, t_0 = time at start of test. The time of occurrence of the delay value is then

$$t_D = t_N + D_N$$

where D_N = programmed value of delay in the Nth cycle. The hardware must calculate the t_D values before the beginning of the current test cycle, and then generate the digital value of the time of the delay event by adding the delay value D_N to the most recent cycle start time t_N. The accuracy of the output edge t_D is then determined only by the interpolator, since the integer cycle portion of t_D is derived from a crystal oscillator and has negligible error. The block diagram of the time synthesizer is similar to that of the Digiphase, except the VCO is replaced by a fixed-frequency reference derived from a crystal oscillator. The same calculated fractional component in the time register is used to program an analog delay interpolator (where a full-scale value is one reference clock) instead of the TAC. The update rate of the calculations can now be synchronized to integer cycles of the fixed frequency reference, which helps minimize analog circuit pulling, since the only critical asynchronous edges are contained within the delay generator. The analog delay is typically a tapped lumped element delay line, or a ramp circuit (similar to an oscilloscope sweep circuit) in which a DAC is used to provide a ramp trigger voltage proportional to the time delay from the start of the ramp. An 8-bit resolution in the DAC will provide 25 psec resolution for a 6.4 nsec clock period. A tapped delay line resolution is typically limited to 6 bits for a 31.25 psec resolution at a 2.0 nsec clock period. The tapped delay line is typically integrated in CMOS or ECL as a programmable number of gate delays. To provide resolution of less than one gate propagation delay, fixed capacitive loading of gate outputs is utilized. Since the analog delay value usually changes on every cycle (even for fixed delay program values), the linearity error of the delay will be converted into highly systematic jitter.

For agile timing (timing which can change from cycle to cycle), it is necessary to track time from the beginning of the test sequence, since each test cycle can have a variable length and is not required to begin on a reference cycle. However, in some applications all test cycles can be constrained to begin on a cycle boundary of a fixed frequency reference. This reference can be generated by a nonagile frequency synthesizer for flexibility and resolution. Since all cycle times are multiples of the period of the synthesizer frequency, there are no fractional cycle components in the time t_N, and a divide-by-N counter can by used to determine the duration of P_n for the current test cycle (N). The delay t_D consists of a fixed number of reference cycles and a constant value for the delay vernier. This offers lower opportunity for jitter to occur, since the time relationships of edges are stationary. The test cycle duration can be programmed from test cycle to test cycle, with a resolution equal to that of the synthesizer cycle time. The maximum synthesizer frequency is typically adjusted to be 4–5 times that of the fastest DUT test cycle rate in order to limit the percentage bandwidth it must provide to ultimately cover an octave range at the fastest DUT test cycle rate. Even so, this range is typically 25%, which requires the analog delay circuits to be frequency independent if recalibration is to be avoided with changes in cycle time. Therefore, even though low jitter results, due to stationary-in-time analog delay values, the accuracy of the resultant edge may not be improved.

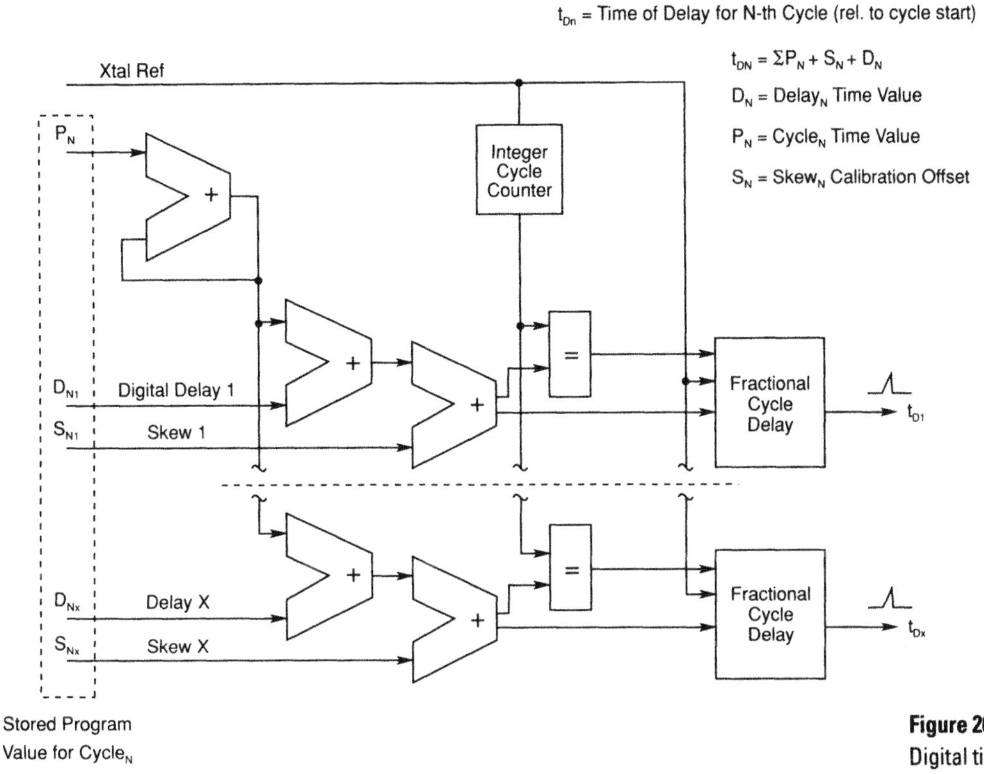

t_{Dn} = Time of Delay for N-th Cycle (rel. to cycle start)
$t_{DN} = \Sigma P_N + S_N + D_N$
D_N = Delay$_N$ Time Value
P_N = Cycle$_N$ Time Value
S_N = Skew$_N$ Calibration Offset
Xtal Ref
P_N
Integer Cycle Counter
D_{N1}
Digital Delay 1
S_{N1}
Skew 1
Fractional Cycle Delay
t_{D1}
D_{Nx}
Delay X
S_{Nx}
Skew X
Fractional Cycle Delay
t_{Dx}
Stored Program Value for Cycle$_N$

Figure 26-11. Digital time synthesis.

Lightning Empiricism

To digress from the analysis above, it is helpful to carry around a few numbers and relationships which keep coming up over and over again. While a SPICE analysis can supply the same information much more accurately, it is also expedient and in reality necessary for analog designers to be able to obtain similar results without computer assistance. Such a case may occur when diagnosing an apparent anomalous behavior online or in investigating the most promising of many design alternatives. In the generation of analog designers before SPICE, this was the only alternative, and quite often the very best designers were also most proficient in what Philbrick made famous as "lightning empiricism." What is meant by this term is the ability without assistance to determine a numerical value or solve a circuit design problem with acceptable error in minimum time. It would probably surprise many analog engineers who have never been required to exercise this skill how fast and accurate the neural network computer on their shoulders can be. Those fortunate enough to have been mentored by the pre-SPICE generation know how embarrassing it was to be stuck on a design problem and have a lightning empiricist walk by, pause for a second, point out the solution, and continue on. In response to this drill it was incumbent upon new designers to also develop this skill, which required carrying around a few numbers and relationships common to the analog world. Such elementary tools as Bode plots, *s*-plane analysis, and circuit analysis were already presumed.

- Numbers (to nearest 1% digit):

Square root of decimal digits 2–10 and decade values 10–90.

- Decibel Conversion:

The voltage ($20 \log_{10}$) decibel values for decimal numbers 1–10, noting that small decibel values are linear (0.1 dB = 1.16%, 0.01 dB = 0.115 %). The decimal factors for decibel values from 1–20.

Example: Converting between binary bits and decimal numbers using the approximation 2 -> 6 dB (6.0206 dB, more precisely): 6 dB $\times$ number of bits is decibel value of the binary power of 2.

Example: How many average detections between single errors with a 2^{-64} error rate? 64 bits $\times$ 6 dB/bit = 384 dB = + 4 dB + 380 dB = 1.6×10^{19}

A precise answer using a calculator is $2^{64} = 1.845 \times 10^{19}$. A +15% scale factor error may be quite acceptable if it's a long walk back to that calculator.

- General:

$k = 1.38 \times 10^{-23}$ (Boltzmann's constant, joule/K) $k_1 = k/q = 8.62 \times 10^{-5}$
(Boltzmann's constant, eV/K) "room" T = 300 K (adjusted for convenience to 27 °C) $q = 1.6 \times 10^{-19}$ (coulombs per electron)
8.7 dB/neper (for converting time constants to percent)
$kT/q = k_1 T = 26$ mV at T = 302 K, or 25 mV at T = 290 K
thermal noise floor (0 dBm ref. = 1 mW) = –174 dBm/Hz
noise resistance = $(60\ \Omega \times e_n{}^2)$ per hertz, given: e_n in nV/(Hz)$^{0.5}$

- Bipolar junction devices:

3 mV/dB = 26 mV/neper = 60 mV/decade (V_{be} or V_j change per unit change in I_c or I_j)

TC of offset voltage for two junctions:
$= (V_{j1} - V_{j2})/\text{Tabs}$ (V_{j2} = reference voltage)
$= \Delta V/300 = 0.33\%\ \Delta V/°C$
= 1 µV/°C for each 0.3 mV of voltage offset
TC of junction voltage = $V_j/°C = -V_j/\text{Tabs}$
$= -0.33\%\ V_j$ (for Tabs = 300 K)
$= -2$ mV/°C (for V_j = 0.60 V)
TC of transistor base current = $(\Delta I_b/I_b)/°C \approx -1\%/°C$
(approx., for low level operation)

- Junction FETs:

gm = $2I_d/(V_p - V_g)$ where V_p = pinch-off voltage
I_d = drain current (in saturation) V_g = gate voltage
TC of $I_d = (I_d/I_d)/°C \approx + 0.7\%/°C$
TC of $V_g = (V_g/°C) \approx -2.2$ mV/°C
Zero TC bias point:
gm (–2.2 mV/°C) + (0.7% I_d)/°C = 0
$I_d/gm = (2.2 \times 10^{-3})/(7 \times 10^{-3}) = 0.315$ V
$V_p - V_g = 2I_d/gm = 0.63$ V

- Settling time of DC amplifier with zeros in the loop gain transfer:

There is a step function settling time residue at the time constant of the zero with a fractional amplitude equal to the reciprocal of the loop gain at the zero frequency.

For example: Amplifier loop gain is rolled off with two poles close to the origin and a zero placed at a time constant one hundred times that of the 0 dB loop gain time constant. Loop gain at the zero time constant is 40 dB. This will result in a 1% residue settling error at the time constant of the zero.

If there are pole zero pairs which do not quite cancel in the loop gain transfer, then the percentage of the zero not cancelled will settle on the zero time constant as above.

References

Best, R., "Phase-Locked Loops: Theory, Design, and Applications," McGraw-Hill, 1984, pp. 221–229, (Fractional-N Phase-Lock Loop).

Blachowicz, L., "Dial Any Channel to 500 MHz," *Electronics,* May 2, 1966, pp. 60–69.

Braymer, N., "Frequency Synthesizer," U.S. Patent No. 3,555,446, filed Jan. 1, 1969, granted Jan. 12, 1971.

Halford, D., "Phase Noise in RF Amplifiers and Frequency Multipliers," memo to J. Barnes, NBS Boulder, October 1967, and personal communication at NBS Frequency and Time Stability Seminar, 1967.

ibid

"Flicker Noise of Phase in RF Amplifiers and Frequency Multipliers: Characterization, Cause, and Cure," Proc. 22nd Annual Symposium on Frequency Control, April 1968, pp. 340–341.

Gillette, G., "Digiphase Synthesizer," Proc. 23rd Annual Symposium on Frequency Control, May 1969, pp. 201–210.

ibid

"The Digiphase Synthesizer," *Frequency Technology,* August 1969, pp. 3–7.

ibid

"Frequency Synthesizer System," U.S. Patent No. 3,582,810, filed May 5, 1969, granted June 1, 1971.

ibid

"A Simple Technique for Analog Tuning of Frequency Synthesizers," Proc. *IEEE* Trans. Instrumentation & Measurement, June 1990, p. 550.

Leeson, D.B., "A Simple Model of Feedback Oscillator Noise Spectrum," Proc. *IEEE,* Feb. 1966, (Short Note).

Mannassewitsch, V., "Frequency Synthesizers," John Wiley, 1987, pp. 43–48, 501–503, (Fractional-N Phase-Locked Loop).

Miller, B., and R. Conley, "A Multiple Modulator Fractional Divider," Proc. 44th Annual Frequency Control Symposium, 1990.

Tykulsky, A., "Spectral Measurements of Oscillators," Proc. *IEEE,* Feb. 1966, p. 306.

27. Some Practical Aspects of SPICE Modeling for Analog Circuits

There are several circuit analysis programs available for computer simulations but the most useful, and also the one most widely used, is SPICE (*S*imulation *P*rogram with *I*ntegrated *C*ircuit *E*mphasis). The SPICE program was developed at the University of California, Berkeley, in 1975 [1] and has had additional refinements by the Berkeley staff and graduate students since that date. Most of the present-day versions of SPICE are based on the Berkeley 2G.6 release, although more detailed updating (mostly of MOS device models) is available with more current SPICE 3 versions. Since SPICE was developed with government grants, it is a public domain program and is available to the user without cost or licensing.

The basic Berkeley SPICE program has always suffered from several problems. The program was (and still is) poorly documented for use by the uninitiated; it takes considerable trial and error skill to obtain good simulation results. Also, convergence has always been a problem, particularly with transient analysis as well as any analysis involving JFETs. The graphical printouts and curves are also quite rudimentary, although more recent 3B versions have improved on graphing with the inclusion of a built-in postprocessor program (Nutmeg).

Most users of SPICE really prefer to utilize one of the several commercial versions available from a number of companies in the private sector. Although generally expensive, most versions have been adapted for use on the personal computer (PC). Some of the more widely used programs available are ALLSPICE, HSPICE, HPSPICE, IS-SPICE, PSPICE, RADSPICE, and ZSPICE, as well as others designated by some predescriptor form as—SPICE. I personally prefer the PSPICE[1] version, as it has excellent documentation along with much improved convergence properties. Further, from an academic viewpoint the PSPICE program is particularly advantageous as a teaching aid, since a student version (adequate for up to 10 transistors) is available from MicroSim Corporation at no cost to the student. All SPICE programs allow DC, AC, and transient analysis of a circuit but also include nonlinear simulations as well as noise, distortion, sensitivity, and transmission line calculations. Of particular usefulness in the design of analog integrated circuits (ICs) is a Monte Carlo statistical simulation program, which allows the circuit designer to evaluate how tolerance spreads on resistors, capacitors, and transistors will affect circuit performance if "*n*" units of the chip are constructed. The Monte Carlo subroutine is available with PSPICE, as well as several other of the commercial SPICE programs. Table 27-1 indicates the types of analysis available with SPICE and PSPICE, while Table 27-2 provides a listing of the various active and passive elements that can be included in a SPICE program.

At the onset it should be stated that simulation results are *only as good* as the

---

1. Registered trademark. Available from MicroSim Corporation, Irving, Calif.

Table 27-1 SPICE Analysis

Analysis Type	Statement
AC	.AC
DC (includes sweeping for device transfer characteristics)	.DC
Fourier analysis	.FOUR
Monte Carlo statistical analysis (PSPICE)	.MC
Noise (.AC analysis must be specified)	.NOISE
DC bias points for devices	.OP
Output plotting	.PLOT
Output printing	.PRINT
Sensitivity	.SENS
Parametric sweep (PSPICE)	.STEP
Analysis at different temperatures	.TEMP
Transfer function (small-signal)	.TF
Transient	.TRAN
Analog behavioral modeling (PSPICE)	VALUE={<....>}
Worst-case sensitivity (PSPICE)	.WCASE

fidelity of the device model and the correlation between the model parameters and the actual device. It is absolutely essential that the SPICE user employ the best data available in the program. Fortunately, most of the semiconductor IC companies have developed the SPICE models for their processes in sufficient detail to allow good correlation between simulated chip performance and final device results.

An analog circuit designer can usually estimate the performance of a circuit with hand calculations, and for simple linear circuits with a few transistors hand calculations may be sufficient. However, for more complex circuits it may be impossible to readily predict performance, particularly if nonlinear behavior is obtained as the transistors move from cutoff through active to saturation conditions. Usually, our hand calculations can only assume small-signal operation, whereas computer analysis can allow a very complete simulation of total operation, including nonlinearities. Of *essential importance* is the use of SPICE to estimate circuit performance *before* the circuit is ever assembled on the bench or on the IC fabrication line. Computer analysis is quite inexpensive, particularly when compared with the cost of reworking an assembled printed circuit board, and certainly very inexpensive compared with the cost of redesigning an IC chip. Conceptually the design of a multitransistor circuit, either analog or digital, should really involve *three* fundamental steps, which in order of procedure should be (1) a basic design concept using hand calculations, (2) circuit simulation via SPICE, and (3) a prototype assembly of the circuit on the bench. For applications such as a large monolithic IC circuit, it may be that a bench prototype is impossible with a "working silicon" requirement for the initial IC fabrication being imposed. This latter case absolutely requires very careful SPICE simulation with both Monte Carlo analysis, as well as simulations with not only typical, but worst-case models for the circuit elements as well.

This chapter is not intended to be a thorough description of SPICE programming. Instead, since this book is primarily intended for a knowledgeable technical audience, it is assumed at the onset that the reader is already familiar with the fundamentals of SPICE usage. For a more complete description of SPICE format, the *SPICE User's Guide* is available as part of the Berkeley program, as well as the *Circuit Analysis Guide for PSPICE*, available from MicroSim Corporation. There are also several good texts devoted to the use of SPICE, or PSPICE [2–7].

Following are some quite practical considerations involving the use of SPICE, as

Table 27-2 SPICE Elements

Type of Element	Example
Passive Elements	
Resistor	RXYZ
Inductor	L2
Capacitor	CBYPASS
Transformer (inductive coupling)	KXFMR
Transmission line	T1
Lossy RC transmission line (SPICE3B)	URC
Independent voltage source	VINPUT
Independent current source	I12
Voltage-controlled voltage source (VCVS)	EA2C
Voltage-controlled current source (VCIS)	GM1
Current-controlled voltage source (ICVS)	HZT
Current-controlled current source (ICIS)	F42
Active Elements	
Diode	D1N4004
Bipolar transistor	Q2N3904
Junction FET	JPROC92
MOSFET	MOSN1
Gallium-arsenide FET (MESFET)	Bxxx(PSPICE); Zxxx(SPICE3)
Digital simulator (PSPICE)	Nxxx; Oxx; Uxxx
Voltage-controlled switch	SW1
Current-controlled switch	WZT
Subcircuit	Xxxx and .SUBCKT
Pole-zero network analysis (SPICE3)	.PZ

well as several examples of modeling to demonstrate the versatility (and often frustrations) of analog circuit simulation.

Some General Requirements

In analysis for both DC and transient solutions, an iterative process is used to converge to a solution of all voltages and currents. The convergence algorithm is satisfied when both of the following conditions apply:

1. Kirchoff's current law sums to zero at each node to within a tolerance of 0.1% or 1×10^{-12} A, whichever is *larger*. Thus, if a JFET (or MOSFET) has a gate current of 0.5 pA, the value predicted by SPICE may be incorrect. This limitation can be removed by using the .OPTIONS statement for ABSTOL, with a statement such as ABSTOL = 1E-14 requiring a current accuracy of 0.01 pA. Similarly, the statement RELTOL = 1E-4 would require the relative accuracy of current summation to be 0.01% rather than 0.1%. Another problem that can be encountered in correctly simulating picoamp gate currents of FETs is the SPICE allocation of a fixed conductance (GMIN) between nodes in a circuit, equal to 1×10^{12} mhos (i.e., a fixed resistor between nodes of $1 \times 10^{12} \, \Omega$). Thus, if the drain of a JFET is at +10 V, and the source of the JFET is at +1 V, with the gate at 0 V, then the SPICE program will obtain a gate leakage current of $10 \text{ V}/10^{12} \, \Omega + 1 \text{ V}/10^{12} \, \Omega = 11$ pA, irrespective of any leakage through

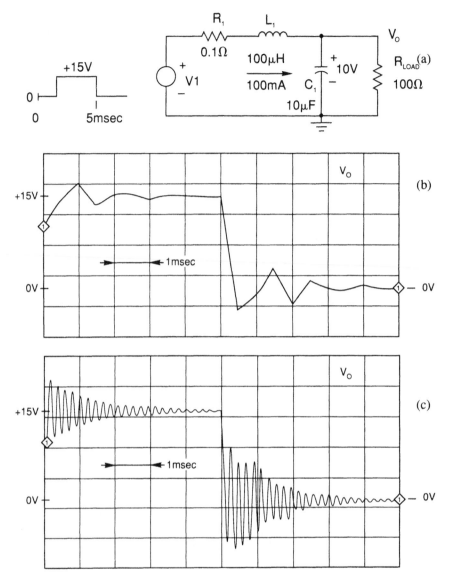

Figure 27-1 (a–c).
(a) Example of a
RLC circuit.
(b) SPICE tran-
sient output with
an internal time-
step of 200 μsec.
(c) SPICE tran-
sient output with
timestep
= 50 μsec.

the gate-source and gate-drain reverse-biased diodes of the FET. To cure this
problem, it would be wise to change GMIN = 1E-14 for such a circuit.

2. Kirchoff's voltage law is correct, so that node voltages converge to within a
tolerance of ± 0.1% or 1 μV, whichever is larger. The voltage accuracy can
also be changed by using the .OPTIONS statement for VNTOL, as well as a
relative accuracy change using ABSTOL.

In some cases the SPICE analysis may fail to reach convergence (this is not as
big a problem with PSPICE, fortunately). In such a case the NODESET statement
can be used to force the circuit to converge to a particular set of conditions. Another
useful choice in nonconverging circuits is to relax the relative accuracy of the solution, by defining the .OPTIONS statement with a reduced relative tolerance
(RELTOL). Normally, RELTOL of 0.1% (0.001) is used. Reducing to RELTOL =
0.01 (1%) may hasten convergence. Also, when DC convergence is not obtained,
SPICE will print out the voltages at each node before convergence ceased. By examining the node voltages it will be apparent which nodes are causing the most

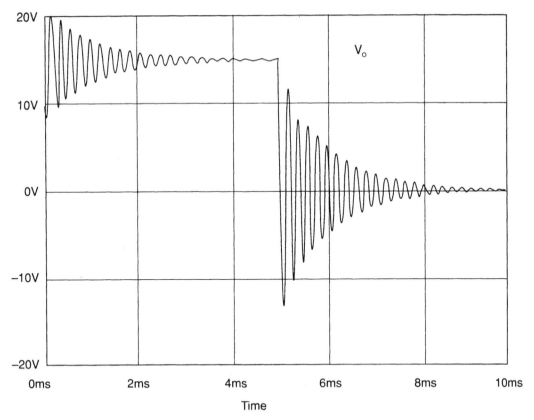

V_o

Figure 27-1 (d).
(d) PSPICE output
for TSTEP =
200 μsec, but
illustrating the
inclusion of a
variable internal
timestep.

difficulty; thus, adding a .NODESET statement for those particular node voltages should aid convergence.

Most problems involving SPICE are usually encountered with transient analysis. This is because an iterative technique is used to evaluate the next point in the time response, with the default ceiling on the internal time step in SPICE being the *smaller* of TSTEP, TSTOP/50, or TMAX, whereas in PSPICE it is the smaller of TSTOP/50 or TMAX, as referred to the general .TRAN control statement,

$$\text{.TRAN TSTEP TSTOP [TSTART [TMAX]]} \qquad (1)$$

Here, TSTEP is the printing or plotting increment, TSTOP is the final time, and (optional) TSTART is the initial output printing time with (optional) TMAX allowing a default ceiling for the iteration time step. As an example of how simulation accuracy is constrained by too large an internal time step, consider the case of the RLC series circuit of Figure 27-1. This circuit is indicative of the output filter network for a switching voltage regulator. The circuit has a resonant frequency of $1/2\pi\sqrt{LC} \approx 5\text{kHz}$, and a load Q of $R_{\text{Load}}/\sqrt{LC} \approx 32$. Suppose we apply an input 15 V pulse of 5 msec width, with a 10 μsec time delay, with 1 μsec rise and fall times. If we want a reasonable printing interval of 0.5 msec, the .TRAN statement might be

$$\text{.TRAN 0.5E - 3 10MS UIC} \qquad (2)$$

where UIC would refer to "use initial conditions" specified for C_1 and L_1 (i.e., $V_{C1(t=0)} = 10$ V, $I_{L1(t=0)} = 100$ mA. Thus, using a Berkeley SPICE program the minimum internal time step iteration would be the smaller of TSTEP (0.5 msec) and

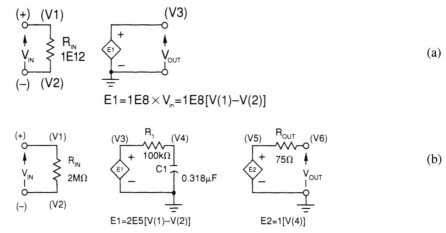

Figure 27-2. Some simple macromodels for op amps. (a) A near-ideal op amp simulation. (b) A basic linear model for a 741 op amp. (c) The Boyle macro-model.

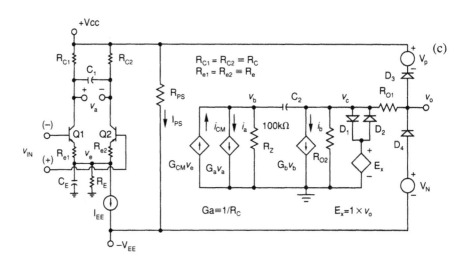

TSTOP/50 (10 msec/50 = 0.2 msec), or 0.2 msec. An observation of the resulting printout in Figure 27-1(b) indicates that insufficient resolution is available to identify the 5 kHz ringing that must occur in this underdamped circuit. However, note that if the time step is reduced to 50 μsec, we begin to see the true ringing response, as shown in Figure 27-1(c). In the PSPICE program a second-order polynomial interpolation is used to determine the internal time step, with a much reduced step used for the fine structure during fast moving transients, and a larger time step (up to TSTOP/50, or TMAX) for time periods where the response is relatively slow changing. For example, the PSPICE output plot from .PROBE (the postprocessor available with PSPICE) for the RLC circuit is shown in Figure 27-1(d), indicating very good resolution during the sudden changes at the output.

Modeling Operational Amplifiers and Comparators

Although SPICE will reasonably model individual components (as shown in Table 27-2), one must create a subcircuit (.SUBCKT) to model an op amp or comparator. Of course, it would be possible to model the actual op amp by specifying models for each transistor, resistor, and capacitor in the circuit (i.e., an exact *micro-model*) *if* the correct models were known. However, only the manufacturer really has complete models, and these are usually proprietary to that particular company. Moreover, the number of nodes required would be large, leading to extensive CPU time required for analysis. Instead it is usually preferable to obtain an overall *macromodel* for the entire circuit, essentially a "black box" approach for the op amp that correctly simulates the internal performance, as well as the input and outut impedances. The primary goal of any macromodel is indeed to obtain equivalent performance but with much reduced complexity and corresponding simulation time.

There are many op amp (or comparator) models that one could use. For example, if a near ideal op amp were desired, the model of Figure 27-2(a) would clearly approximate a circuit with near infinite input resistance ($R_{in} = 1 \times 10^{12}\ \Omega$), zero output resistance, and a near infinite open-loop gain ($A_{OL} = 100$ million) that is not frequency dependent. Only three nodes, plus a ground node, are required. For simple linear circuit analysis, the model for a 741 op amp would extend the simple circuit of Figure 27-2(a) to include a finite open-loop gain (typically $= 2 \times 10^5$), typical input resistance of 2 MΩ, output resistance of 75 Ω, and a single pole equal to the gain-bandwidth product divided by the open-loop gain, or approximately 1 MHz/$2 \times 10^5 = 5$ Hz. The 5 Hz pole is obtained by a simple RC network, or 5 Hz $= 1/2\pi R_1 C_1$. Note now, however, that six nodes plus ground are required.

For an inclusion of nonlinear effects such as input and output voltage limitations, input bias currents, and output slew-rate effects and current limiting, more complex modeling is required. There are two basic approaches to obtaining a more accurate macromodel for opertional amplifiers.

The Boyle Macromodel

The most widely used macromodel is that due to Boyle *et al.* [8], which was introduced in 1974 and is shown in Figure 27-2(c). A later addition by Krajewska and Holmes [9] extended the Boyle model to allow the inclusion of either JFETs or MOSFETs as front-end devices. Several manufacturers have adopted the Boyle model and, in fact, have available on disk SPICE subcircuits for many of their op amps and comparators.[2] Most commercial versions of SPICE also include an internal modeling program that allows the user to enter parameter specifications and obtain a modified Boyle .SUBCKT that can then be used in circuit analysis.

In the model of Figure 27-2(c) the op amp is modeled by a first stage using near ideal transistors Q_1 and Q_2; this stage obtains the desired input bias (and offset) currents of the op amp, DC offset voltage, as well as R_{in} and input common-mode specifications. The inclusion of C_1 allows a second open-loop pole, while C_E allows a second-order slewing effect to occur. For convenience, the first-stage voltage gain is chosen as unity. The intermediate stage is coupled by the main compensation capacitor C_2, which determines both the dominant open-loop pole and the slew-rate for the amplifier. The output stage comprises the output resistance ($R_{O1} + R_{O2}$), as well as short-circuit current limitation (via D_1, D_2, and $E_x = 1.0 \times v_O$). The output maximum voltage range is defined by V_P and D_3 for V_{out} (max.-positive), and V_N and D_4 for V_{out} (max.-negative). The typical quiescent power supply current of the amplifier is obtained with R_{Ps}, or $I_{Ps} = (|V_{CC}| + |V_{EE}|)/R_{Ps}$. The overall model requires 16 nodes plus ground.

2. Several SPICE op amp models are currently available on disks from Analog Devices, Burr-Brown, Comlinear, Harris, Linear Technology, Precision Monolithics, and Texas Instruments.

As an example of the use of the Boyle model, consider the following typical specifications for an LM318 op amp, with ±15 V power supplies:

Small-signal gain (A_{OL}) = 110 dB (3.16×10^5)	Phase-margin (ϕ_M) = 45°
CMRR = 100 dB	Slew-rate (SR) = +115, –80 V/μsec
PSRR = 80 dB(neg.), 85 dB(pos.)	Common-mode input range (V_{ICMR}) = ±11.5 V
$Z_{in} \approx 3 \text{M}\Omega \mid\mid 3 \text{ pF}^3$	Output voltage swing $[V_{O(max)}]$ = ±13 V
Input bias current (I_B) = 150 nA	SC Current (I_{SC}) = ±21 mA
Input offset current (I_{OS}) = 30 nA	Power supply current (I_{Ps}) = 5.1 mA
$R_{out} \approx 100 \ \Omega^3$	Input noise voltage (ENV) $\approx$ 11 nV/$\sqrt{\text{Hz}}$
Offset voltage (V_{OS}) = ±4 mV	ENV corner frequency (f_{bv}) = 100 Hz
Gain-bandwidth (GB) = 18 MHz	Input noise current (ENI) $\approx 2qI_B$ = 0.22 pA / $\sqrt{\text{Hz}}^3$
	ENI corner frequency (f_{bi}) $\approx$ 10 kHz^3

In the op amp configuration of Figure 27-2(c) there are several arbitrary choices made. The differential input stage (which could also be pnp, n-, or p-channel JFETs, or MOSFETs) is balanced with the criteria that $R_{C1} = R_{C2} \equiv R_C$ and $R_{e1} = R_{e2} \equiv R_e$. Further, for convenience the first stage voltage gain is unity, while the voltage-controlled current source (VCIS) G_a is defined by a unity transfer of current $i_a = G_a v_a \equiv (1/R_C)v_a$. Thus, if Q_1 and Q_2 are in linear operation with small-signal currents $|i_{C1}| = |i_{C2}| \equiv |i_C|$ then $|v_a| = 2|i_C|R_C$, or $|i_a| = 2|i_C|$. The voltage gain across the integration capacitor is large since most of the open-loop gain is obtained with the VCIS dependent-generator G_b. Hence when large-signal slewing occurs, the collector current of Q_1 (or Q_2) approaches the current source value I_{EE}, or $i_a = 2I_C \approx I_{EE}$, which is then available to charge capacitor C_2, or for npn inputs the positive slew-rate defines the collector current of Q_1 (and Q_2) as

$$SR^+ = \frac{\Delta V_0}{\Delta t} = \frac{i_{a(max)}}{C_2} = \frac{2I_{C1}}{C_2} \approx \frac{I_{EE}}{C_2} \tag{3}$$

or effectively

$$I_{C1} = I_{C2} = \frac{C_2 SR^+}{2} \tag{4}$$

The negative output slew rate SR^- would be smaller due to charge storage in C_E or

$$SR^- = \frac{2I_{C1}}{C_2 + C_E} \tag{5}$$

For the case of $SR^- > SR^+$ one should use pnp input devices, with Eqs. (3) and (5) reversed. Krajewska [9] suggests a diode in series with C_E for p-channel JFET input op amps to obtain $SR^- > SR^+$.

For convenience, Boyle chose the reverse saturation current (I_{S1}) of Q_1 as 8×10^{-16} A. The offset voltage (V_{OS}) is then defined by the difference of the base-emitter voltages, $V_{OS} = V_{be1} - V_{be2}$. But the V_{be} voltage is related by the Ebers-Moll relationship (the Early voltages of Q_1 and Q_2 are ∞)

---

3. Estimates only. A 10 kHz 1/f noise corner-frequency is indicative of this early (1970s) op amp process.

$$I_c = I_S \exp(V_{be}/V_T) \tag{6}$$

where V_T is the thermal voltage $V_T = kT/q$. Thus the offset voltage defines I_{S2} by

$$V_{OS} = V_{be1} - V_{be2}$$

$$= V_T \ln(I_{S1}/I_{S2})$$

or

$$I_{S2} = I_{S2} \exp(V_{OS}/V_T) \tag{7}$$

The input bias current (I_B) and the input offset current (I_{OS}) define the base currents, and thus the beta terms (β, or h_{FE}) by

$$I_{B1} \equiv I_B + \frac{I_{OS}}{2}, \qquad I_{B2} \equiv I_B - \frac{I_{OS}}{2}$$

$$\beta_1 = \frac{I_{C1}}{I_{B1}}, \qquad \beta_2 = \frac{I_{C2}(= I_{C1})}{I_{B2}} \tag{8}$$

The input stage resistors are defined by the gain bandwidth product and the assumption of unity voltage gain for the first stage. The open-loop f_{-3dB} dominant corner frequency is basically at the interstage node v_b and given by the Miller effect term,

$$f_{-3dB} \approx \frac{1}{2\pi R_2 C_2(1 + G_b R_{O2})} \approx \frac{1}{2\pi R_2 C_2 G_b R_{O2}} \tag{9}$$

But the DC open-loop voltage gain is (with a unity voltage gain for the input stage) related to the gain-bandwidth product GB

$$A_{OL} \approx 1 \times G_a R_2 \times G_b R_{O2} = \frac{GB}{f_{-3dB}} \tag{10}$$

Thus combining Eqs. (9) and (10) with $G_a = 1/R_C$ gives

$$R_C = R_{C1} = R_{C2} = \frac{1}{2\pi(GB)C_2} \tag{11}$$

In a similar fashion the unity voltage gain requirement defines the added emitter resistor R_e as

$$R_C = R_{e1} = R_{e2} = \frac{\beta_1 + \beta_2}{2 + \beta_1 + \beta_2}\left[R_C - \frac{V_T}{I_C}\right] \tag{12}$$

Since the collector currents of Q_1 and Q_2 are now known, the current source I_{EE} is defined by

$$I_{EE} = \frac{I_{C1}}{\alpha_1} = \frac{I_{C2}}{\alpha_2} = \left(\frac{1+\beta_1}{\beta_1} + \frac{1+\beta_2}{\beta_2}\right)I_C \tag{13}$$

The resistance R_E is used to simulate the output impedance of the constant current source furnishing I_{EE}. Boyle [8] suggests a value equal to the Early voltage divided by I_{EE}, or

$$R_E \approx \frac{V_A}{I_{EE}} \approx \frac{200\,\text{V}}{I_{EE}} \tag{14}$$

The remaining capacitor C_1 is used to reflect the phase margin for the op amp, by providing a second pole in the open-loop response due to C_1, R_{C1}, and R_{C2} as

$$f_{P2} = \frac{1}{2\pi(2R_c)C_1} \tag{15}$$

with the added phase shift $\Delta\phi$ at $f = GB$ defined by

$$\Delta\phi = \tan^{-1}(GB/f_{p2}) = \tan^{-1}\left[\frac{\frac{1}{2\pi R_c C_2}}{\frac{1}{2\pi R_c C_1}}\right]$$

$$= \tan^{-1}(2C_1/C_2) = 90° - \phi_M \tag{16}$$

Hence, C_1 is defined by

$$C_1 = \frac{C_2}{2}\tan(90° - \phi_M) \tag{17}$$

Since the input stage voltage gain is unity, then from Eq. (10) the value of G_b is known

$$G_b = \frac{A_{OL}}{G_a R_2 R_{O2}} = \frac{A_{OL}R_c}{(100\text{k}\Omega)R_{O2}} \tag{18}$$

The output resistance for the op amp is the sum of R_{O1} and R_{O2}. At higher frequencies R_{O2} is shorted by C_2, thus one should choose R_{O1} as the high-frequency value while R_{O2} is equal to

$$R_{O2} = R_{out} - R_{O1} \tag{19}$$

Usually, however, a choice is made for R_{O1} based on the voltage drop $I_{SC}R_{O1}$.

In the model an attempt is made to include common-mode effects by the inclusion of the VCIS generator $G_{cm}\,v_e$. Since $R_E \gg R_{e1} + 1/g_{m1}$, then the signal v_e is the same as the common-mode input voltage v_{inCM}. The common-mode signal at v_b is thus

$$v_{bCM} \approx (G_{CM}R_2)v_{inCM}$$

but since the differential-mode and common-mode gains from v_b to v_o are identical, the common-mode rejection-ratio (CMRR) defined by the ratio of open-loop differential gain to common-mode gain is

$$\text{CMRR} = \left| \frac{A_{\text{OL}}}{A_{\text{CM}}} \right| = \frac{1(v_b/v_a)(v_o/v_b)}{1(v_{b\text{CM}}/v_{\text{inCM}})(v_o/v_b)}$$

$$= \frac{G_a R_2}{G_{\text{CM}} R_2} = \frac{(1/R_c)}{G_{\text{CM}}}$$

Then the G_{CM} value is

$$G_{\text{CM}} = \frac{1}{(\text{CMRR})R_c} \tag{20}$$

Current limiting in the output stage is provided by R_{O1}, diodes D_1 and D_2, and the VCVS E_x, with $E_x = 1.0 \, (v_o)$. Thus the positive output short-circuit current will produce a voltage drop across R_{O1}, and if that is equated to the conduction voltage of diode D_1, then

$$V_{D1} = I_{sc}{}^+ R_{O1} = V_T \ln(I_x/I_{SD1}) \tag{21}$$

and similarly for the negative output short-circuit current

$$V_{D2} = I_{sc}{}^- R_{O1} = V_T \ln(I_x/I_{SD2}) \tag{22}$$

The choice of I_x is somewhat arbitrary, since a tenfold increase in I_{D1} or I_{D2} is obtained with only 60 mV change in the value of $I_{SC}R_{O1}$. The current i_b furnished by the VCIS $(G_b \, v_b)$ has a maximum overdriven input value of approximately

$$I_{\text{max}} = I_x + I_{SC} \approx I_{EE} R_2 G_b \tag{23}$$

Thus the maximum current through either D_1 or D_2 is really limited to

$$I_x \approx I_{EE} R_2 G_b - I_{SC}$$

$$= I_{EE} \left(\frac{A_{\text{OL}} R_c}{R_{O2}} \right) - I_{SC} \tag{24}$$

In practice, a reasonable approximation to current limiting is achieved if I_x and I_{SC} are approximately equal, hence the choice of reverse saturation currents for diodes D_1 and D_2 are thus based on Eqs. (21) and (22) as

$$I_{SD1} \approx I_{sc}{}^+ \exp\left(-I_{sc}{}^+ R_{O1}/V_T\right)$$

$$I_{SD2} \approx I_{sc}{}^- \exp\left(-I_{sc}{}^- R_{O1}/V_T\right) \tag{25}$$

Diodes D_3 and D_4, along with fixed voltage sources V_P and V_N determine output voltage limiting as

$$V_{o(\text{max})}{}^+ = V_{CC} - V_P + V_{D3}$$

$$-\left(V_{o(\text{max})}{}^-\right) = -\left(V_{EE} - V_N + V_{D4}\right) \tag{26}$$

Now that the appropriate equations have been obtained, it is merely a matter of inserting specifications for the op amp to obtain the SPICE model. Hence, the following calculations are obtained at $T = 300$ K ($V_T = 0.02586$) using a dominant node integration capacitance of 28 pF, obtained from the LM318 schematic:

$$I_{C1} = I_{C2} = \frac{C_2(SR^+)}{2} \approx \frac{28pF(115V/\mu \sec)}{2} = 1.61mA$$

$$SR^- = \frac{2I_{C1}}{C_2 + C_E}, \text{ so } C_E = 12.25pF$$

Let $I_{SQ1} = 8 \times 10^{-16} A$, so $I_{SQ2} = 8 \times 10^{-16} \exp(4 \times 10^{-3}/0.02586) = 9.34 \times 10^{-16}$

$$I_{B1} = I_B + \frac{I_{OS}}{2} = 165 \text{ nA}; \quad I_{B2} = I_B - \frac{I_{OS}}{2} = 135 \text{ nA}$$

$$\beta_1 = I_{C1}/I_{B1} = 1.61mA/165nA = 9758; \quad \beta_2 = 11,926$$

$$R_{C1} = R_{C2} = \frac{1}{2\pi(GB)C_2} = \frac{0.159}{18MHz \times 28pF} = 315.48\Omega$$

$$G_a = 1/R_C = 3.17 \text{ milliSiemens (millimhos)}$$

$$R_{e1} = R_{e2} = \left(\frac{\beta_1 + \beta_2}{2 + \beta_1 + \beta_2}\right)\left(R_C - \frac{V_T}{I_C}\right) = 299.4$$

$$I_{EE} = 3.22 \text{ mA}, \quad R_E \approx 200/3.22 \text{ mA} = 62k\Omega$$

$$C_1 = \frac{C_2}{2}\tan(90° - \phi_M) = 14pF$$

$$R_{O1} + R_{O2} \approx 100\Omega; \text{ let } I_{SC}R_{O1} \approx 0.7V, \text{ so } R_{O1} = 0.7/21mA = 33\Omega$$

$$R_2 = 100k\Omega; \ G_b = \frac{A_{OL}R_C}{R_2R_{O2}} = \frac{3.16 \times 10^5(315.48)}{(1 \times 10^5)(67\Omega)} = 14.88 \text{ Siemens(mhos)}$$

$$G_{cm} = \frac{1}{CMRR(R_C)} = \frac{1}{(1 \times 10^5)315.48} = 31.7 \times 10^{-9} \text{Siemens(mhos)}$$

$$I_{SD1} = 21mA \exp(-21mA \times 33\Omega/V_T) = 48.3 \times 10^{-15} A = I_{SD2}$$

Since $V_{o(max)} = \pm13V$, then $V_P = V_N \approx 15V - 13V + V_D \approx 2.7V$

$$R_{PS} = 30V/5.1mA = 5.88k\Omega$$

The complete circuit diagram is shown in Figure 27-3 with nodes defined. The 3 pF capacitor is added across the input to give the correct input capacitance. Also, the input resistance for the basic Boyle model is too high, namely

$$R_{IN} \approx \left(1 + \beta_1\right)\left(R_{e1} + \frac{1}{g_{m1}}\right) + \left(1 + \beta_2\right)\left(R_{e2} + \frac{1}{g_{m2}}\right) \approx 6.84 M\Omega$$

Thus the input is paralleled with a 5.34 MΩ resistor to obtain the desired value of $R_{IN} = 3$ MΩ. The SPICE .SUBCKT description for Figure 27-3 is as follows:

```
.SUBCKT LM318      2      3      4      6      7
*                 -in    +in   -VEE   out   +VCC
*Device Char: Aol=3.16E5, CMRR=100dB, Zin=3meg//3pF, Ib=150nA,
*Ios=30nA
*Vos=4mV, GB=18MHz, SR=+115,-80 V/usec, PM=45deg, Output=+-13V,
*SC current=+-21mA, Rout ~ 100 ohms, Ips=5.1mA. ENI~0.22pA/rthz,
*ENI(fb)~10Hz. Total nodes=16 (18 if diode limiters are added).
*- - -this is the Boyle model- - - - - -
RC1   9   7   315.48
RC2   10  7   315.48
C1    9   10  14pF
RE1   12  11  299.4
RE2   13  11  299.4
CIN   2   3   3PF
RADDIN  2   3   5.34MEG
CE 11   0   12.25PF
RE   11  0   62K
IEE   11  4   3.22MA
RPS   7   4   5.88K
GCM   (0 14)  (11 0)   31.7NMHOS
GA    (14 0)  (9 10)   3.17MMHOS
R2    14  0   100K
C2    15  14  28PF
GB    (15 0)  (14 0)   14.88
RO2   15  0   67
RO1   15  6   33
VP    7   18  2.7
VN    17  4   2.7
*
D1    15  16  DA
D2    16  15  DA
D3    6   18  DA
D4    17  6   DA
Q1    9   2   12  QN1
Q2    10  3   13  QN2
.MODEL   DA   D(IS=48.3FA)
.MODEL   QN1  NPN(IS=8E-16   BF=9758   KF=3.2E-15   AF=1)
.MODEL   QN2  NPN(IS=9.34E-16  BF=11926  KF=3.2E-15  AF=1)
.ENDS   LM318
```

The circuit of Figure 27-3 does not adequately include several important features of the LM318 device. The model is inherently limited to two poles by the choice of circuit design. The PSRR is not included, nor is the input common-mode range; the circuit of Figure 27-3 has a positive input common-mode value of approximately $+15 \text{ V} - I_C R_C \approx +14.5$ V (the actual spec should be +11.5 V), while the negative

Some Practical Aspects of SPICE Modeling for Analog Circuits

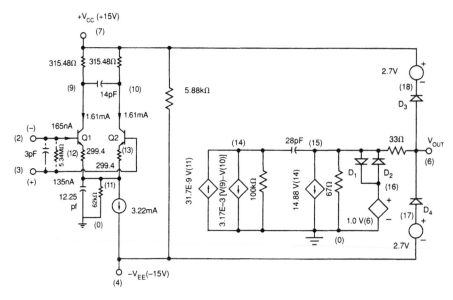

Figure 27-3.
Resulting Boyle
macromodel for
the LM318.

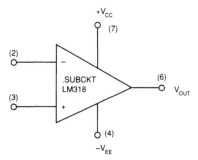

input common-mode voltage is not −11.5 V (as in the LM318 specification) but instead is dependent on both the 62 kΩ (R_E) resistor to ground and the essentially infinite impedance of the 3.22 mA (I_{EE}) constant current source. In reality, the inclusion of the R_E resistor is *not correct*, as it gives rise to an error current produced by the common-mode voltage between the emitter terminal [node (11) in Figure 27-3] and ground. Actually, the modeling of the output impedance of the constant current source driving the emitters of Q_1 and Q_2 should be represented by a resistance in parallel with I_{EE}, and between the emitter terminal and −V_{EE}. This will also require a decrease in the I_{EE} current, since current would now flow through R_E as well; alternatively, one could of course replace R_E by a series combination of R_E and a large capacitor to eliminate the DC current in R_E.

Although the equivalent noise current (ENI) is correct for the Boyle model, one must carefully define the SPICE parameters *AF* and *KF* to correctly obtain the 1/*f* noise corner frequency f_{bi}. The power spectral density for the base current shot noise in SPICE is

$$S_i(f) = i_b^2 \equiv 2qI_B + \frac{KF(I_B)^{AF}}{f}$$

$$= 2qI_B \left[1 + \frac{KF(I_B)^{AF-1}}{2qf} \right]$$

$$\equiv 2qI_{B}\left(1 + \frac{f_{bi}}{f}\right) \tag{27}$$

where the definition can be made that (for AF = 1.0)

$$KF \equiv 2qf_{bi} \tag{28}$$

Thus, in Figure 27-3 the shot noise current sources in parallel with the base to ground for Q_1 and Q_2 would be (in rms amps-squared per Hertz.)

$$i_{b1}^{2} = 2q(165\text{nA})\left(1 + \frac{10\text{kHz}}{f}\right)$$

$$i_{b2}^{2} = 2q(135\text{nA})\left(1 + \frac{10\text{kHz}}{f}\right) \tag{29}$$

where $AF = 1.0$ and $KF = 2q\,(10\text{ kHz}) = 3.2 \times 10^{-15}$. The 5.34 M$\Omega$ parallel resistor will also add noise, which can be treated as a thermal noise current source in parallel with the input of value $i_R^2 = 4kT/5.34\text{M}\Omega$, $or\ i_R = 0.056\text{pA}/\sqrt{\text{Hz}}$, which is really negligible compared to the $\sim 0.22\text{pA}/\sqrt{\text{Hz}}$ noise of i_{b1} and i_{b2}.

The equivalent noise voltage (ENV) of the actual op amp is approximately $10\text{nV}/\sqrt{\text{Hz}}$ with a noise corner (f_{bv}) near 100 Hz. Because the choice of I_{C1} and I_{C2} are based on gain-bandwidth considerations, and the arbitrary choice of R_{E1} and R_{E2} is made to obtain an input stage voltage gain of unity, there can be no real correlation between actual ENV and the ENV of the model of Figure 27-3. It is possible to add series resistance to both inputs in Figure 27-3 to increase the ENV of the circuit, but one must be careful that the added voltage drops produced by I_{B1} and I_{B2} do not change the offset voltage (V_{OS}) for the circuit.

A "Circuits Approach" Macromodel

Many of the deficiencies of the Boyle model are limited by the choice of an input stage voltage gain of unity and the use of transistors in the input stage. A more fundamental circuits approach can be used, where transistors are eliminated, if a model is formed by using mostly passive components along with both fixed and dependent voltage and current sources. The author has found the model of Figure 27-4 to be quite useful for not only bipolar processes, but JFET and MOSFET op amps as well.[4] Further, the model easily allows an extension to multiple poles and zeroes, such as are required for modeling more complex op amps such as the OP27 and OP37.

In the circuit of Figure 27-4 since the offset voltage, common-mode rejection ratio, and power-supply rejection ratio can be defined in terms of an equivalent input voltage source, this effect is obtained by the series connection of sources V_{OS}, ECMRR, and EPSRR, which are defined for the earlier example of the LM318 as

$$V_{OS} = 4\text{mV}$$

---

4. This model is described in more detail in Reference [10], Appendix D.

$$ECMRR = \frac{V_{IN}(common - mode)}{CMRR(= 100dB)} = \frac{1}{10^5}\left[\frac{V(8,0) + V(3,0)}{2}\right]$$

$$EPSRR = \frac{\Delta V_{CC}}{PSRR(+)} + \frac{\Delta V_{EE}}{PSRR(-)}$$

$$= \frac{[+15V - V(7,0)]}{85dB(= 1.8 \times 10^4)} - \frac{[-15V - V(4,0)]}{80dB(= 10^4)} \tag{30}$$

Both *ECMRR* and *EPSRR* can be obtained in SPICE as a second-degree poly-nominal *VCVS*.

The input bias currents of the op amp are obtained with diodes DN1 and DN2 (for npn or PFET inputs), or with diodes DP1 and DP2 (for pnp or NFET inputs). For the previous example, since $I_B^- = 165$ nA and $I_B^+ = 135$ nA, the choice of the reverse saturation currents is made as $I_{S-DN1} = 165$ nA, $I_{S-DN2} = 135$ nA. Just as previously, the ENI values will also be correct if the SPICE parameters are chosen with $AF = 1.0$ and $KF = 3.2$ E-15. The thermal noise current of R_{IN} will produce a slight increase in ENI. The I_S currents of DP1 and DP2 are chosen $\ll 150$ nA, or say $\sim 10^{-15}$ A.

The input circuit models the open-loop input resistance and capacitance by R_{IN} (3 MΩ for the LM318) and C_{IN} (~3 pF), although the inclusion of the common-mode input resistance and capacitance can be added as well using the RCM1, CCM1 (as well as RCM2, CCM2) network. If used, these components should be chosen as RCM1 = RCM2 = 2 R_{CM}, CCM1 = CCM2 = C_{CM}/2.

The input common-mode voltage range for the LM318 is obtained with the

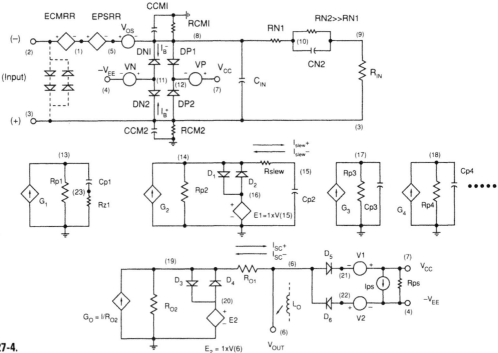

Figure 27-4.
A "circuits approach" macromodel.

diodes and the *VN* and *VP* sources. If *VN* is chosen as ~3.7 V then diode DN1 or DN2 would conduct when either input node is more negative than the V_{ICMR} value of −11.5 V. Similarly, if either input node is more positive than +11.5 V, then either diode DP1 or DP2 will conduct, provided that *VP* is chosen as ~4.2 V.

The equivalent noise voltage of the op amp can easily be added in the input circuit as a series resistance, whose noise is equal to that of the ENV in the "white" (or, midband) noise region. Since $ENV = 11nV/\sqrt{Hz}$, then the choice of RN1 for the LM318 is

$$4kT(RN1) = 1.66 \times 10^{-20}(RN1) = (ENV)^2 = 1.21 \times 10^{-16}$$

or

$$RN1 = 7.3k\Omega \tag{31}$$

Further, it is possible to provide some reasonable simulation of the increase of ENV at low frequencies by the use of the *RN2, CN2* network. This network provides an increase in ENV at low frequencies (noise resistance equal to $RN2 + RN1 \approx RN2$), but at high frequencies reduces to a noise resistance $\approx RN1$. Solving for the noise obtained across R_{IN} due to *RN1* and *RN2* (with the requirement that $R_{IN} \gg RN2 \gg RN1$) leads to a choice of *RN2* and CN2 based on the corner frequency value f_{bv}. A selection for RN2 and CN2 is based on choosing two values of ENV from the manufacturer's input noise curve at frequencies $f = f_{bv}$ and $f = f_{bv}/M$, where $M \gg 1$. Substituting into the ENV expression leads to the following equations for *RN2* and *CN2*:

$$RN2 = RN1(M + 1)$$

$$CN2 = \frac{\sqrt{M}}{2\pi(RN2)f_{bv}} \tag{32}$$

Thus, using the LM318 data for $M = 10, 100,$ and 1000 would give

$$M = 10: \qquad RN2 = 80.3k\Omega, \quad CN2 = 62.6nF$$

$$M = 100: \qquad RN2 = 737k\Omega, \quad CN2 = 21.6nF$$

$$M = 1000: \qquad RN2 = 7.3M\Omega, \quad CN2 = 6.9nF$$

The best simulation of the low-frequency increase in the ENV occurs with the largest value of *M*; however, *RN2* is in series with R_{IN} of the op amp. Hence, for this example using $RN2 = 80.3$ kΩ, $CN2 = 62.6$ nF will obtain a reasonable approximate to the input noise only to $f_{bv}/10$ (≈ 10 Hz), and yet will only attenuate the DC gain by the ratio of *RN2* to *RIN*, or about 2.6%. If an accurate representation of ENV down to 1 Hz were desired ($f_{bv}/100$), then the 737 kΩ requirement for *RN2* would have resulted in an attenuation of the DC input signal by 20%, which would then require that the overall DC gain be later increased by this factor. For FET op amps, since $R_{IN} \sim 10^{12}$ Ω, one could reasonably represent the low-frequency increase of ENV down to frequencies $\ll f_{bv}$ without a loss of DC gain.

The rest of the circuits in Figure 27-4 model the open loop poles (and also zeroes) and set the slew-rate and output limiting. From a closer observation of the manufacturer's data sheets for the LM318, it is observed that the open loop frequency response has a pole-zero pair (due to loss of gain in one side of the input stage current-mirror load) with a pole $f_P \approx 200$ kHz followed by a zero at $f_Z \approx 400$ kHz. Thus, the gain function initially decreases with a gain-bandwidth slope of

2×18 MHz = 36 MHz, so the dominant open-loop pole is really at 36 MHz/110 dB = 114 Hz. The pole-zero pair is accomplished in Figure 27-4 by $(R_{P1} + R_{Z1}) C_{P1}$ and $R_{Z1} C_{P1}$, respectively, while the dominant pole is obtained with $R_{P2} C_{P2}$. Further, the actual phase response of the LM318 shows a very steep slope past the $f_{GB} = 16$ MHz crossing, which indicates several more poles in the response. As an approximation, the $R_{P3} C_{P3}$ and $R_{P4} C_{P4}$ networks are added in Figure 27-4, with each arbitrarily adding an additional phase shift of $(90° - \phi_M)/2$, or $22.5°$ phase lag each at $f = 16$ MHz, or the pole frequencies are at $f_{P3} = f_{P4} = 16$ MHz/tan $(22.5°) \approx 39$ MHz.

The slew-rate for the LM318 of +115, −80 V/μsec is obtained with the D_1, D_2, E_1, and RSLEW elements. Since the slew-rate is determined by the maximum current that will charge the dominant node capacitance, the current into (or out of) C_{P2} in Figure 27-4 is limited by the requirement that $I \times$ RSLEW = V_{D1} (or V_{D2}). Hence, if we arbitrarily choose $C_{P2} = 100$ pF [(Note this requires $R_{P2} = 1/2\pi(100$ pF)(114 Hz) = 13.85 MΩ)], then the maximum slew-rate limited currents through RSLEW will be

$$I^+ = \left(SR^+\right)100 \text{pf} = 115 \times 10^6 \times 100 \text{pF} = 11.5 \text{mA}$$

$$I^- = \left(SR^-\right)100 \text{pf} = 80 \times 10^6 \times 100 \text{pF} = 8 \text{mA} \tag{33}$$

Thus, allowing ~ 0.8 V drop across RSLEW at 11.5 mA requires RSLEW = 69.6 Ω, and using a clamping current in $D_1 \approx I^+$ (just as we did in the Boyle model) obtains the required reverse saturation current values for D_1 and D_2 as

$$(11.5 \text{mA})69.6\Omega = \frac{kT}{q} \ln\left(\frac{11.5 \text{mA}}{I_{SD1}}\right)$$

$$(8 \text{mA})69.6\Omega = \frac{kT}{q} \ln\left(\frac{8 \text{mA}}{I_{SD2}}\right) \tag{34}$$

or, solving, gives $I_{SD1} = 4.16 \times 10^{-16}$ A and $I_{SD2} = 3.57$ pA.

The last circuit in Figure 27-4 provides a short-circuit current limit of ±21 mA using D_3, D_4, E_2, and R_{O1}, identical to the circuits in the Boyle model. If the choices of Figure 27-3 are used, with $R_{O1} = 33$ Ω and $R_{O2} = 67$ Ω, then as in Eq. (28) we require that $I_{SD3} = I_{SD4} = 48.3 \times 10^{-15}$ A. Similarly the D5, D6, V1, and V2 elements in Figure 27-4 limit the output voltage swing to ± 13 V, just as in the Boyle model. Hence using values from the Boyle model, we require V1 = V2 $\approx$ 2.7 V. There is an added current source in Figure 27-4 at the output, so that the sum of I_{PS} + $[(V_{CC} - V_{EE})/R_{PS}]$ more correctly defines the power supply current. These component values can be obtained by taking the slope of the manufacturer's power supply current vs. supply voltage curve at the desired operating point (here, ±V = ±15 V). The values obtained for the LM318 are $I_{PS} = 4.84$ mA, $R_{PS} = 125$ kΩ.

The model of Figure 27-4 also allows the inclusion of diode limiters at the input, as shown by the "dashed" diodes, and the possibility of an increasing Z_{out} with frequency, which very often occurs with high-frequency op amps. The latter feature is obtained by adding an inductor (L_O in Figure 27-4), or possibly a parallel LC network in series with the output terminal.

The circuits approach model of Figure 27-4 requires a total of 23 nodes, where the Boyle model of Figure 27-3 required 16 nodes. The increase of 7 nodes is attributed to the added poles and zeroes, as well as the inclusion of input stage voltage limiting and noise representation. A SPICE format .SUBCKT for the circuits approach model is shown below.

```
.SUBCKT LM318        2       3       4       6       7
*                    -in     +in     -VEE    out     +VCC
*Device Char: Ao1=3.16e5, CMRR=100dB, Zin=3meg//2pf,
*Zin(cm)~200meg//2pf,
*Ib=150nA, Ios=30nA, Vos=4mV, GB=18MHz, SR=+115,-80V/usec,
*PM = 45deg,
*Output drive=+-13V, SC current=+-21mA, Rout~100 ohms, Ips=5.1mA,
*Input CM range~+-11.5V, ENV=11nV/rthz, ENV(fb)~100Hz,
*ENI~0.22pA/rthz,
*ENI(fb)~10kHz. Open-loop poles est. at 114 Hz, ~200kHz, ~39MHz (2), zero
*est. at ~400kHz. Total nodes=23 (25 if diode limiters are added to input).
*- - - - - - -this is a circuits approach model- - - - - - -
*- - - -INPUT STAGE- - - - - -
ECMRR 2 1 POLY(2) (8 0) (3 0) (0 5U 5U)
EPSRR 1 5 POLY(2) (7 0) (4 0) (2.33333M -55.55555U 0.1M)
VDS   5   8   4MV
CCM1  8   0   1PF
CCM2  3   0   1PF
RCM1  8   0   400MEG
RCM2  3   0   400MEG
CIN   8   3   2PF
RN1   8   10  7.3K
RN2   10  9   80.3K
CN2   10  9   62.6NF
RIN   9   3   3MEG
DN1   11  8   DN1
DN2   11  3   DN2
DP1   8   12  DA
DP2   3   12  DA
VN    11  4   3.7
VP    7   12  4.2
*- - - -SECOND STAGE GIVES fp1=200kHz, fz1=400kHz, gain=1.0
G1    (0 13)  (3 9)   0.01
RP1   13  0   100
RZ1   23  0   100
CP1   13  23  3.975NF
*- - - -Third stage gives A01=3.16e5, dom.pole at 114 Hz, & SR limits- - -
G2    (0 14)  (13 0)  22.65MMHOS
RP2   14  0   13.95MEG
D1    14  16  DB
D2    16  14  DC
E1    (16 0)  (15 0)  1.0
RSLEW  14  15  69.6
CP2   15  0   100PF
*- - - -4th & 5th stages have Av=1.0, poles at 39Mhz- - -
G3    (0 17)  (15 0)  0.01
RP3   17  0   100
CP3   17  0   40.77PF
G4    (0 18)  (17 0)  0.01
RP4   18  0   100
CP4   18  0   40.77PF
*- - -Output stage has Av=1.0, ISC limits, & voltage limiting- - - -
```

```
GO    (0 19)  (18 0)   14.9254MMHOS
RO2   19   0   67
D3    19   20   DD
D4    20   19   DD
E2    (20 0)  (6 0)   1.0
RO1   19   6   33
D5    6   21   DA
D6    22   6   DA
V1    7   21   2.7
V2    22   4   2.7
IPS   7   4   4.84MA
RPS   7   4   125K
*-------MODELS--------
.MODEL DA D(IS=1E-15 RS=100)
.MODEL DB D(IS=0.416FA)
.MODEL DC D(IS=3.57PA)
.MODEL DD D(IS=48.3FA)
.MODEL DE D(IS=1FA RS=10)
.MODEL DN1 D(IS=165NA RS=100 KF=3.2F AF=1)
.MODEL DN2 D(IS=135NA RS=100 KF=3.2F AF=1)
*-----------------
.ENDS LM318
```

Comparisons with Manufacturer's Data

A comparison is made in Figure 27-5 between the manufacturer's data and the two SPICE models for the open-loop gain magnitude and phase. The circuits model with the added zero, and multiple poles, more closely approximates the true device characteristics. Figure 27-6(a) compares the actual ENV for the LM318 with the models. It is apparent that the circuits model is reasonably close to the device published data, whereas the Boyle model obtains much too low a value of ENV. In Figure 27-6(b) a comparison is made for the large-signal slew-rate limited pulse response, for a closed-loop gain of +1.0 (a unity-gain follower connection). Both the Boyle model and the circuit's model very closely approximate the negative slew-rate limited response. For the positive SR-limited edge of the pulse, only the circuits model produces a good approximation to the actual device performance. It is noted, however, that the circuit's model has a rather slow recovery after the sharp excursions of the pulse, due primarily to the clamping of the input by diodes DN1 and DP1, as well as the charge on capacitor C_{P2}.

Modeling Current Feedback Operational Amplifiers

The primary limitation to achieving a high frequency response in a typical operational amplifier is the internal dominant node capacitance and its associated large parallel resistance. If the amplifier topology could be changed, however, so that current amplification using relatively low impedances could be realized, then the dominant time constant for the amplifier would decrease, with a corresponding increase in the overall frequency response. Several recent op amps operating at significantly higher currents, with better bipolar processes [device gain-bandwidth products (f_T) > 2 GHz for both npn's and pnp's are now achievable] can realize a significant reduction in both dominant-node capacitance and resistance, with the

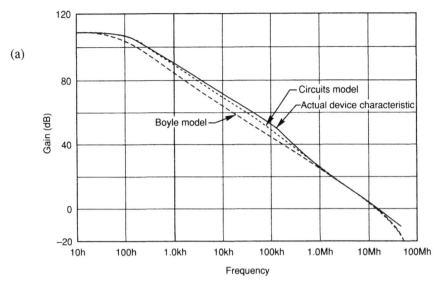

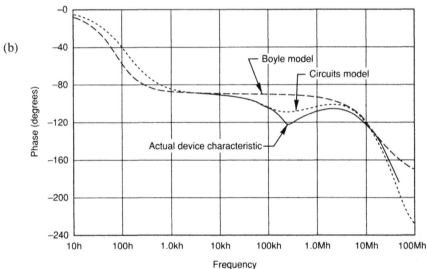

Figure 27-5. Comparison between manufacturer's data and the SPICE models. (a) Open-loop gain magnitude (dB). (b) Open-loop phase.

result that op amp gain-bandwidth products > 200 MHz are possible for a unity-gain connection. Although a high-frequency bandwidth is possible at unity gain, when the gain is increased the closed-loop bandwidth must correspondingly decrease. This can easily be seen by assuming a single-pole op amp connected in a noninverting connection with feedback resistors R_F and R_1, as indicated in Figure 27-7(a). If one assumes a large open-loop voltage gain ($A_{OL} \gg 1$), with a high input impedance (R_{IN} between the positive and negative inputs is large), and a single open-loop pole set by a dominant node resistance and parallel capacitance ($\omega_{dom} = 1/R_{dom} C_{dom}$), then it is relatively simple to show that the closed-loop noninverting gain (A_{CL}^+) reduces to [**10**]:

$$A_{CL}^+ = A_{CL}^+(\text{ideal}) \left\{ \frac{1}{1 + sR_{dom} C_{dom} \left[\dfrac{A_{CL}^+(\text{ideal})}{A_{OL}} \right]} \right\} \qquad (35)$$

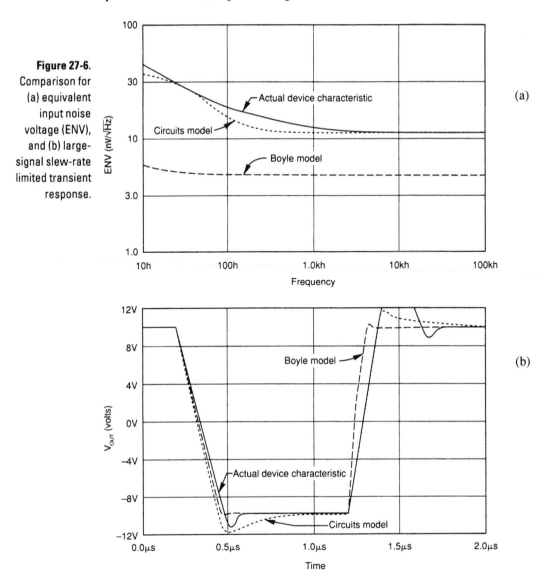

Figure 27-6. Comparison for (a) equivalent input noise voltage (ENV), and (b) large-signal slew-rate limited transient response.

where $A_{CL}{}^+(ideal = (R_F + R_1)/R_1$. So the closed-loop -3 dB bandwidth is

$$f_{-3dB(CL)} = \frac{1}{2\pi R_{dom} C_{dom} \left[\dfrac{A_{CL}{}^+(ideal)}{A_{OL}} \right]}$$

(36)

and thus the bandwidth in a standard op amp *varies inversely as the closed-loop gain*. Hence, an op amp with GB = 200 MHz would have $f_{-3dB(CL)}$ = 200 MHz for $A_{CL}{}^+ = 1$, 20 MHz for $A_{CL}{}^+ = 10$, etc. Of course, in practical terms there is always more than one pole in the loop, plus the addition of input and output capacitance so that Eq. (36) is quite idealized.

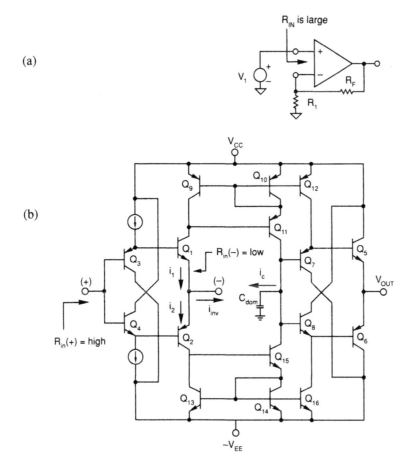

(a)

(b)

Figure 27-7.
(a) Noninverting
gain connection
for a standard op
amp, having R_{IN}
very large.
(b) Simplified
circuit diagram of
a current-feed-
back amplifier.
(Circuit reprinted
with permission
of Comlinear
Corporation.)

The Comlinear Current-Feedback Circuit

The use of current feedback for operational amplifiers has been pioneered by Comlinear Corporation[5]. These amplifiers have quite different input impedances than a standard op amp. The input impedance looking into the noninverting input terminal is high (typically, several hundred kΩ), while the input impedance at the inverting input terminal is very low (ideally, 0 Ω). A simplified circuit diagram for a current-feedback (CF) amplifier is shown in Figure 27-7(b). In this figure it is apparent that a large noninverting input resistance is achieved by Q_3 and Q_4 in parallel, each connected as an emitter-follower. Note also that the voltage gain from the input bases of Q_3–Q_4 to the inverting input terminal is close to unity (~+1.0). However, the input resistance looking-into the inverting input terminal is quite low, since the emitters of Q_1 and Q_2 are in parallel, so $R_{in}^- \approx r_{e1} \parallel r_{e2}$. Since Q_1 and Q_2 act as a common-base connection for currents i_1 and i_2 ($i_{inv} = i_1 - i_2$), then the collector currents i_{C1} and i_{C2} are approximately i_1 and i_2, also. The connection of Q_9–Q_{11} (and Q_{13}–Q_{15}) forms a Wilson current mirror; thus, if the emitter-base areas of Q_9

5. Comlinear Coporation, Fort Collins, Colorado 80525.

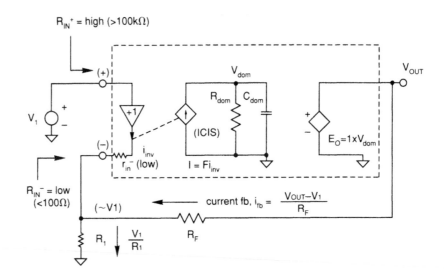

Figure 27-8.
A block diagram
representation of
the current
feedback
operational
amplifier.

$R_{IN}^+ = $ high (>100kΩ)

V_{dom}

V_{OUT}

(+)

R_{dom} C_{dom}

V_1 +

+1

(ICIS)

$E_0 = 1 \times V_{dom}$

(−)

i_{inv}

r_{in}^- (low)

$I = Fi_{inv}$

$R_{IN}^- = $ low
(<100Ω)

(~V1)

current fb, $i_{fb} = \dfrac{V_{OUT} - V_1}{R_F}$

R_1 $\dfrac{V_1}{R_1}$

R_F

and Q_{10} (as well as Q_{13} and Q_{14}) are equal, then the currents i_{C11} are i_{C15} are also approximately equal to i_1 and i_2. Thus, the current charging the dominant-node capacitance C_{dom} is the inverting input error current,

$$i_c = i_1 - i_2 = i_{inv} \tag{37}$$

The voltage produced at the dominant node is then amplified by the unity voltage-gain diamond driver (Q_5–Q_8)) to appear at the output. Note also that the diamond driver obtains a large current amplification between the dominant node and V_{out} (approaching β^2); thus, the output transistors have the capability of furnishing large currents to the load.

An analysis of the CF topology can be made with the help of the simplified model of Figure 27-8. The unity-gain input buffer forces the inverting input voltage to follow the noninverting input, so the voltage across R_1 is equal to V_1. The inverting input current is then related to the feedback currents as

$$i_{inv} = \frac{V_1}{R_1} - \frac{V_{out} - V_1}{R_F} \tag{38}$$

The current-controlled current-source (ICIS) at the dominant node is dependent on i_{inv} as

$$I = Fi_{inv} \tag{39}$$

and the resulting voltage v_{dom} is thus related to V_{out} as

$$V_{out} = v_{dom} = Fi_{inv}\left(\frac{R_{dom}}{1 + sR_{dom}C_{dom}}\right) \tag{40}$$

If Eq. (38) is substituted into Eq. (40) and the result solved for the closed-loop non-inverting gain expression, we have

$$A_{CL}^+(s) = \frac{V_{out}}{V_1} = \left(\frac{R_F + R_1}{R_1}\right)\left[\frac{1}{1 + s\left(\dfrac{R_F C_{dom}}{F}\right)}\right] \tag{41}$$

or the closed-loop −3dB bandwidth for the CF amplifier gives (normally, F = 1)

$$f_{-3dB(CL)} = \frac{1}{2\pi R_F C_{dom}} \tag{42}$$

If Eq. (42) is compared to Eq. (36) for a standard op amp, it is apparent that the ideal CF op amp has a *fixed bandwidth, irrespective of the closed-loop gain!* The bandwidth is dependent only on the value of the feedback resistor (R_F) and the dominant node capacitance. For example, if the dominant node capacitance in Figure 27-7(b) is 3 pF (due to the base-collector capacitances of Q_{11}, Q_7, Q_{15}, and Q_8, as well as the substrate capacitance), then for a 200 MHz bandwidth a feedback resistance of $R_F \approx 250\ \Omega$ would be required.

Another interesting feature of the CF topology is that the large-signal slew-rate rise-time is basically the same as the small-signal rise-time. To see this, suppose in Figure 27-8 that a positive step voltage ΔV_1 is applied. From Eq. (38) the initial change in the inverting current is

$$\Delta i_{inv}(t = 0^+) \approx \Delta V_1 \left(\frac{1}{R_1} + \frac{1}{R_F} \right) \tag{43}$$

since V_{out} has not yet responded at $t = 0^+$. The ICIS thus charges the dominant node capacitance with an equal current ($F = 1$), or the initial slew-rate limited charging of C_{dom} is

$$SR \equiv \frac{\Delta V_{out}}{\Delta t} = \frac{\Delta v_{dom}}{\Delta t} = \frac{\Delta I}{C_{dom}} \approx \frac{\Delta V_1}{C_{dom}} \left(\frac{R_1 + R_F}{R_1 R_F} \right) \tag{44}$$

or, in the final analysis, since the gain relation is $\Delta V_{out}/\Delta V_1 = (R_1 + R_F)/R_1$, then Eq. (44) reduces to

$$SR \approx \frac{\Delta V_{out}}{R_F C_{dom}} \tag{45}$$

or the response is essentially the same as that of the small-signal model, namely an exponential response with a time constant of $\tau = R_F C_{dom}$. The settling time of the response is also enhanced, since the time required for the output signal to settle to within 1% of its final value would be given by

$$\left(1 - e^{-Kt/\tau} \right) = 0.01$$

$$\tag{46}$$

$$K = 4.6\tau$$

or only 4.6 time constants would be required. Using the earlier example of a 200 MHz CF op amp, with $R_F = 250\ \Omega$ and $C_{dom} \approx 3$ pF, then the small-signal 10–90% rise-time should be approximately $2.2 R_F C_{dom} = 1.7$ nsec, while the 1% settling time should be approximately 3.5 nsec.

In actuality, the CF amplifier has a finite inverting input resistance (typically, 20–60 Ω), as well as several high frequency poles in addition to the dominant pole. Thus the simple calculations of the previous paragraphs are somewhat optimistic, and both the bandwidth and slew-rate are somewhat dependent on the closed-loop gain.

Figure 29.9 (a).
A circuits
macromodel for
current-feedback
op amps (repro-
duced with
permission from
Comlinear
Corporation, Fort
Collins, Colorado
80525).

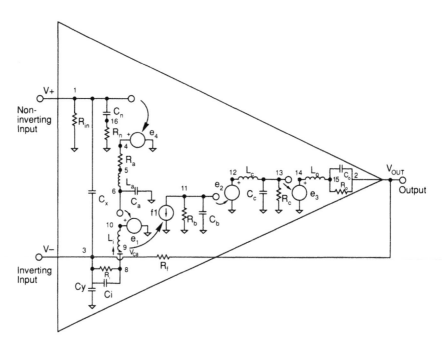

Due to the basic differences between the standard op amp and the current-feedback op amp, the Boyle model is not applicable in developing a SPICE
macromodel. Comlinear Corporation does provide a basic circuits-approach model
for most of its CF op amps, as indicated by the basic topology of Figure 27-9(a)
[11]. In this circuit the noninverting input impedance is obtained by R_{in}, C_n, and R_n,
while the inverting input impedance is modeled by R_i, C_i, L_i, and C_y. The unity
voltage-gain transfer between $V+$ and $V-$ is obtained with dependent voltage
sources $e_4 = 1.0$ and $e_1 \approx 1.0$. The finite gain-bandwidth of the input buffer [Q_1–Q_4 in
Figure 27-7(b)] is approximated by the RLC network of R_a, L_a, and C_a. The unity
current transfer of the inverting input error current is obtained with f_1, and the dominant-node time-constant $R_b C_b$. Two higher-order poles of the op amp are obtained
with L_C, R_C and C_C, while the output impedance is modeled by L_o, R_o, and C_o. Both
e_2 and e_3 are unity voltage-gain transfer sources.

A Circuits Approach SPICE Model for CF Op Amps

Although the model of Figure 27-9(a) well represents the frequency dependence of
the CF op amp, it does not include DC offsets, as well as noise or saturation limits.
However, by adopting the results of the circuits approach model of Figure 27-4, an
improved circuits model is obtained, as shown in Figure 27-9(b). In Figure 27-9(b)
the *CMRR, PSRR* and V_{OS} values are obtained identically to those of Figure 27-4.
The input common-mode voltage range is also similar, determined by the diode
clamping network of $-V_{EE}$, V_N, and D_{N1} and D_{N2} for $V_{INCM(-)}$, and by V_{CC}, V_P, and
D_{P1} and D_{P2} for $V_{INCM(+)}$. The output short-circuit current limits, as well as the
maximum output voltage range, are determined by the identical output circuit of
Figure 27-4. The ENV of the CF op amp is determined in the midband noise region
by the sum of the thermal noise of R_{N1} and R_{INI} in the Figure 27-9(b) model as

$$\text{ENV}^2 = 4kT(R_{N1} + R_{INI}) \tag{47}$$

The ENI is quite different in the CF topology, with the ENI at the inverting input
$>>$ ENI at the noninverting input. In Figure 27-9(b) we can model (ENI)$_{noninv}$ by

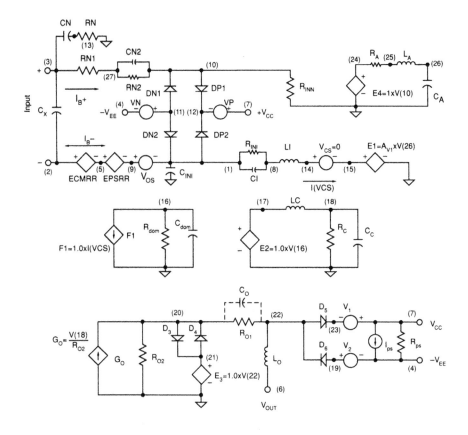

Figure 27-9 (b).
An improved
circuits macro-
model to include
noise, voltage
and current
offsets, short-
circuit current
limits, and input
and output
voltage limits.

the shot noise of the reverse leakage current of D_{N1} and D_{P1},[6] as well as the $4kT/R_{INN}$ thermal noise of R_{INN}. Normally, the noninverting input resistance is large, so the $4kT/R_{INN}$ term can be neglected. However, we also wish to model the noninverting input DC bias current (I_B^+) by the reverse current of D_{N1} and D_{P1} as well. Thus, two equations are required, based on the reverse saturation current (I_S) of each diode,

$$I_B^+ = I_{S-D_{N1}} - I_{S-DP1} \tag{48}$$

$$2q\left(I_{S-D_{N1}} + I_{S-DP1}\right) = (ENI)_{noninv}^2 \tag{49}$$

with a resulting solution

$$I_{S-DP1} = \frac{(ENI)_{noninv}^2}{4q} - \frac{I_B^+}{2} \tag{50}$$

$$I_{S-DN1} = I_B^+ + I_{S-DP1}$$

The low-frequency corner for the ENI can be obtained as in the standard op amp, by the choice of the SPICE parameter $KF = 2q f_{bi}$.

---

6. Since the reverse leakage currents of D_{N1} and D_{P1} (also, D_{N2} and D_{P2}) are temperature dependent, the ENI values obtained by this model are only valid at room temperature (=27°C). For other temperatures one should adjust the I_S values as needed.

The ENI for the inverting input of the op amp can generally be modeled in the same way as for the $(ENI)_{noninv}$, using Eqs. (48)–(50), although the resulting reverse saturation currents of DN2 and DP2 may be much larger than the values for DN1 and DP1, since typically $(ENI)_{inv} \gg (ENI)_{noninv}$ for a CF op amp. One point of interest concerns the I_B^- value. Note in Figure 27-9(b) that the polarity of I_B^- is indicated as *either* into or out of the inverting input. The I_B^- bias current is truly an "error current," equal to the DC unbalance of the emitter currents of input transistors Q_1 and Q_2 of Figure 27-7(b), and thus the net direction and magnitude of I_B^- will vary from unit to unit. As a matter of a worst-case approach, it is suggested that I_B^- be chosen as opposite to the direction of I_B^+ in the model of Figure 27-9(b).

The low-frequency corner (f_{bv}) for the ENV can be approximated somewhat similar to Figure 27-4, by the use of the $R_{N2} C_{N2}$ network of Figure 27-9(b). Since one must require that $(R_{N1} + R_{N2}) \ll R_{INN}$ so as not to attenuate the low frequency gain, then R_{N2} cannot be as large as the values typically employed in the circuits model for a standard op amp. However, by moving the location of the $R_{N1}, R_{N2} - C_{N2}$ network from its position in Figure 27-4, to the *input* in Figure 27-9, then we now have an equivalent *increase* in the low-frequency ENV produced by the voltage drop across $R_{N1} + Z_{N2}$ by I_{S-DN1} and I_{S-DN2} as

$$(ENV)^2_{low-freq.} \approx 4kT(R_{N1} + R_{N2}) + (R_{N1} + R_{N2})^2 2q(I_{S-DN1} + I_{S-DP1})\left(1 + \frac{f_{bi}}{f}\right) \quad (51)$$

However, notice also that this change in network topology will change the net DC input offset voltage, since now in Figure 27-9(b) the resulting DC offset voltage will be

$$\text{DC input offset voltage} = I_B^+(R_{N1} + R_{N2}) - V_{os} \quad (52)$$

so we will have to change the value of V_{OS} accordingly.

As an example of the use of the macromodel of Figure 27-7(b), consider a Comlinear CLC400 current-feedback op amp whose typical data sheet parameters are, for a closed-loop gain of +2, with $R_F = 250\ \Omega$ and $R_{Load} = 100\ \Omega$:

$V_{CC} = +5\ V, -V_{EE} = -5\ V$	$V_{INCM} = \pm 2.1\ V$
Bandwidth = 200 MHz	$V_{out-max} = \pm 3.5\ V$
R_{IN} (noninv) = 200 kΩ	ENV = 2.7 nV/$\sqrt{Hz}$; $f_{bv} \approx 40$ kHz
SR = 700 V/μsec	$(ENI)_{inv} = 16$ pA/$\sqrt{Hz}$; $f_{bi} \approx 45$ kHz
$V_{OS} = \pm 2$ mV	$(ENI)_{noninv} = 2.8$ pA/$\sqrt{Hz}$; $f_{bi} \approx 45$ kHz
$I_B^+ = 10\ \mu A$	$\tau_r(0.5\ V\ step) = 1.6$ nsec, 0% O.S.
$I_B^- = \pm 10\ \mu A$	$I_{SC} = \pm 70$ mA
PSRR = 51 dB	$\pm 1 \le A_{CL} \le \pm 8$
CMRR = 53 dB	

Further, from Comlinear Application Note OA-09, the model parameters for the CLC400 are:[7]

$R_{INN} = 200$ kΩ	$E_1 = 0.9957$
$R_N = 0.01\ \Omega$	$E_2 = E_3 = E_4 = 1$
$C_N = 5.5$ pF	$V_{CS} = 0$

---

7. Data reprinted with permission of Comlinear Corporation.

$$C_x = 0.91 \text{ pF}$$
$$C_{INI} = 1.8 \text{ pF}$$
$$R_A = 1 \ \Omega$$
$$L_A = 11 \text{ pH}$$
$$C_A = 180 \text{ pF}$$
$$R_{INI} = 59 \Omega$$
$$C_I = 5.3 \text{ pF}$$
$$L_I = 33 \text{ nH}$$

$$R_{dom} = 125 \text{ k}\Omega$$
$$C_{dom} = 3.9 \text{ pF}$$
$$L_C = 62 \text{ pH}$$
$$C_C = 470 \text{ pF}$$
$$R_C = 0.29 \Omega$$
$$R_{O1} + R_{O2} = 7.3 \ \Omega$$
$$L_o = 13 \text{ nH}$$

The other necessary parameters are determined for the circuit of Figure 27-9(b) as follows:

$$ECMRR = \frac{V(10) + V(1)}{CMRR} = \frac{V(10) + V(1)}{446.7}$$

$$EPSRR = \frac{5V - V(7)}{PSRR^+} - \frac{[-5V - V(4)]}{PSRR^-} = 28.18 \times 10^{-3} - \frac{V(7)}{354.8} + \frac{V(4)}{354.8}$$

$$R_{NI} = \frac{\left(2.7 nV/\sqrt{Hz}\right)^2}{4kT} - R_{INI} = 380\Omega$$

$$I_{S-DP1} = \frac{\left(2.8 pA/\sqrt{Hz}\right)^2}{4q} - \frac{10\mu A}{2} = 7.25\mu A$$

$$I_{S-DN1} = 7.25\mu A + 10\mu A = 17.25\mu A$$

$$I_{S-DN2} = \frac{\left(16 pA/\sqrt{Hz}\right)^2}{4q} - \frac{10\mu A}{2} = 395\mu A$$

$$I_{S-DP2} = 10\mu A + 395\mu A = 405\mu A$$

Since V_{INCM} = $\pm 2.1 \text{V}$, then for ~ 1mA clamp current we need

$$V_N = V_P = 5V - (2.1V - V_D) \approx 2.9V$$

Choose R_{O2} = 1 – ohm, so $R_{O1} = 7.3 - 1 = 6.3\Omega$. Since $I_{SC} = 70\text{mA}$,

then $I_{SD3} = I_{SD4} \approx 70\text{mA}/\exp(70\text{mA} \times 6.3\Omega/kT/q) = 2.75\text{mA}$

Since $V_{out-max}$ = $\pm 3.5 \text{V}$, and letting $I_{SD5} = I_{SD6} = 10^{-14} A$,

then $V_1 = V_2 \approx 5 - (3.5V - 0.7V) = 2.2V$

The choice of resistor R_{N2} is somewhat arbitrary. Using the manufacturer's noise data we can observe that the noise corner frequency for the ENV is ~ 40 kHz, with the noise at 100 Hz approximately equal to 52 nV/$\sqrt{Hz}$). Thus, using Eq (51) leads to a choice of $R_{N2} \approx 490 \ \Omega$. Then, since R_{N2} and R_{N1} are comparable, we can choose C_{N2} for a break frequency at ~ f_{bv}, or let $C_{N2} \approx 8.2$ nF. The resulting SPICE .SUBCKT model for the CLC400 is as follows:

```
.SUBCKT CLC400      2      3      4      6      7
*                  -in    +in   -VEE   out   VCC
*OFFSET ADJ @ PIN 1
*CURRENT FEEDBACK, Comlinear, op amp. GB=200MHz (at Acl=2),
*Rin=200K (noninv)
*and 59 ohm (inv). Rout=7.3(est.). En=2.7nV/rthz, In(inv)=16pA/rthz
*and (noninv)=2.8pA/rthz. IB(+)=10uA=IB(-). SR=700 V/usec(for Av=+2)
```

```
*CMRR=53DB, PSRR=51 dB. Vos=2mV. OS=0%, typ, for Acl=2.
*Iout(max)=+-70mA.
*VinCMR=+-2.1V, VoutR=+-3.5V (max) for RL=100.
*RECOMMENDED RF=250-ohm and RL (typ)=100-ohm. Gain range
*suggested is+-1 to
*+-8. Suggested series R is 33-ohm(10pF load), 40(20pF) and 30(50pF). Use
*the CLC401 for larger gains. - - - -TOTAL NODES=27- - - -NOTE:
*NOISE CURRENTS are
*ONLY VALID AT ROOM TEMPERATURE (~ 27 degC)
*- - - -input section, including noise and limiting - - - - - - - - - -
CN    3   13   5.5PF
RN    13   0   0.01
CX    3   2   0.91PF
RN1   3   27   380 ; models Rn(midband), along with RINI
RN2   27   10   490
CN2   27   10   8.2NF
VOS   9   1   10.7MV ; adjusts Vos to balance Ib(+) x (RN1+RN2) voltage drop
ECMRR   2   5   POLY(2)   (10 0)   (1 0)   0   2.239M   2.239M
EPSRR   5   9   POLY(2)   (7 0)   (4 0)   28.184M -2.8184M +2.8184M
DN1   11   10   DA
.MODEL DA D(IS=17.25U   KF=14.4F   RS=10) ; models
*Inoise=2.8PA/RTHZ,Fbi=45KHZ
VN    11   4   2.9
DP1   10   12   DB
.MODEL   DB   D(IS=7.25U   KF=14.4F   RS=10);   total IN= DA+DB
VP    7   12   2.9
RINN   10   0   200K ; OL input resis. for noninv input
CINI   1   0   1.8PF
DN2   11   1   DC
DP2   1   12   DD
.MODEL DC D(IS=395UA   KF=14.4F   RS=10)
.MODEL DD D(IS=405UA   KF=14.4F   RS=10)
RINI   1   8   59 ; models Rin for inv. input
CI    1   8   5.3PF
LI    8   14   33NH ; gives XL=RINI @ 300MHZ
VCS   14   15 ; basically, an 'ammeter' to monitor current
E1    (15 0)   (26 0)   0.9957
*- - - -INTERMEDIATE NETWORK FOR PHASE LOSS THROUGH Av=1
*- - -SECOND-ORDER POLES @ 883MHZ AND 3570MHZ- - -
E4    (24 0)   (10 0)   1.0
RA    24   25   1
LA    25   26   11PH
CA    26   0   180PF
*- - — -MODEL FOR DOMINANT POLE AT 316KHZ- - - -
F1 (16 0) VCS 1.0
RDOM 16 0 125K
CDOM 16 0 3.9PF
*- - - -ADDITIONAL SECOND-ORDER POLES @ 748MHZ AND 937MHZ- -
E2    (17 0)   (16 0)   1.0
LC    17   18   62PH
CC    18   0   470PF
```

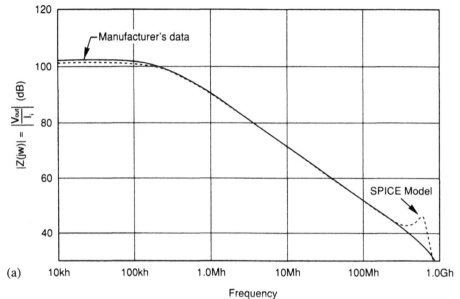

Figure 27-10.
Comparison of
the open-loop
transimpedance
gain, Z(jw) =
V_{out}/I_i for (a)
magnitude and
(b) phase
response, for the
CLC400 op amp.

(a)

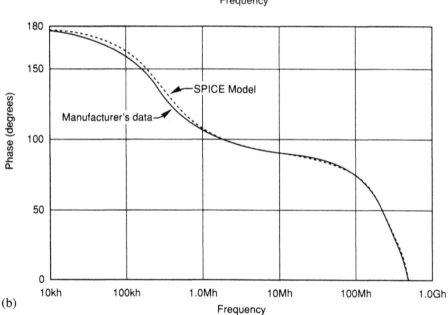

(b)

```
RC   18   0   0.29
*----OUTPUT CIRCUIT, AND LIMITING-----
GO   (0 20)   (18 0)   1.0
RO2   20   0   1.0
D3   20   21   DSC
D4   21   20   DSC
.MODEL DSC D(IS=2.75E-9)
E3   (21 0)   (22 0)   1.0
RO1   20   22   6.3
LOUT   22   6   13NH ; Gives XL=Rout @ 90MHZ
D5   22   23   DE
D6   19   22   DE
```

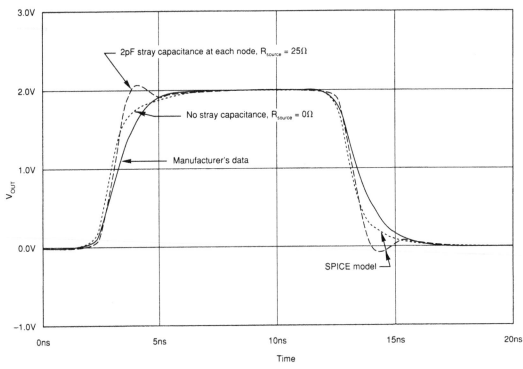

Figure 27-11.
The pulse re-
sponse for the
CLC400 op amp
for $A_{CL} = 2$, for
$V_{out} = 2$ V, peak-
to-peak.

```
.MODEL   DE D(IS=1E-14 RS=10)
V1   7   23   2.2
V2   19   4   2.2
RPS   7   4   666 ; Simulates total P.S. current, for +-5V operation.
.ENDS CLC400
```

In Figures 27-10 through 27-12 a comparison is made between the SPICE model of Figure 27-9(b) and the manufacturer's data for the CLC400 op amp. Figure 27-10 indicates the open-loop transimpedance gain obtained by driving the inverting input with a current source I_i, and plotting the magnitude and phase of $Z(j\omega) = V_{out}/I_i$. Both the magnitude and phase of $Z(j\omega)$ are well represented by the SPICE circuits model, except for a very high frequency peaking in the model's magnitude response at ~ 700 MHz.

The comparison between actual and simulated responses for the transient response of a gain of ± 2 V/V circuit is compared in Figure 27-11. The manufacturer's data and the SPICE simulation are reasonably close, with no overshoot indicated, for negligible stray capacitance. Also shown is a SPICE simulation for the case of 2 pF stray capacitance assumed at each input and output node, and the source represented by a 50 Ω output pulse generator driving a terminated 50 Ω input resistance at the noninverting input. The ringing observed in the model is also representative of the response actually observed for this case.

Comparisons between manufacturer's curves and simulation results for noise is indicated in Figure 27-12. The SPICE model has a mid-frequency ENV of ~ 2.9 nV/$\sqrt{Hz}$ versus 2.7 nV/$\sqrt{Hz}$ from published data. The corner frequency for the model is quite accurate, as is the $1/f$ low-frequency noise. The ENI comparisons are shown in

Figure 27-12(b), where the inverting input equivalent noise current $(ENI)_{inv}$ is modeled precisely as the actual manufacturer's data. The noise current at the noninverting input, $(ENI)_{noninv}$, is in good agreement with actual data and simulation results for $f < 10$ MHz; however, at higher frequencies the SPICE model predicts an increase in ENI.

Simulation results were also obtained for the loop transmission $T(j\omega)$ (loop gain) for the range of closed-loop gain recommended for the CLC400 op amp, namely A_{CL} from ± 1 to ±8. For a gain of +1 the DC loop transmission was $T_0 = 51.5$ dB, with a loop crossing $(|T(j\omega)| = 1)$ at a frequency of 126 MHz, with a phase margin of 67°. For a gain of +8, the value of T_0 decreased to 43.9 dB, with a reduction in the crossing frequency to 51.3 MHz, and a corresponding improvement in the phase margin to 79.4°. Although the frequency bandwidth does change by ~60%, for a gain range from +1 to +8, this is still a significant improvement over a standard configuration 126 MHz single-pole op amp, whose bandwidth would change from 126 MHz $(A_{CL} = +1)$ to ~ 16 MHz $(A_{CL} = +8)$.

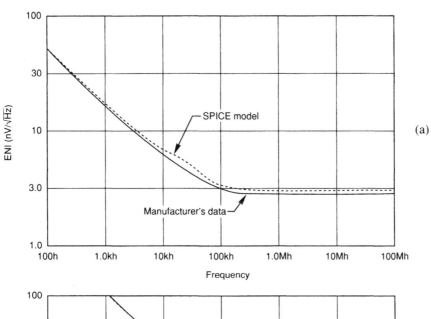

(a)

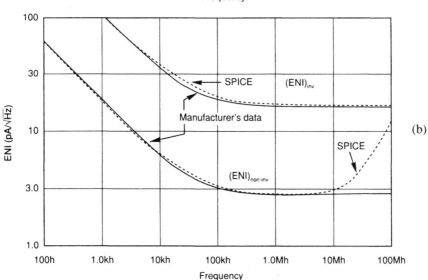

(b)

Figure 27-12.
Comparison of the ENV and ENI for the CLC400 op amp. (a) ENV. (b) ENI.

References

1. L.W. Nagel, *SPICE 2: A Computer Program to Simulate Semiconductor Circuits*, ERL Report ERL-M520, University of California, Berkeley, 1975.

2. P. Antognetti and G. Massobrio [eds]., *Semiconductor Device Modeling with SPICE* (McGraw-Hill, New York, 1988).

3. P.W. Tuinenga, *SPICE, A Guide to Circuit Simulation and Analysis Using PSPICE* (Prentice-Hall, Englewood Cliffs, N.J., 1988).

4. W. Banzhaf, *Computer-Aided Analysis Using SPICE* (Prentice-Hall, Englewood Cliffs, N.J., 1989).

5. L. G. Meares and C. H. Hymowitz, *Simulating with SPICE* (Intusoft, San Pedro, Calif. 1988).

6. M. R. Rashid, *SPICE for Circuits and Electronics Using PSPICE* (Prentice-Hall, Englewood Cliffs, N.J., 1990).

7. E.J. Kennedy, *Semiconductor Devices and Circuits: Theory, Design, and Applications*, Appendix C- SPICE programming Fundamentals and Applications [Holt, Rinehart and Winston, Philadelphia, 1992 (Tentative)].

8. G. R. Boyle *et al.*, "Macromodeling of Integrated Circuit Operational Amplifiers," *IEEE J. S.S. Circuits*, vol. SC-9, No. 6, pp. 353-363, Dec. 1974.

9. G. Krajewska and F. E. Holmes, "Macromodeling of FET/Bipolar Operational Amplifiers," *IEEE J. S.S. Circuits*, vol. SC-14, No. 6, Dec. 1979.

10. E. J. Kennedy, *Operational Amplifier Circuits, Theory and Applications* (Holt, Rinehart and Winston, Philadelphia, 1988).

11. *1989 Data Book*, published by Comlinear Corporation, Fort Collins, Colorado 80525.

Robert J. Matthys

28. Design of Crystal Oscillator Circuits

This chapter is primarily about the circuits used in crystal oscillators, and is only incidentally about the crystals used in them.

Circuit Characteristics

Crystal oscillator circuits are linear analog circuits with carefully controlled overload properties. Both the linear and overload properties are important. The linear properties control the gain and phase shift, and the overload properties control the wave shape and oscillation amplitude. There are many oscillator circuits—some are simple, some are complex. Some contain 90° phase shift networks, while others don't.

Why are there so many circuit types? The primary reason is that an oscillator's circuit design is dominated by the wide variation in a crystal's internal resistance with frequency. To drive the crystal, the circuit's impedance level has to somehow match into the crystal's internal series resistance R_s, which can vary from 200 KΩ at 1 kHz down to 10 Ω at 20 MHz. A circuit that works well into a 10 Ω load is considerably different from one that works into a 200 KΩ load. Whatever the crystal's internal resistance is, the circuit's impedance level must be shifted up or down to drive into it. Some circuits work best at low frequencies. Others work best at high frequencies. There is no universal oscillator circuit.

There are certain crystal characteristics that affect the design of an oscillator circuit. The most important is the internal resistance of the crystal, which varies widely with frequency. Figure 28-1 shows the maximum crystal resistance vs. frequency, taken from various crystal specifications. Another characteristic is that the maximum power into a crystal has to be limited, to minimize frequency drift from heating effects. Figure 28-2 shows the maximum power into a crystal as a function of the oscillation frequency, again taken from various specifications. And Figure 28-3 [from Reference 1] shows the maximum voltage across the crystal at series resonance, in order to stay within the crystal's power limit.

Frequency stability is important in an oscillator. A good circuit will contribute less than half of the short-term (less than a few seconds) drift, and only a small part of the long-term (weeks and months) drift. In a good circuit, most of the frequency drift, both short and long-term, comes from the crystal. But in a poor circuit the frequency drift caused by the circuit itself will far exceed that of the crystal. In particular, the short-term instability of a poor circuit with low in-circuit Q can exceed that of the crystal by more than an order of magnitude. The best short-term stability is in a bridge circuit, which can even reduce the crystal's short-term instabilities. Reducing the magnitude of the circuit's short-term frequency drift so that it is less than that of the crystal forms a large part of the circuit design effort.

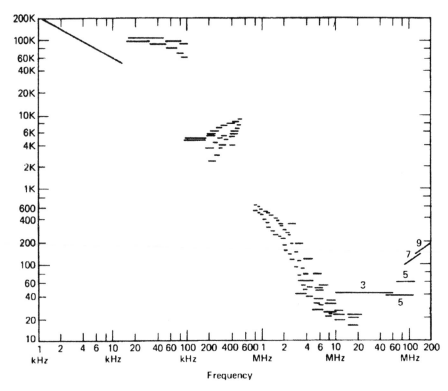

Long-term circuit drift comes mostly from phase changes in the 90° phase shift networks used in some circuits. Stable components must be used in these networks to avoid long-term frequency drift. The frequency trimming capacitor can also cause long-term drift if it's not mechanically and thermally stable.

Most oscillator circuits contain only one transistor. A minority use two transistors, and only a few circuits use three or four transistors. Both sine and square wave outputs are available from most crystal oscillators. On those few oscillators with sine wave drive on the crystal, only a sine wave output is available without adding wave form squaring circuitry.

Design Basics

For a circuit to oscillate, it needs positive feedback and a loop gain greater than one. Its impedance level has to match into the crystal's internal resistance, as mentioned before. And it must not degrade the crystal's internal Q too much. It needs good wave form in both the linear and overload modes, and must have no spurious oscillations or parasitics. It also must have enough loop gain to oscillate. A circuit that does all of these things will have good short-term stability.

A crystal is electrically equivalent to a narrow bandpass filter. The oscillation frequency wanders some about the center of the passband, with the size of the wander depending on how fast the phase changes away from the center frequency. Since the width of the passband is inversely proportional to Q, the highest Q will give the highest frequency stability. The resistive circuit load seen by a crystal, "looking out" into the circuit, causes oscillation losses that add to the internal losses in the crystal, and hence the crystal's in-circuit Q is normally lower than the crystal's internal Q. The trick is to minimize these resistive circuit losses and thereby maxi-

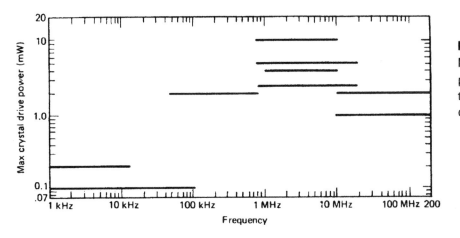

Figure 28-2.
Maximum crystal power dissipation as a function of frequency.

mize the crystal's in-circuit Q and the oscillator's short-term frequency stability. A 2-to-1 maximum reduction of the crystal's internal Q when put in the oscillator circuit is an acceptable target.

Some circuits have a higher in-circuit Q than others. And a high in-circuit Q is easier to obtain with high resistance crystals than it is with low resistance crystals. At 10–20 MHz, a crystal's internal series resistance R_s is typically 10 Ω. The circuit resistance seen by a crystal is usually the sum of the circuit's source resistance driving the crystal plus the circuit's load resistance on the crystal's output. To keep the in-circuit Q reduction to just 2-to-1 requires circuit source and output load resistances of 5 Ω each. It can take two emitter-followers cascaded in series to get such a low source resistance.

An alternative at these low resistances is to operate the crystal at one of its harmonic frequencies (third, fifth, etc.), and take advantage of the fact that the crystal's internal resistance increases with the order of the harmonic. Then the circuit would drive into an internal crystal resistance of 40 (third harmonic) to 60 Ω (fifth harmonic), which is much easier than driving into 10 Ω. The higher impedance level of harmonic operation also means a lower supply current, which is helpful if low power is a consideration.

As shown in Figure 28-1, the internal resistance of a crystal decreases with frequency, assuming fundamental oscillation. Matching a circuit's impedance level up or down to the resistance of the crystal is helped by the fact that the impedance level of amplifier stages in general also decrease with frequency, to combat the increasing effect of stray capacitance to ground and across components. Somewhere between

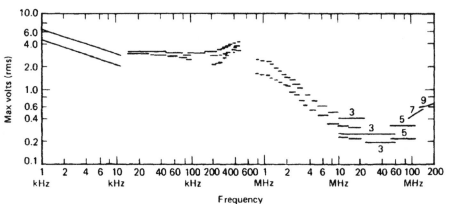

Figure 28-3.
Maximum crystal-drive voltage at series resonance as a function of frequency. (From Reference 1, ©1983, reprinted with permission.)

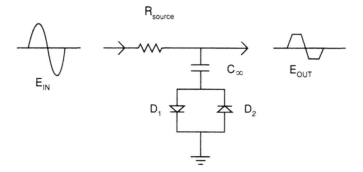

Figure 28-4.
Diode amplitude
limiter.

10 and 100 MHz, the shunt impedance of stray capacitance to ground becomes so low that it's very difficult to get any gain at all, and it then becomes necessary to tune out the stray capacitance with a shunt inductor to get reasonable gain at frequencies above this point. Unfortunately, the gain of an amplifier stage also decreases as the circuit's impedance level goes down. This makes it harder at the higher frequencies to get enough gain for a circuit to oscillate, without adding more amplifier stages.

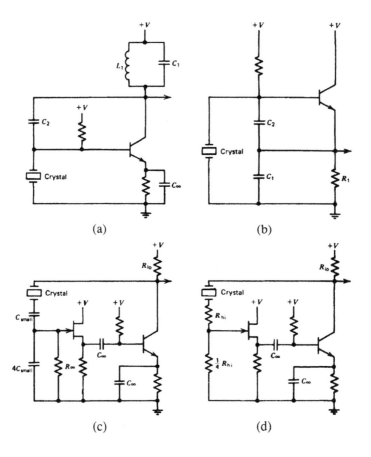

Figure 28-5.
Parallel-resonant circuits (high load impedance): (a) Miller—poor circuit; poor frequency stability. (b) Colpitts— good circuit; fair frequency stability. Circuit is far more complex than it appears to be; widely used. (c) Low capacitance load—works reasonably well; fair frequency stability. (d) High resistance load—works reasonably well; poor frequency stability. (©1983, reprinted with permission.)

The oscillation amplitude can easily exceed a crystal's maximum drive level, particularly at frequencies above 3 MHz. Two paralleled diodes with reversed polarity, as shown in Figure 28-4, make a good amplitude limiter to reduce the drive level down to a more reasonable level. They also provide a good overload wave form in the process. Reducing the power supply voltage also will reduce the drive level. Some circuits have poor overload characteristics, and adding a diode amplitude limiter of the form shown in Figure 28-4 is sometimes the only way to get a good wave form out of those circuits.

Most oscillator circuits work best with high gain transistors. The higher gain can be traded off with a lower output load resistance on the crystal, resulting in a higher in-circuit Q. At frequencies above 1 MHz, the Miller effect becomes significant, so minimizing a transistor's feedback capacitance between collector and base is important at those frequencies. Bipolar transistors are more useful than FETs in oscillator circuits, as the bipolar will give five to ten times more gain than a FET will. Some useful high gain transistors are MCM3960, MPSA18, and MM6515 (all Motorola). A useful high transconductance FET is J309 (Siliconix).

In laying out a circuit board, the normal advice to use a ground plane has not been of much help with oscillator circuits. The purpose of a ground plane is to reduce the inductance of a ground lead between two points that are some distance apart and also

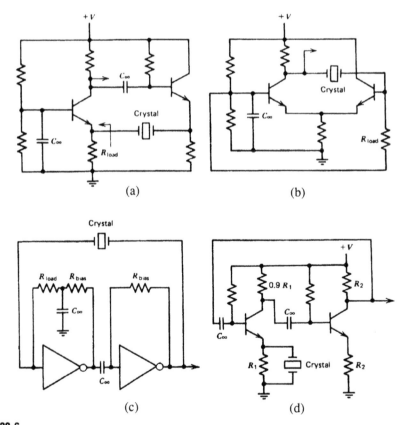

Figure 28-6.
Series-resonant circuits (low load impedance): (a) Common Base—works very well; good circuit; good frequency stability. (b) Common Base, low frequency—works very well; good circuit. Provides high crystal load resistance needed at low frequencies; good frequency stability. (c) Two-inverters-IC—works fairly well; fair frequency stability. With TTL, oscillates spuriously when crystal is removed; widely used. (d) Emitter coupled—works fairly well; good frequency stability. (©1983, reprinted with permission.)

Design of Crystal Oscillator Circuits

Figure 28-7.
More series-resonant circuits (low load impedance); (a) Pierce—very close to series resonance. One of the best circuits; very good frequency stability, best overall design; widely used. (b) Pierce-IC—close to series resonance. Good circuit; good frequency stability; widely used. ©1983, reprinted with permission.

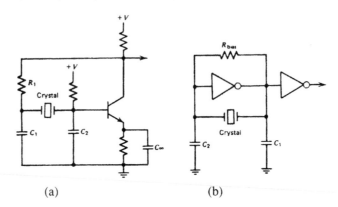

(a)   (b)

provide some shielding. Shielding is necessary with oscillator circuits. But keeping the leads short at the higher oscillation frequencies is even more important. At 100 MHz, the inductance of the lead wires can be ignored if the component leads and PCB traces are all kept to 3/16 in. maximum length.

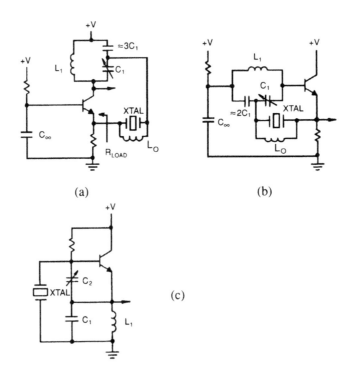

Figure 28-8.
Basic harmonic circuits. (a) Butler common base—operates at or near series resonance. Fair to poor circuit design. Has parasitics, touchy to tune. Fair frequency stability. (b) Butler emitter follower—operates at or near series resonance. Good circuit design. No parasitics, easy to tune. Good frequency stability. (c) Colpitts harmonic—operates 30–200 ppm above series resonance. Physically simple, but analytically complex. Fair frequency stability. (©1987, reprinted with permission.)

Figure 28-9.
Basic harmonic circuits, continued. (a) Pierce harmonic—operates 10–40 ppm above series resonance. Good circuit design. Good to very good frequency stability. (b) Emitter coupled harmonic—operates at or near series resonance. Circuit somewhat complex. Very good frequency stability. (©1987, reprinted with permission.)

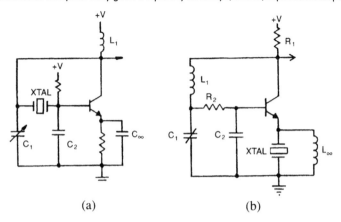

(a) (b)

Oscillator Circuits

Oscillator circuits for the traditional type of quartz crystal can be divided into three main categories: fundamental, harmonic, and bridge. There are some special crystal categories that won't be covered here—such as tuning fork crystals for wristwatches, SC three-frequency crystals with a short thermal response time, and flanged fundamental crystals at UHF frequencies up to 500 MHz.

(a)

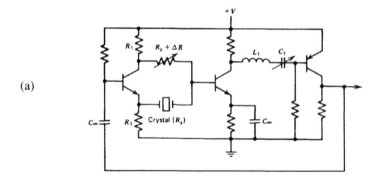

(b)

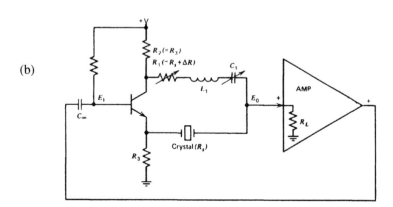

Figure 28-10.
Bridge circuits:
(a) Modified Meacham—One of the best circuits; best short-term frequency stability. Complex circuit, difficult to design, requires tuned amplifier.
(b) RLC Half-bridge—as above except easier to design, uses untuned amplifier.
(From Reference 1, ©1983, reprinted with permission.)

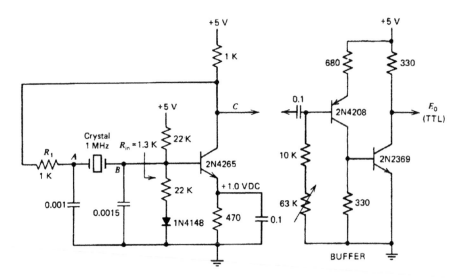

Fundamental circuits are used at frequencies up to about 20 MHz, and harmonic circuits above 20 MHz. The harmonic circuits contain an LC tank (or network) to select the desired harmonic. The bridge circuits are more complex but are the only circuits in which the in-circuit Q can be higher than the crystal's internal Q. They do this by using negative feedback to cancel out part of the crystal's internal resistance without cancelling out the crystal's inductive or capacitive reactance.

There's not enough space here to cover all the oscillator circuits in detail. So the basic schematics of the principal circuit types given in References 1 and 2 are presented in Figures 28-5 to 28-10 as typical examples. Figures 28-5 to 28-10 also include a short mention of the primary characteristics of each circuit. For actual working circuits of each, with component values and detailed information, the reader is referred to References 1, 2, and 3.

Which Circuit Should You Use?

Which circuit you should use depends on what the internal series resistance R_s of your crystal is, the oscillation frequency, and what your needs are with respect to cost, complexity, and frequency stability. To start with, you should pick one that will give a source and output load resistance for the crystal that's equal to or less than the crystal's internal resistance R_s.

A few suggestions on circuit selection might be in order. For a fundamental oscillator at a frequency less than 20 MHz, the Pierce circuit works well and is also an easy circuit to design [1]. Figure 28-11 shows an example of a working Pierce circuit at 1 MHz from Reference 1. Its in-circuit Q is 90% of the crystal's internal Q. For frequencies between 20 MHz and 100 MHz, the emitter-coupled harmonic circuit is the best one (Refs. 2 and 3). Figure 28-12 shows an example of a working emitter-coupled harmonic circuit at 100 MHz from Reference 3. It has the best short-term frequency stability, i.e., the lowest phase noise of any harmonic circuit. Its in-circuit Q is 80% of the crystal's internal Q. Both the Pierce and the emitter-coupled harmonic circuits need stable components in their phase shift networks.

For frequencies above 100 MHz, the Butler emitter-follower circuit is recommended. And, for a frequency standard, the higher short-term stability of a bridge circuit is a good choice. Figure 28-13 shows an example of a working bridge circuit

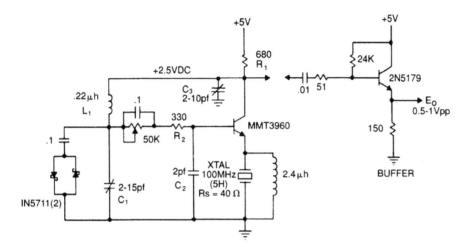

Figure 28-12.
Emitter Coupled
Harmonic at 100
MHz. (©1987,
reprinted with
permission.)

at 10 MHz from Reference 1. In this circuit, the crystal's in-circuit Q is two times higher than its internal Q.

Circuit Loading Effect on Crystal Q

Making the circuit's load resistance R_c (the sum of the crystal's source and output load resistances) on the crystal very small will give a high in-circuit Q. But this is only part of what's really going on in the oscillator circuit. What's not apparent is that a high circuit load resistance R_c on the crystal will also give a high in-circuit Q, and that intermediate values of load resistance R_c will give (surprise!) a very low

Figure 28-13.
RLC Half-bridge
at 10 MHz.
(©1983, reprinted
with permission.)

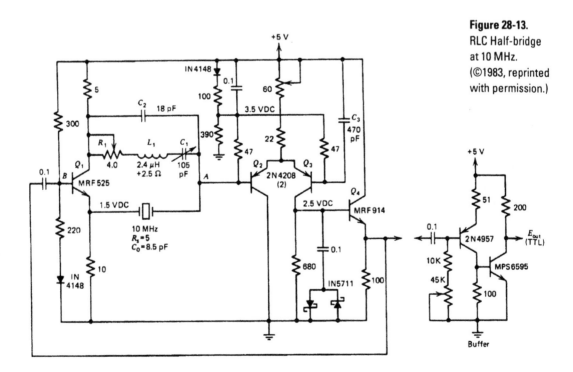

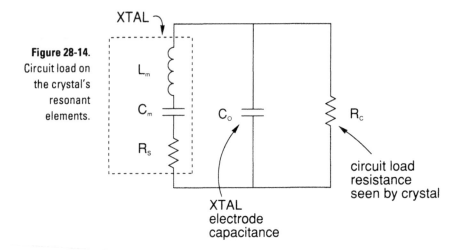

Figure 28-14.
Circuit load on the crystal's resonant elements.

XTAL

L_m

C_m

R_s

C_o

R_c

circuit load resistance seen by crystal

XTAL electrode capacitance

in-circuit Q. These effects are due to (1) the shunt capacitance C_o across the crystal, and (2) what happens when a parallel RC network is converted to its equivalent series RC network.

When a crystal oscillates, energy is transferred back and forth between its motional inductance, L_m, and motional capacitance C_m at the rate of the oscillation frequency. Because of the piezo-electric coupling to the crystal's terminal electrodes, part of this oscillatory energy exchange flows as a current in the crystal's external electrical circuit, i.e., out of one crystal terminal and back into the other, through whatever external impedance is connected between these two terminals. Any resistance in this external current path will absorb oscillation energy and decrease the crystal's Q.

Figure 28-14 shows the crystal and its external circuit load resistance R_c. As mentioned before, this circuit load resistance R_c on the crystal is the sum of the circuit's source resistance driving the crystal and the crystal's output load resistance. Figure 28-14 also shows that the capacitance C_o between the crystal's electrodes forms a second shunt path between the crystal's terminals. Thus part of the crystal's external current will go through the electrode capacitance C_o and part will go through the circuit's load resistance R_c on the crystal.

If the load resistance R_c in Figure 28-14 is $0\ \Omega$, all of the external current will go through R_c. With no external resistance losses, a high-Q circuit condition exists, and

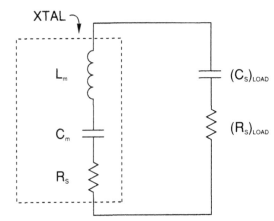

Figure 28-15.
Equivalent series load on the crystal's resonant elements.

XTAL

L_m

C_m

R_s

$(C_s)_{LOAD}$

$(R_s)_{LOAD}$

the in-circuit Q is equal to the crystal's internal Q. If the circuit's load resistance R_c on the crystal is infinite, all of the external current will go through the electrode capacitance C_o. C_o will store part of the energy but won't lose it, as it theoretically has no resistive losses. Again, with no external losses, a high-Q circuit condition exists, and the in-circuit Q is again equal to the crystal's internal Q. But if the circuit load resistance R_c is equal to the reactive impedance of the electrode capacitance C_o, then half the external current will go through the load resistance R_c and half through the electrode capacitance C_o. With external resistive losses, a low-Q circuit condition exists, and the in-circuit Q is then less than the crystal's internal Q.

The action can be made clearer by converting the parallel resistive (R_c) and capacitive (C_o) loads on the crystal in Figure 28-14 into an equivalent series $(R_s\ C_s)_{load}$ network, as shown in Figure 28-15. The equivalent series load resistance $(R_s)_{load}$ in Figure 28-15 represents the true load resistance external to the crystal that needs to be minimized for maximum in-circuit Q. The ratio $[(R_s)_{load} + (R_s)_{crystal}]/(R_s)_{crystal}$ gives the reduction in the crystal's Q when it is placed in the oscillator circuit. The ratio is also a good measure of how good a circuit is, by how much or how little the crystal's internal Q is degraded by the circuit. It can be shown mathematically that the lowest in-circuit Q (and the highest circuit losses) occur when the circuit load resistance R_c on the crystal is equal to the reactive impedance of the parallel electrode capacitance C_o. For example, assume that the circuit load resistance R_c on the crystal in Figure 28-14 is 39 kΩ, and is equal to the reactance of the 4 pF electrode capacitance C_o at 1 MHz. This gives in Figure 28-15 an equivalent series load resistance $(R_s)_{load}$ on the crystal of 28.1 kΩ. If the crystal's internal resistance R_c is 300 Ω, which is a typical value at 1 MHz, then the in-circuit Q is (28,100 + 300)/300 or 95 times lower than the crystal's internal Q. Obviously, this would not be a good circuit design.

The effect of the circuit's load resistance R_c on in-circuit Q is quite broad. The in-circuit Q will be at least a factor of two lower than the crystal's internal Q if the circuit's load resistance R_c on the crystal is within two orders of magnitude (either larger or smaller) of the reactance of the electrode capacitance C_o. If the crystal's in-circuit Q is not to be degraded by more than 30% (1.3$\times$), then the circuit's load resistance R_c must be at least three orders of magnitude larger or smaller than the reactance of the electrode capacitance C_o. This effect has been verified by both calculation and experiment.

The equivalent series load capacitance (C_s) load shown in Figure 28-15 is in series with the crystal's internal motional capacitance C_m and reduces the total net series capacitance in the loop. This raises the crystal's LC resonant frequency slightly and is the basis behind the common practice of putting a variable capacitor in series with the crystal for frequency tuning purposes.

Sometimes, as in the case of the Pierce circuit, the circuit load on the crystal is both resistive and capacitive. To determine the effect on Q, the impedance of the circuit's resistive and capacitive load network on the crystal is combined with the parallel electrode capacitance C_o to form an equivalent series $(R_s\ C_s)_{load}$ network, as shown in Figure 28-15. The equivalent series resistance $(R_s)_{load}$ again determines how much the crystal's internal Q is degraded in the circuit.

Almost all oscillator circuits have medium to low source and output load resistances on the crystal. There are only a few circuits with high crystal source and output load resistances. Because of this, the duality of obtaining a high in-circuit Q with either a low or high circuit load resistance R_c on the crystal is seldom mentioned.

In summary, the crystal's electrode capacitance C_o has to be included as part of the circuit load on the crystal when calculating the crystal's Q reduction by the loading of the oscillator circuit. This may seem obvious after being explained, but without explanation, it's not.

Testing and Optimizing

After the circuit type has been selected, a test circuit needs to be built and checked out experimentally for a clean overload wave form, proper source and output load resistances on the crystal, transistor current level, absence of spurious oscillations and parasitics, and calculation of the in-circuit Q. Looking at the circuit wave forms with an oscilloscope is the only way you can check (1) for proper circuit operation, and (2) on the overload characteristics which control the oscillation amplitude and most of the crystal's drive wave form. Because of this, very little of an oscillator's circuit design can be done analytically. Most of the design effort occurs at the test bench.

An oscilloscope lets the circuit tell you what it wants and needs for good operation. That may sound peculiar, but a lot of what happens when optimizing a circuit is intuitive and depends on how much experience the designer has in reading what the circuit is saying on the oscilloscope.

To optimize the circuit, several things can be done:

1. Vary the impedance level of one part of the circuit with respect to another, looking for a better match into the rest of the circuit, a larger oscillation amplitude, a better wave form, or a reduced loading effect of one component on another.
2. Try different L/C ratios in the LC tank or network (if there is one), looking for a larger signal amplitude, a better wave form, or a better match into the drive transistor or the crystal load.
3. Vary the R/C ratio in the R_c phase shifting network (if there is one), to minimize loading effects on the preceding or following parts of the circuit.
4. Increase or decrease the transistor current, to get a more symmetric wave form or a larger undistorted signal amplitude.

If any of these changes increases the amplitude of oscillation or improves the wave form, it usually is an improvement, and more change in the same direction should be tried.

Sometimes a circuit won't oscillate, and that can be very discouraging. It usually means there's not enough loop gain. To find out where the gain is being lost, it's helpful to break the circuit at some convenient point and insert a test frequency from a signal generator. The test frequency is usually outside the narrow bandpass of the crystal, so the crystal is temporarily replaced with a resistor whose value is approximately equal to the crystal's internal series resistance R_s. Then relative signal levels and phase shifts can be measured at various points in the circuit. Finally, the relative signal levels are converted into localized voltage gains and losses around the oscillator loop.

In trying a new UHF circuit, it's frequently helpful to try it first at a lower frequency, such as 1 MHz, where lead lengths are not as important and the 'scope will show more detail in the wave forms. It's easier then to get a feel for what is and isn't important in the circuit, and what you can and can't do about improving it.

Any spurious oscillations or parasitics should be eliminated. If present, they usually show up as oscillatory spikes in the wave form (assuming a wideband scope), or as sharp peaks or dips in the transistor's DC current when an LC tank is tuned. The Butler common base circuit is particularly prone to these when its load is tapped down on the capacitive side of its LC tank.

Inductors are used in several of the oscillator circuits and are inductive only over a limited frequency range. Like any component, they have a shunt wiring capacitance across them. At frequencies above the resonant frequency of the inductance with its shunt capacitance, the inductor stops being inductive and becomes capacitive, its reactance being equal to that of the shunt winding capacitance. Inductors at high frequencies are usually single layer solenoids with a shunt capacitance of about 2 pF. This shunt capacitance of 2 pF provides an upper limit to the maximum inductance you can get at any given frequency, and is a very real limit in circuit design. Any inductor used must be checked to be sure it is being used below its resonant frequency, where it will behave inductively rather than capacitively.

If the inductor must be stable, it has to be of the air core type. Stable air core inductors are available as spiral metal films on glass cylinders. The various iron and other powdered core materials used to increase a coil's inductance are not linear, so the inductance of coils using them will vary 2% to 20%, depending on temperature and signal amplitude. In addition, all of these core materials (except air) have an upper frequency limit (different with each type) where they become quite lossy, and are essentially useless at frequencies above that point. As a rule of thumb, the higher this loss limiting frequency is, the lower is the inductance improvement provided by the core material. Unfortunately, inductors are not marked with this core loss limit, so any cored inductors out of the junk box are of questionable use unless their core limit frequency can be identified.

Oscilloscope probe grounding is important if the wave forms shown on the 'scope are to bear any relation to what's going on in the circuit. At oscillator frequencies up to 1 MHz, a separate ground wire 3 ft long from the 'scope case to anywhere on the test circuit ground is sufficient. At 10 MHz, a 6 in. ground wire from the back end of the probe handle to circuit ground is sufficient. For oscillators at 20 MHz and above, a ½ in. maximum length ground wire from the front tip of the probe to a circuit ground point within ½ in. maximum of the specific test point being observed is needed. Some probe manufacturers sell short ground clips for convenient use at the probe tip.

At 10 MHz and above, the typical 'scope probe with 10 to 15 pF of input capacitance is almost useless for oscillator work. The low shunt impedance of the probe capacitance at these frequencies changes the circuit's phase shifts, reduces gain, and eliminates the high frequency detail in any wave forms observed. Some circuits will quit oscillating because of the heavy shunting effect of the probe's capacitance. Oscilloscope probes of 1 or 2 pF input capacitance and with at least 1 MΩ input resistance, such as the Tektronix P6201 (10×) or P6202A, are a necessity for oscillator work at frequencies above 10 MHz. Probes with different cable lengths or from different manufacturers also have different internal time delays. This has to be allowed for when checking the phase shift between two circuit points, if probes of different lengths or from different manufacturers are used.

Oscilloscope bandwidth is also important. In looking at a 100 MHz wave form on a 200 MHz oscilloscope, it must be remembered that however badly the circuit wave form is actually distorted, it will only show up on the scope as a slightly warped sine wave. Being able to see the third harmonic of the oscillator frequency on the 'scope is sort of a minimum capability, if the 'scope wave form is to have much meaning. Trying out a UHF oscillator circuit at a lower frequency is also helpful in determining what the characteristic wave forms should look like in a particular circuit.

Electrical Properties of Oscillator Crystals

As an electrical circuit element, an oscillator crystal acts as a series LC tank of very high Q and makes an excellent very narrow bandpass filter. The crystal is usually made of quartz, because quartz is piezoelectric and has low losses and low thermal

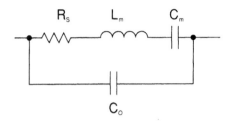

Figure 28-16.
Equivalent circuit
for a crystal.

expansion. Quartz crystals are piezoelectrically excited, and oscillation creates an acoustic standing wave of energy in the crystal, with the resonant frequency controlled by the mechanical dimensions of the crystal. Crystals are available from 1 kHz to 500 MHz. The AT cut is used in the most popular frequency range, 1 to 100 MHz.

Crystals can oscillate at frequencies higher than their fundamental resonance frequency, which are called overtones. The overtones of the AT cut are at the odd harmonics of their fundamental frequency. In some other crystal cuts, the overtone frequencies are not harmonically related to the fundamental frequency. Operating the crystal in a vacuum container rather than in one that's gas filled will give a three times improvement in Q but only at frequencies below 30 MHz [4]. Crystals are normally operated at their fundamental resonance if the frequency is below 20 MHz. Above 20 MHz, they are usually operated at one of their overtone frequencies.

The equivalent electrical circuit for a crystal is shown in Figure 28-16. The crystal's resonant motional components are L_m and C_m, with R_s representing the crystal's internal losses. R_s is the crystal resistance that the oscillator circuit has to match into, when the crystal oscillates at series resonance. C_o is the capacitance between the drive electrodes, which are located on opposite sides of the crystal. The electrode capacitance C_o should be considered part of the external circuit load on the crystal's resonant elements L_m and C_m.

There is no such thing as a series resonant or parallel resonant crystal. The crystal is the same in both cases. The difference refers to whether a crystal is looking into a low impedance load (series resonant) or a high impedance load (parallel resonant). In practice, all crystals oscillate at or very near to their series resonant frequency. The phrase "parallel resonant" is misleading but has been traditionally used in the industry to describe crystal operation with a high impedance load on it.

At frequencies above 50 MHz or so, the shunt impedance of the crystal's electrode capacitance C_o becomes so low that it tends to short out the crystal's resonant elements L_m and C_m from the rest of the circuit. This can be eliminated by parallel resonating the electrode capacitance with a shunt inductance, as can be seen in some of the harmonic circuits in Figure 28-8. Not all oscillator circuits above 50 MHz require such a shunt inductor.

Crystals are manufactured to frequency tolerances of 0.001% to 0.02%. If the oscillator must be set to an exact frequency, the oscillation frequency can be trimmed by adding to or subtracting from the phase shift at any point around the oscillator circuit. Adding a phase lag will lower the oscillation frequency. Subtracting a phase lag will raise it. The most common trimming method is to insert a small 30 to 50 pF variable capacitor in series with the crystal, but this approach only allows raising the crystal frequency and not lowering it.

References

1. R.J. Matthys, *Crystal Oscillator Circuits*, Wiley & Sons Interscience, 1983. Revised 2nd Ed. Krieger Publishing Co., 1991.

2. R.J. Matthys, "Survey of VHF Crystal Oscillator Circuits," *Proceedings of RF Design Expo,* Feb. 1987, pp. 371–381.

3. R.J. Matthys, "High Performance VHF Crystal Oscillator Circuit," *RF Design*, March 1988, pp. 31–38.

4. H.E. Bommel, W.P. Mason, and A.W. Warner, "Dislocations, Relaxations, and Anelasticity of Crystal Quartz," *Phys. Rev.,* V 102 n 1, April 1, 1956, pp. 64–71.

5. B. Parzen, *Design of Crystal and Other Harmonic Oscillators*, Wiley & Sons Interscience, 1983.

29. A Tale of Voltage-to-Frequency Converters

Ancient History

Once upon a time, there weren't any voltage-to-frequency converters (V/F converters) or voltage-controlled oscillators (VCOs). I couldn't tell you exactly when that was, but back in the 1950s and 1960s, very few people ever heard about an oscillator whose frequency could be controlled by a voltage. In those days, when you wanted to change an oscillator's frequency, you changed a pot or a resistor or a capacitor, or maybe an inductor.

I checked up on this, because I spent a couple hours searching in the old M.I.T. *Radiation Lab Series*, published in 1949. There were no oscillators or multivibrators whose frequency could be controlled by a voltage—no VCOs, as far as you could learn by looking at the "Bible" of that time. (See also the last section, A Final Note.) It's true that FM radio transmitters used frequency modulated oscillators back as early as the 1930s, and these were modulated by voltages, but they only covered a relatively narrow frequency; when I refer to a VCO, I am talking about oscillators whose frequency could be controlled over a range of 10:1 or 100:1 or 1000:1 or more. In general, this type of oscillator is expected to have a pulsed or square-wave output, *not* a sine wave.

Less Ancient History

In 1961, when I graduated from M.I.T. and joined up with George A. Philbrick Researches (127–129 Clarendon Street, Boston 16, Massachusetts), I joined a company that made operational amplifiers, analog multipliers, integrators, and all sorts of analog computing modules. And just about everything was made with vacuum tubes. There were applications notes and applications manuals to tell you how to apply operational amplifiers (in those days we would never say "op amp"). And there was the big old Palimpsest, a sort of collection of old stories on things you could do with operational amplifiers and analog computing equipment. But, there were no digital voltmeters, and no voltage-to-frequency converters.

About 1963, I became aware of a high-performance operational amplifier—the 6043 project. This chopper-stabilized amplifier had been designed by Bruce Seddon, one of our senior engineers and one of our vice-presidents. The customer for this amplifier was DYMEC, a subsidiary of Hewlett-Packard, and this amplifier was to be the front end of a voltage-to-frequency converter instrument. The amplifiers were functioning pretty well, and they were meeting just about every specification, but they had a problem with undesired and unpleasant little offset shifts and jumps.

As the project went on, more and more engineers were looking over Bruce's shoulder, trying to help him solve this problem of the jumpy offset. Some people suspected that the amplifier stages might be rectifying out signals that were caused by—?—perhaps—the transmitter of a local taxicab company? I do not know if it was ever resolved what was the cause of this drift or wobble—it was hard to resolve, just a few dozen microvolts—but, I have the impression that the amplifier project was not a success, and we never went into full production. If there was any circuitry of the actual V/F converter, that was never discussed at Philbrick—that was proprietary within DYMEC, and it was only our job to make an operational amplifier with low offset.

Of course, this amplifier had several vacuum tubes and a chopper. The first amplifier tubes, 6CW4 nuvistors, ran with their 6 V heaters connected in series, so as to run on a 12 V DC bus and avoid the noises inherent in using 60 cps heaters. Then there were two more gain stages, DC-coupled. There was a chopper, Airpax 172 or Bristol or similar, and a chopper amplifier based on a 12AX7. The whole circuit was similar to the Philbrick "USA-3" amplifier; it did, however, run on ± 150 V instead of ± 300 V. Dan McKenna, the senior technician who worked on the project, said he always suspected that it was the blame of the heaters that were connected in series, because the data sheet on the 6CW4s said, "do not connect heaters in series." But I realized later, connecting two heaters in series was surely okay; putting 10 or 20 tubes' heaters in series, as in a TV set, was the procedure that was prohibited. So, even though Dan grouched about it, this series stacking of two heaters was probably quite wise, not risky, as it would force the tubes to run at about the same number of watts in each heater.

Anyhow, that is ancient history, a design with vacuum tubes. Even then, even though we were trying to design part of a V/F Converter, we didn't have one in our lab. We didn't have a digital voltmeter (DVM)—we had the old Fluke 805 differential voltmeters. Now, these meters have many elements of accuracy that good DVMs have these days—good resolution and stability—but they were big and heavy and slow. If you wanted to read 5.032 V, for example, you could do it, but it was a tedious deal. You had to turn the first knob to 5, and then the next one to 0, and the next one to 3, and then turn up the gain and look at the analog meter to see if the residue looked like a "2." That was how you learned the voltage was 5.032 V. If you have ever spent a few hours twisting the knobs of one of those old Fluke meters, you may remember it with nostalgia, but you must admit, it was awfully boring.

When the first DVMs came, from HP and from Non-Linear Systems (NLS), they were slow and (in the case of the NLS) clunky and noisy, and they did not have excellent accuracy or features compared to the old Fluke differential meters. But they sure were faster and easier.

And there was another way to do a DVM—you could buy a voltage-to-frequency converter from DYMEC and feed its output frequency into an "event counter" (an EPUT meter—Events Per Unit Time—from Beckman Berkeley), and its neon-discharge display tubes would glow and tell you what the frequency was, in terms of pulses per unit of time, and that was supposed to be linearly proportional to the input voltage.

This new DYMEC V/F converter had several solid-state circuits and tricky pulse generators. To this day I do not know how those proprietary, secret pulse circuits were supposed to work. It had a pulse generator based on some special pulse transformers and blocking-oscillator circuits. The instrument had a pretty good temperature coefficient (TC), *but*, only because it had a little oven to hold all the transistorized circuits (which *really* had a rotten TC) at a constant temperature of +65 °C. Consequently, when you turned it on, you had to wait at least half an hour before the accuracy would finally settle out to its ultimate value—waiting for the oven to settle its temperature. It could handle a full-scale voltage of + and –1.0 V. It works,

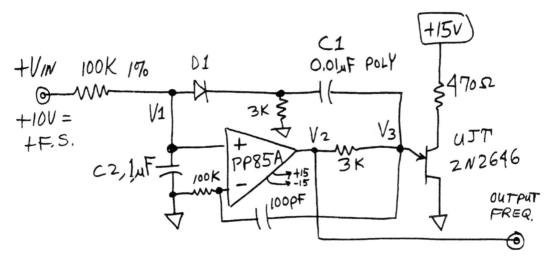

Figure 29-1.
The Swedish
voltage-to-
frequency
converter circuit
(drawn from
memory,
approximate
schematic).

pretty well. It was claimed, originally, to have better than 0.02% of linearity. I measured it once, and it had some nonlinear errors around 0.024%—not bad, but apparently something had drifted slightly out of spec. It cost $1600, back when $1600 would buy you a Volkswagen. I still have one of the DYMEC V/F Converters, Model DY-2211B, and the book on it.

Now let's move up to about 1967. We engineers at Philbrick were working mostly on solid-state operational amplifiers—amplifiers made of 6 or 8 or 10 discrete transistors. The integrated circuit amplifiers were arriving, but most of them were pretty crude in performance and features.

One day Bill Bernardi, one of the senior applications engineers, told me that a customer in Sweden had made a linear voltage-to-frequency converter using a PP85A (one of our standard operational amplifiers) and a UniJunction Transistor (UJT). And the nonlinearity was, he said, about 0.1%. When I heard this, I got *very* curious, because everybody knows that UJTs are the crudest, dumbest, most imprecise oscillator you can find. Just about every student learned that a UJT looks very cute because it can oscillate with a minimum amount of external parts, *but* it's an awfully junky circuit. You could gold-plate the sow's ear, and it was still a junky circuit. So when I heard that a UJT was involved with a V/F converter of very good linearity, I was impressed, but I was suspicious, and I looked into what they were doing. I didn't know anything about V/F converters, but I was curious. I found that the PP85A was used as a comparator, and the UJT was mostly used to provide some "negative resistance" or positive feedback, to make a pulse whose amplitude or width are not critical and thus did not hurt the accuracy of the V/F converter. Ah, but how is that? How is there a V/F converter that uses one simple comparator and a crude UJT pulser, and no other obvious precision components, and yet provides a 0.1% linear V/F converter?

As near as I can recall and reconstruct, the circuit was basically that in Figure 29-1. The principle of operation is that when the current from V_{in} causes the main integrating capacitor C2 to rise up to 0 V, and the op amp starts to swing its output positive, it triggers the UJT, which then puts out a crude pulse which kicks the minus input of the comparator; and the output *also* kicks a certain amount of charge through a charge-dispensing capacitor, C1, back to the integrating capacitor, to reset it. This amount of charge must be constant and invariant of anything, especially invariant of the repetition rate. If you can get that, you get excellent linearity. Apparently the Swedish engineers had stumbled onto this crude but functional circuit.

Now that I understood the principles, I figured out that there was room for a good

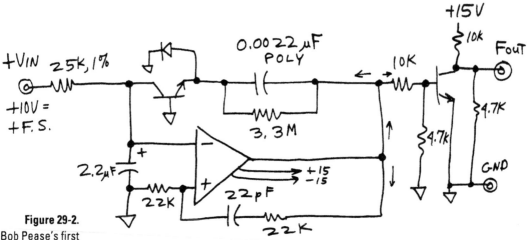

Figure 29-2.
Bob Pease's first
attempt at
designing a
voltage-to-
frequency
converter
(A1-Philbrick T82
AH ≅ Amelco
805BE. Full scale
is 10 kHz. Good
linearity, poor
PSRR and TC).

bit of improvement. I started fooling around with some breadboards. I got the full-scale frequency up to 10 kHz (the Swedish circuit worked well up to just 1 kHz, which is not nearly as useful as 10 kHz), and got the nonlinearity down to 0.03%. And I invented a scheme so the operational amplifier's output could be capacitively coupled back to its positive input, causing enough regeneration or positive feedback, that the UJT was no longer needed. I used an Amelco 805BE integrated-circuit operational amplifier as the comparator. Now, the whole thing would fit into a 1.5 in. square package, just 0.5 in. high—a small epoxy-potted module that was rather smaller than the PP85A amplifier and associated parts as shown in Figure 29-2. We built up a prototype and we tested it, and it worked pretty well. We potted it in our usual hard black epoxy and shipped it to a customer in Connecticut—a customer of Larry Plante, who was our Sales Engineer for that region. Also, I sent in a patent application to our patent attorneys. I forget exactly who it was—was it Mr. X in New York, or Mr. Y in Waltham? No matter.

That must have been a busy year, because by the time I got off the other hot projects I was set to work on, for a number of high-priority customers, I realized I had not heard anything from this customer in Connecticut. I got in touch with Larry Plante. All he knew was, the customer didn't like it. Worse yet, a whole year had elapsed since I had sent the part in interstate commerce, and the patent attorney had done nothing, so the patent application was now worthless. I was quite cross, and I read the riot act to these attorneys. Then I set in at the work-bench with a vengeance.

I realized the old circuit had depended on the power supply stability for its gain factor, so it had no power supply rejection, or, to be more nearly correct, a PSRR of about 0 dB. I added a zener in a bridge, to give a PSRR of perhaps 45 dB. (Note, that was the first time I had ever seen that zener bridge (see Figure 29-3)—was I one of the earliest inventors of that circuit? It is a *neat* and useful circuit.) I added improved features around the amplifier, to improve the start-up and the accuracy. I replaced the (sole-sourced) 805BE with the more popular and inexpensive LM301A. Refer to Figure 29-3; a description of how it works is provided nearby. I gave it to my technician, Dick Robie, to oven it and graph out the temperature coefficient (TC) from the temperature data. That night, on the way out the door, I asked Dick what had he seen for the TC. He replied, about zero. I asked, "Whatd'ya mean, zero? You mean, down near 100 parts per million per degrees C?" He replied, "Oh, much better than 100—less than 10 ppm per °C." I was shocked. How could it be that good? The next day, I figured out the fortuitous situation: of course,

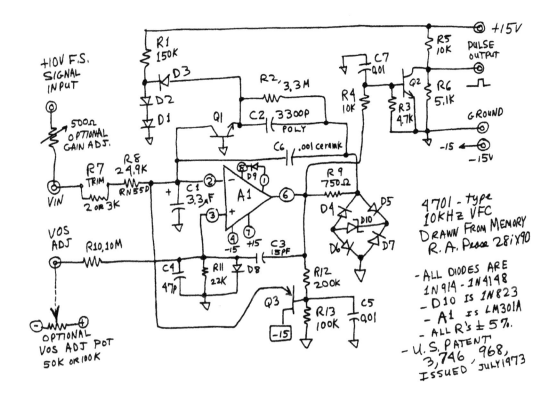

for the "charge-dispensing capacitor, C2," I had used one of the best capacitors in the house, the most precise and stable ones, which were polystyrenes with a TC of −110 ppm/°C. This TC was just about cancelled out by the TC of the entire group of diodes in the rest of the circuit. Namely, the driven end of the capacitor moves about 12.4 V p-p, plus the V_f of four diodes. These four diodes run rich at about 6 mA, and have a TC of about −2.0 mV/°C. The p-p voltage of these four diodes is approximately cancelled by that of the other four V_{be}'s at the other end of the poly capacitor, but those diodes run at about 0.1 mA, and their TC is about −2.3 mV/degree. The difference of these is about 4×0.3 mV/°C, or 1.2 mV/°C, which is about big enough to cancel out the −110 ppm/°C of the capacitor. Now, there were several things I could have done to fix it if the TCs had not come out even—I could have used 3 diodes, or 5, or 4 ½ or 3 ½, but, if 4 was the right answer, I'd go with it.

I got my boss, Dave Ludwig, to approve the layout of a printed-circuit board, and I think Wayne Norwood was the guy who laid it out. We built up a few dozen and evaluated them thoroughly. I wrote up a draft of a data sheet, and I negotiated with Skip Osgood and Bill Bernardi to get it printed properly. I got some test equipment going, and a silkscreen, and we called it the 4701. We were in the voltage-to-frequency converter business.

I don't recall exactly how we got these V/F converters to be so popular. I can't recall how we found so many customers, or how we got out publicity. I asked Frank Goodenough—now the senior editor for analog circuits at *Electronic Design* magazine—I knew he had been involved. He recalled how he had gotten involved: He had looked at some of the characteristics of this 4701, and he suspected that a good V/F converter might be useful at the Foxboro company, the big process-control experts. Indeed, he did find some interest there. They were very interested—but they never bought very many, because Foxboro was very concerned about buying only parts available from multiple sources.

Figure 29-3.
The legendary 4701 voltage to frequency converter, designed and drawn by Bob Pease. (All diodes are in 914-1N4148; DIO is IN823; A1 is LM301A; all R's ±5%.)

The 4701 became popular, with many customers at many companies. It became profitable. It became good business. If I had all the time and space in the world, I would tell you how the 4701 led to the 4702 (a 10 kHz frequency-to-voltage converter, using the same charge-dispensing principles) and the 4703 (100 kHz full scale V/F converter) and the 4705 and 4707 (1 MHz and 5 MHz V/F converters). Also, the 4709, 4711, 4721, and 4715 V/F converters, and 4704, 4706, 4708, 4710, and 4722 frequency-to-voltage converters. Some of these had a moderate TC of 150 or 44 ppm/°C, but some were well-trimmed and guaranteed to 6 ppm/°C—as good as the best DVMs of the day.

But it all started with that crummy little 4701—and the principle that one cheap operational amplifier driving a little charge-dispenser could make a very linear data converter. This came from an understanding that you could build an oscillator with a UJT and an operational amplifier to help improve the linearity, and *then* throw out the UJT! I was able to do that because I was receptive to the concepts that would make a good V/F Converter, even though I had never seen a V/F Converter! I was able to make accurate measurements, to throw in precision components—zener references, capacitors, resistors—and to get inexpensive IC amplifiers, and to optimize the whole little system. What is the underlying principle?

I like to think of the words of Branch Rickey, manager of the St. Louis Cardinals in the 1920s (and later manager of the Brooklyn Dodgers). One of his favorite sayings was "Good luck is the residue of design." When he referred to his baseball teams, he meant that his players were able to handle any play, field the ball, throw to the right base, etc., no matter how surprising the bounce of the ball might be. If his players had learned the fundamentals, if they were prepared for any situation, they could take advantage of game situations and *win*. Other people might say his players were just lucky, but he knew darned well that he had trained them and drilled them so they would instinctively know to do the right thing.

I, too, was in a situation where I was prepared to take advantage of the opportunity, and I didn't drop the ball when it came my way. (Well, I fumbled it for a while but then I got it right.) We built up that V/F Converter business to about one-tenth of all the business at Teledyne Philbrick, and if you look at the schematic of that little V/F converter, you can tell it was pretty profitable when we sold it for $59. But as I mentioned in another chapter, when a guy had spent his $59, he really got a lot of satisfaction. We had some very loyal customers, because the performance per dollar was right, even if the parts list would have looked pretty skimpy, if the epoxy were not so opaque and obscuring. As with other circuits like the P2, the list of parts looked like not much of a big deal, but the way you put them together was what made the value to the customer. To some extent, this is like the integrated circuit business, where the price of an IC often is related to the value it provides to the customer and is not related at all to the cost of manufacturing.

The V/F Converter business eventually became just popular enough that other analog-circuit manufacturers decided to get into the business. Several competitors such as Dynamics Measurements Corp., Intech, and a couple others started making units that competed. But these companies were all followers—none of them was making or selling V/F converter modules until the 4701 came along and popularized the concept. Finally Raytheon started making an integrated-circuit V/F converter— the RC4151. But it did not use charge-dispensing techniques—it used a timer, similar to the 555. It was inexpensive but did not have any guaranteed TC, and only poor linearity (0.15%).

In 1976 I left Teledyne Philbrick and went to work for National Semiconductor. Bob Dobkin put me to work designing a V/F converter integrated circuit that would have all the advantages of previous designs but avoid the disadvantages. I came up with a design that was nothing like a 4701—nothing like any of the Philbrick V/F converters. This LM331 would run on any power supply from 4 to 40 V (whereas the Philbrick ones needed ± 15 V, and ± 12 to ± 18 V was about as wide as they would accept). It's been pretty popular, and to this day, 13 years after it was introduced, the LM331 is still getting designed in to new circuits. People tell me they like the way it is versatile and offers good precision.

At Philbrick, after I had refined the circuits of the 4701, I realized that this went far beyond the original patent application I had filed. So, I wrote up a new application to take into account the new schemes, and we filed that one in about 1970, and eventually it issued, in July of 1973. The number of that U.S. Patent is 3,746,968. After 17 years, that patent expired in July 1990, and consequently I don't feel bad at all about talking about the 4701. After all, one of the functions of the patent is to teach the reader how to do something. The patentee holds a monopoly right for 17 years, but 17 years is the limit.

Vignettes—Little Stories...

In those early days of the Philbrick 4701 family, our arch-rival was Analog Devices. We were kind of nervous about how AD would bring out competing modules. Would they steal our circuits? Would they infringe on our patents? Year after year we waited for the shoe to drop. Finally, after just about every other competitor was selling V/F converters and F/V converters, Analog Devices brought out its modules. They did *not* infringe, they did *not* steal our circuits. We also thought they did not have very impressive performance or features, as they had designed their circuits to do things the hard way. But mostly, they were late in the marketplace. Later, we found out why:

At Analog Devices, the engineers always designed what the marketing managers told them to. This was rather different from Philbrick, where the engineers often designed things that marketing people could barely comprehend—but on a good day, these unrequested products made a lot of good business, and a lot of friends, and good profits, too. But the marketing people at AD had looked at the marketplace, and they decided there was "no market for V/F or F/V converters." Year after year, they decreed that, and V/F converters were studied, but no products were introduced. Finally, several of the AD salesmen presented evidence that even though the marketing people could prove there was "no market" for V/F converters, there really were *customers* for V/F converters, and they twisted the marketers' arms until they finally agreed to get into that business.

That reminds me of a funny story. When we had just three V/F converter products at Philbrick, and I was designing several more, one of the marketing managers at Philbrick—I recall that it was Maurice Klapfish, who had recently arrived from Analog Devices—decided to commission a market survey about V/F and F/V converters. The engineers were summoned to hear the report and conclusions of the survey. This fellow had asked many people, "Is there any market for good V/F Converters?" In each case, the answer was, "no, no significant market." We asked this fellow, whom had he asked? He said that he asked all the manufacturers who now made V/F converter instruments. Well, what would you expect *them* to say? Of course they would not want to encourage competitors!

At this point, I stood up and asked if he realized that Philbrick's entire building—the complete facilities and every concrete block in every wall—had been paid for by the profits on a product (the Philbrick P2 amplifier, see Chapter 9) that nobody had ever asked for, and that marketing never said they wanted, and in fact marketing had never even planned or specified or run a market survey on it—it was just thrust into the marketplace by engineers—did he realize that? The market-survey fellow said, no, he did not believe that. Well, of course I walked out of the meeting.

After a while, our management wisely decided that a few new V/F converter products might just be okay, if the customers continued to show good acceptance of these V/F converter products. I won't say that marketing people never approved of the designs, and features, and specs. Let's just say that I put on my marketing hat, and decided what kind of features a customer might like to buy, in a marketplace where nothing of this sort had ever been seen before.

Frank Goodenough reminded me that at one sales reps' meeting in Chicago, where all the 50-odd reps were gathered to hear the Philbrick marketing experts explain what good new things we had to sell, Frank had stood up to tell the reps about the new V/F converters and F/V converters. For some reason, the Marketing Manager, John Benson, had not gotten his full agreement on how the V/F converter line was to be promoted, so just when Frank (who, if you have ever met him, is a *very* enthusiastic guy) was beginning to say glowing things about the V/F converters, John told him to "shut up and sit down and not talk about those things." Well, if you want to make sure a person pays a lot of attention to what a speaker is saying, you just tell the speaker to "shut up and sit down." Frank did what he was told, but he sure got the reps' interest that day!

In 1988, I was interviewed for a biographical page in *EDN* magazine. When the biographical material was shown to me at the time it was ready for publication, it stated that I had designed a V/F converter using vacuum tubes, and thus I had proved that one could have made good voltage-to-frequency converters back in the 1950s and 1940s. However, even though I had been threatening for 17 years to design and build such a machine, I had never actually done it. I could either tell the editors to drop out that phrase, or I would have to actually design such a machine. I was a little rusty in my art of designing with tubes, but I took an old hi-fi amplifier and rewired a group of tubes—some 6SN7s and 6SL7s and an old OB2 reference tube, and with a minimum of tweaking, I got a good V/F converter running with the same basic design as the 4701. The linearity was down near 0.08% on the first try. So, if had a time machine, I could go back to the year 1940 and be a hero by inventing V/F converters and F/V converters and digital voltmeters using V/F converters. Unfortunately, in the year 1940, I was not even 1 year old, and if anybody was going to invent V/F converters in 1940, it was going to be somebody other than me! Still, in concept, there was nothing to prevent this design from being marvelously useful, back 40 or even 50 or 60 years ago. The V/F converter could have teamed up with the frequency counter or "EPUT Meter," to make digital voltmeters and analog-to-digital converters, 20 or 30 years before the first DVMs were sold commercially. . .

In 1971 I was invited to do a lecture tour, talking about Philbrick's new products in a 14-day tour of Japan. I talked about many other products as well as the V/F converters, with a slide show and various lectures and data sheets. One of the sequences I made up was to show the linearity of the model 4705. After we trimmed the gain-adjust pot to deliver 1,000,000 Hz at 10.00000 V input, I showed the test results—a 5-digit DVM sat on top of a frequency counter. For example, when you put in 5.00000 V, the frequency might typically be 500,008 Hz—a pretty good match. I showed the sequence we actually followed in our final test sequence—the input was decreased down to 2 V, 1 V, 0.1 V, 0.010 V, 0.001 V, and 0.0001 V. Finally I showed the last slide where the input was 0.00001 V, and the frequency was 1 Hz—showing a dynamic range of 1 million to 1. To my astonishment, the room full of Japanese engineers burst into spontaneous applause! I must say, this happened in Osaka, where the Japanese often show enthusiasm. In Tokyo, I noticed the engineers in the audience were impressed, but in Tokyo, people are more reserved, and applause did not seem appropriate to them. Still, that applause was the high point of my tour—and perhaps of my entire career!

Notes on "Markets"

Why did the 4701 and its companions become so popular? I think there are several good applications which made for a lot of the popularity, and most of these were not really obvious, so it is not silly to say that marketing people might be fooled.

A major application was analog-to-digital conversion. The 4701 could cover a 10-bit range, from 10 Hz to 10 or 11 kHz, and at the price it was a little better than a comparable 8-bit ADC. Further, you could get the output isolated by 1000 V or more, just by driving the output into an opto-coupler. That made it *very* attractive.

An additional variation on this theme: this ADC could serve as an integrator. Instead of just feeding the output to a 10-bit counter, you could feed it to a 20-bit counter, or to a 16-bit or 24-bit counter with a 10 or 16 or 20-bit prescaler. So, you could integrate signals, just as with a wattmeter, by adding a couple of inexpensive counters, CD4020 or 4040 or similar. I think that made a lot of friends, because in the early 1970s, people were not interested in screwing around with analog computers or analog integrators. They didn't want to reset them, or have anything to do with them. The 4701 let them get the job done and all the user had to add was a couple of DIP ICs in the digital domain. Some of our big orders came because a major Japanese instrument maker was integrating signals related to air pollution, integrating them all day long. . . .

The other major application was isolation of analog signals, over a large AC or DC voltage offset. These days you can run over to Analog Devices or Burr Brown and get some cute isolators that stand off 1000 V or more, *but* those circuits only came along because the 4701 and 4702 pioneered the isolation business. I recall the first order we ever got for 1000 4701 V/F converters *plus* 1000 4702 frequency-to-voltage converters. The customer was in Japan. Many Philbrick people got *really* curious: what was he going to do with them? What industry was he in? Oh, he was in the shipbuilding industry.

But a ship is a solid slab of steel, and why would anybody be interested in isolation in such a case? Finally, some knowledgeable person pointed out, almost everything on a ship might be "grounded," but there are often dozens and hundreds of volts of AC and DC and transients and spikes, between any two "ground" on the ship. Maybe our customers weren't so stupid, after all. Maybe there *was* a "market" for V/F converters! I am still not an expert at "marketing," but I find that when I listen to the customers, as a class, they can figure out a bunch of things I never could imagine. So, even if there is no obvious place for these circuits to go, well, just stand around and see who shows up with money. Shucks—I nearly forgot: the words of Jay Last, one of the founders of Teledyne. He liked to say, "The only valid market survey is a signed purchase order." That's the last word.

How *Does* that Kluge Work?

The best way for me to explain how the 4701-type circuit works is to suggest that you assume that it *really does* work; and after we analyze each section of it, then you will agree that was a reasonable assumption. Figure 29-3 is my schematic of the circuit.

First, let's assume that the negative input of the op amp A1 is a few millivolts more negative than the positive input, and that V_{IN} is some positive voltage. Then rather soon, the voltage at the negative input will cross over and exceed that of the

꒛TELEDYNE PHILBRICK

10kHz
High Performance
Voltage to Frequency
Converters

4701
4713
4725

The 4701, 4713, and 4725 are high performance, low cost, voltage to frequency converters capable of producing a 10Hz to 10kHz output pulse train from a + 10mV to + 10V input signal. Twenty percent overrange, up to 13 bit resolution and low noise feedthrough are some of the inherent features of these general purpose devices. They are available to three different guraranteed nonlinearity specifications: ± 0.1%FS (4713), ± 0.05%FS (4705) and ± 0.015%FS plus ± 0.015% signal (4725). Full scale and offset errors, ± 0.75%FS and ± 0.03%FS respectively, are the same for the three units. Applications include FM telemetry, magnetic tape recording and digital to frequency conversion.

Applications Information

Precalibrated to meet all published specifications, these devices provide the user with optional trimming for applications requiring greater accuracies (see figure below). Input offset voltage is trimmed by applying a 100mV signal to the input terminal and adjusting R2 for a 100Hz output. Full scale is then trimmed by applying 10V to the input terminals and adjusting R1 for a 10kHz output. Repeat above procedure for precise calibration.

FEATURES

- ± 0.008%FS Nonlinearity
- 20% Overrange
- 13 Bit Resolution
- High Noise Rejection
- Low Cost

APPLICATIONS

- FM Telemetry
- Precision Integrators
- Common Mode Voltage Isolation
- Digital to Frequency Conversion

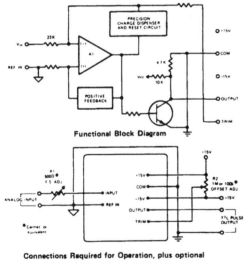

Functional Block Diagram

Connections Required for Operation, plus optional Input Offset and Full Scale Adjustments

꒛TELEDYNE PHILBRICK

Allied Drive @ Rte. 128, Dedham, Massachusetts 02026
Tel: (617) 329-1600.   TWX: (710) 348-6726,   Tlx: 92-4439

7-3

Figure 29-4.
The data sheet for the 4701 family.

positive input. Now, all this time, A1's output has been at the positive limit, near +13 V; but when the inputs cross, the output will rapidly go to −13 V. What are the interesting things happening then?

1. There is positive feedback through C3 = 15 pF, so the positive input is driven a few volts negative. This ensures that there will be a good wide healthy pulse—at least 17 or 21 μsec.
2. The right-hand end of C2 is driven from about +7.5 V to −7.5 V, and the left-hand end of C2 is discharged through the emitter of Q1. That is a specific amount of charge, $Q = C \times \Delta V$. The ΔV is not just the 15 V p-p at the right-hand end of C2, but rather it is that 15 V p-p *less* the 2.5 V p-p excursion at the left-hand end. When this charge is driven into the emitter of Q1, it comes out the collector (well, at least 99.5% of it does) and pulls the voltage on the 3.3 μF down by about 12 mV. All this time, the voltage at the positive input of A1 is much more negative than that, so this step of −12 mV is just a minor jump. But, that charge is the magic element of precision. The size of the 12-mV jump is not important, but the *charge* is.

Note, in most conventional V/F Converters, the charge is dispensed by a timer circuit such as an LM555, which gates a current ON and OFF, so $Q = I \times T$. However, you need several precision resistors, and even then, the timer is subject to drift and noise, so that is a rather crude, drifty, noisy, unstable kind of charge dispenser. In the 4701, the gain depends almost entirely on just three elements—the zener voltage, the capacitor C2, and the input resistor R8. It's true that the diodes enter in, but since the V_fs of D1, D2, D3, and Q1 cancel out the V_fs of D4, 5, 6, and 7, then there is not a lot of error or drift likely there.

3. Now that the output is staying at a negative 13 V and most of the charge on C2 has flowed through Q1, there are two more details going on:
   a. The voltage at pin 3 of A1 is tailing up gradually to be more positive than at the pin 2. After all, pin 2 was pushed down 12 mV. Soon, after about a total of 20 μsec, V_3 will cross V_2 and the output will bang back up to +13 V.
   b. During that time, the current through R2 pulls at the emitter of Q1 and makes sure that Q1's emitter settles at a stable voltage. It makes sure that Q1's emitter voltage does not tail off to some drifty voltage. Even though R2 looks like it will dump in current that would hurt precision, it actually helps the precision.
4. Okay, now finally V_3 crosses V_2 and the output goes positive. Now we have to wait for the current through R8 to pull V_2 up those 12 mV that it was pushed down. That time will of course depend (inversely) on the size of the signal input; the bigger, the faster. That means the time between pulses could be anything between 70 μsec and 9 or 90 msec. Are we forgetting something? Yes. The p-p voltage at the left end of C2 must be stable and constant and invariant of rep rate. But the diodes there might give a long tail—the voltage might settle quite gradually and give a different p-p value at different speeds. By definition, that would hurt the linearity. What's the fix? The current through R2 is the fix. That current flows through D1, D2, and D3, and forces the left end of C2 to settle to within a millivolt or two in just 50 or 60 μsec. Without R2, the linearity goes to pot. Now, it looks *really stupid* to have a circuit like this where the "precision capacitor" C2 has a resistor across it that obviously makes so much "leakage." But that controlled "leakage" turns out to be *exactly* the reason for the precision and excellent linearity. The Swedish design didn't have this, and while their circuit had good linearity at 1 kHz, it

could not be made to work well at 10 kHz. But this basic charge dispenser, when driven with suitably fast transistors, works well at 100, 1000, and even 5,000 kHz.

What else is in the circuit? D9 is needed between pins 1 and 8 of the LM301A to keep it from wasting excessive current in its negative limit. D8 is a good idea to protect the positive input from overdrive in the positive direction. Q3 functions only when you overdrive the input—let's say—pull V_{IN} up to 50 V, and put in so much current that the V/F converter stops. Then it stops with pin 2 of A1 at +1 V, and pin 6 at –13 V. It would never put out another pulse—it would never restart, even if V_{IN} falls to a legal level such as +1 or +5 V—except that after a lag, C5 gets pulled minus, and Q3 turns on and pulls pin 2 so low that operation does start again. In normal operation, Q3 just sits there and does nothing, biased OFF.

C7 acts as a pulse-stretcher. The pulse width at the output of A1 is about 22 µsec. But we had a companion F/V converter, the 4702, that could only accept 20 µsec (or wider) pulses. If A1's output pulse got any narrower than 20, the 4702 would lose accuracy. We solved the problem by putting in C7 so that when A1 makes a 20 µsec pulse, the base of Q2 would be held off a little longer than that, due to the RC lag—about 15 µsec extra. Then a 4701's pulse was always plenty wide enough to drive a 4702.

The little capacitor C6 was added to make the p-p voltage at V_2 a little bigger, so when some LM301's were a little slow, there was now a bigger signal between V_2 and V_3, and the amplifier would not balk. After all, the LM301 is characterized as an operational amplifier, and if some are a little slower than others when you run them as a comparator, you really can't complain. . .

As you can see, the 4701 circuit did get a couple Band-aids®, but not excessively many, and we never really did get stumped or stuck in production. Our customers loved the linearity, the TC was pretty good, and the frequency output would never quit. They figured they really got their money's worth, and I certainly couldn't disagree with a satisfied customer.

A Final Note

Now, in July 1988 I did read a letter, which a Mr. Sidney Bertram of San Luis Obispo, California had written to the *IEEE Spectrum* (July 1988) about how he had worked on frequency-modulated oscillators in 1941. To quote from the letter: "When I joined the sonar group at the University of California's Division of War Research in 1941, I was told about three frequency-modulated oscillators they had had developed under contract—one by Brush Development, one by Bell Laboratories, one by Hewlett-Packard Co. The Hewlett-Packard oscillator, a positive-bias multivibrator, was the simplest and most satisfactory. It became the heart of the subsequent work, and one of my jobs was to give it a reproducible linear voltage-frequency characteristic."

I wrote off to Mr. Bertram and he was very helpful. He sent me a copy of a paper he had written, published in the *Proceedings of the IRE*, February 1948. This paper showed Mr. Bertram's work (in 1941) to take a basic two-tube multivibrator with six resistors and optimize it to a linearity of ± 200 Hz over a range of 37 kHz to 70 kHz, or about 0.6%. Not bad for a 1941 design!

So, we cannot say there were *no* VCOs before 1950, but they were not common knowledge, as you could not find them unless you looked outside the "Bible"—the 3000 pages of the Rad Lab Series.

Dan Sheingold

30. Op Amps and Their Characteristics

This chapter is about two kinds of circuits, *operational-amplifier circuits* and *operational amplifiers*. An operational-amplifier circuit performs one among an essentially infinite variety of defined *operations*, employing one or more amplifying devices known as operational amplifiers, or "op amps."

In this chapter, we shall discuss the kinds of operations the circuits can perform—and the role played by the characteristics of the op amps, and other elements used in the circuits, in determining how well those operations are performed. In order to accomplish this, we shall have to consider the characteristics and behavior—and to some extent the design and construction—of the op amp as a component.

Here are two basic definitions:

An op amp—the *component*—is a high-gain amplifier, usually with symmetrical differential input, designed to perform stably when connected in feedback loops having large amounts of negative feedback. An *ideal op amp* would have infinite gain and bandwidth; used with an appropriate power supply, there would be no limitations on the magnitudes (small or large) of either the signals appearing at or the impedances connected to the output or input terminals.

An operational-amplifier *circuit*, using one or more op amps—plus other circuit components—relies on the high gain of the op amp(s) in order to provide a predetermined functional relationship between the output and one or more inputs. Some simple examples of functions that can be performed by operational amplifiers include multiplying by a precise constant, adding, integrating with respect to time, active filtering, signal generation and wave form shaping, accurate voltage buffering, and voltage-to-current conversion. If ideal op amps were used, the function performed by such a circuit would depend solely on its configuration and the properties of the external components used to establish the function.

Brief History

The operational amplifier circuit is historically inseparable from the idea of negative feedback. One of the first functions of the negative-feedback electronic circuit, invented by H. S. Black and described in AIEE's *Electrical Engineering* in 1932, was to provide stable gain—independent of vacuum-tube properties and other amplifier parameters—in telephone circuits. Stable gain is still one of the most prevalent of op-amp circuit applications. Operational-amplifier circuits first appeared in recognizable form in about 1938, when feedback circuits and electronic analog computing were first employed to simulate dynamic control loops.

The term, *operational amplifier,* and a discussion of its use in analog computers to analyze problems in dynamics, appeared in a landmark paper by Ragazzini *et al* in 1947. The first op amp to appear on the market as a general-purpose functional

Figure 30-1.
Conventional
simplified
representation of
an operational
amplifier.

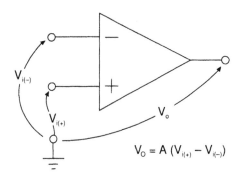

Figure 30-1. Conventional simplified representation of an operational amplifier.

$$V_O = A (V_{i(+)} - V_{i(-)})$$

building block, with differential inputs and today's familiar triangular block diagram, was the Model K2-W, which was introduced in 1952 by George A. Philbrick Researches, Inc.

Solid state op amps employing germanium transistors first appeared commercially in about 1958, high-performance types employing silicon transistors started to appear in 1960, the first low-cost potted modules appeared in 1962, and the first successful IC op amp, the μA709, surfaced in 1964. Since then, the pace has quickened, and many technologies have proliferated to make the op amp the low-cost, universal building block of electronic circuitry, available in wide variety, that it is today.

Block Diagram

Figure 30-1 is the conventional simplified representation of an operational amplifier—a triangular building block with an output terminal, two input terminals, and an external reference terminal, or *common*. Not shown are the power-supply terminals or other connections for offset trimming or external frequency compensation. The *common* terminal serves both as a basic reference terminal for voltages appearing in the op-amp circuit and as a return path to the power supply for currents flowing through elements of the op-amp circuit.

Although it is incomplete, we will use this representation because it is helpful when envisioning and discussing operational-amplifier circuits from a functional point of view. It reduces the clutter caused by repetitive (though essential) wiring. Increased levels of detail will be seen later.

In principle, depending on its circuit configuration, an electronic operational amplifier may accept input signals in the form of either current or voltage, and the output may be generated as either a voltage or a current. However, for the discussions that follow, *operational amplifiers* will be considered to respond to voltage, and to generate voltage outputs. The circuits they are used in, however (op-amp *circuits*), may be designed to respond to voltage or current, or both, and to generate rationally determined voltages, currents, impedances, or power levels.

Gain

Let us assume that the inputs and outputs of the amplifier in Figure 30-1 can swing positively and negatively with respect to common through a range of at least ±10 V. If the voltage at the input labeled "+" increases (i.e., becomes more positive), the

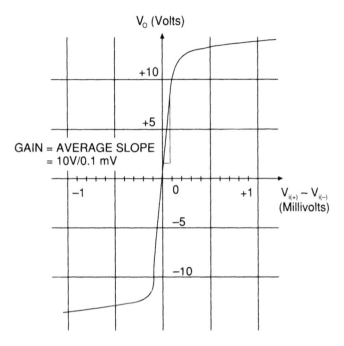

Figure 30-2.
Saturation point for an opertional amplifier.

V$_o$ (Volts)

+10

+5

GAIN = AVERAGE SLOPE
= 10V/0.1 mV

−1   0   +1   V$_{i(+)}$ − V$_{i(-)}$
                                    (Millivolts)

−5

−10

output will tend to increase; if the input labeled "−" increases, the output will tend to decrease (become less positive or more negative). Thus these inputs are *differential*; the output tends to respond to their difference. Let us further assume that the amplifier has an *open-loop gain* of 100,000, or

$$V_o = A_{vol}\left(V_{i^+} - V_{i^-}\right)$$

$$= 100,000\left(V_{i^+} - V_{i^-}\right)$$

This means that if the difference between the inputs changes by as little as 100 μV (10^{-4} V), the output will change by 10 V. Another way of stating this is to note that for the output to change by less than ±10 V, the input difference must change by less than ±100 μV. If the input difference is larger, it may cause the output to become *nonlinear*, and eventually to *saturate*, as shown in Figure 30-2.

The Comparator

One consequence suggested by this property is that the amplifier can be used simply to determine whether one voltage is more positive than the other. If V_i^+ is even 1 mV more positive than V_i^-, the output will swing to its upper limit; if V_i^+ is 1 mV more negative than V_i^-, the output will swing to its lower limit. When functioning in this way, the op amp is acting as a *high-gain comparator*.

The comparator mode has its uses. For example, by making a binary (two-valued) decision, it produces a digital signal based on analog information (that is, it acts as a *one-bit analog-to-digital converter*). It is important to note, though, that this mode of operation represents a very small fraction of the potential applications of the op amp. Of much greater usefulness is the behavior of the op amp in negative-feedback circuits where the output is forced by the amplifier to be always within bounds.

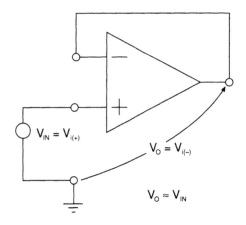

Figure 30-3.
Simplified model
for examining
effects of nega-
tive feedback.

$V_{IN} = V_{i(+)}$

$V_O = V_{i(-)}$

$V_O \approx V_{IN}$

The Follower

Perhaps the easiest way to begin to develop an understanding of how negative feedback around an op amp makes precision operational-amplifier circuits possible is to
consider a circuit in which the only feedback element is a piece of wire. In Figure
30-3, the output is jumpered to the negative input (V_i^-), and an input voltage, V_{in}, is
connected between the positive input, V_i^+, and common. We make the appropriate
assumptions that the impedance from the input terminal to common is extremely
high compared with the source impedance of V_{in}, and the impedance in series with
the output terminal is quite low, compared with the load impedance.

Continuing to use the example of an amplifier with an open-loop gain of 100,000,

$$V_o = 100,000\left(V_{i^+} - V_{i^-}\right)$$

Since $V_i^- = V_o$, we can solve for V_o,

$$V_o = \frac{V_{i^+}}{1 + \dfrac{1}{100,000}}$$

This means that the output voltage must follow the input voltage to within $1/A_{vol}$,
and the circuit performs as a high-accuracy unity-gain follower with 0.001% error.
The reader should note that if there is a tendency for V_o to change, that change will
be applied to the V_i^- terminal, causing the amplifier to oppose and correct for the
change.

To show just how powerful the effect of feedback is, let us consider the same
circuit, but with one modification: that the amplifier is driving a 1,000-Ω load, and
that its output impedance is an outrageous 1,000 Ω (Figure 30-4). Only half the
output voltage, V_o, is being fed back to V_i^-, *but* the voltage applied to the load is still
equal to V_i^-, which must still be very nearly equal to V_i^+. Therefore,

$$V_{out} = (1/2)\left(100,000\left(V_{i^+} - V_{out}\right)\right)$$

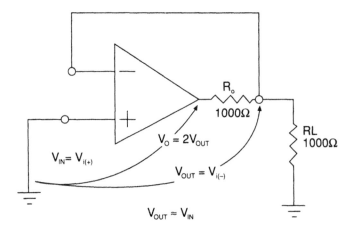

Figure 30-4.
Modification of
the circuit in
Figure 30-3.

and

$$V_{out} = \frac{V_{i^+}}{1 + \dfrac{1}{50,000}}$$

This example shows that the high open-loop gain and negative feedback cause the amplifier to do whatever is necessary to make the voltage at its negative input very nearly equal to the voltage at the positive input, including—in this case—providing an amplifier output voltage, V_o, that is twice as great as the required output voltage, V_{out}.

The effect of the large output impedance, *inside the loop*, is to halve the open-loop gain. The overall effect is to double the extremely small error to only 0.002%; note that the larger the gain, the smaller the error, and the smaller the effects of output impedance. The important point is that the actual value of the open-loop gain is unimportant, as long as it is sufficiently high.

Practical note: This example of extreme loading was chosen to demonstrate in a dramatic way the high gain-stability of op-amp circuits. It should be evident that the amplifier's maximum output span will have to be twice the voltage span applied to the load. There may also be dynamic stability and bandwidth problems in circuits with high series output impedance.

Application Principles

So far, we have demonstrated two inherent principles that can be applied to make the op amp useful: high-precision voltage *comparison* and high-precision voltage *following*. Here is a list of some additional principles of operational-amplifier circuits that lead to high-precision functions that can be performed with single amplifiers:

- Amplification (follower with gain)
- Voltage-to-current conversion
- Voltage null and current balance
- Voltage inversion
- Voltage amplification-attenuation
- Current-to-voltage conversion

- Independent current summing
- Independent voltage summing
- Voltage-current relationship—static or dynamic, linear or nonlinear
- Current-voltage relationship—static or dynamic, linear or nonlinear
- Transadmittance functions
- Inverse functions (apparent reversal of causality)
- Differential operations
- Operations—linear and nonlinear—involving positive feedback

Each of these principles is a potential problem-solving tool for the circuit designer, suggesting a multitude of specific embodiments. Each stems directly from the following statements, which express the underlying basic principles of ideal differential-input op amps:

1. When negative feedback is applied to an ideal differential-input operational amplifier, the differential input voltage must approach zero.
2. No current flows into either input terminal of the ideal differential-input op amp.
3. In addition, noise and DC offsets do not exist; and there are no constraints on input voltage range (from very small to very large), output voltage range, output current range, energy dissipation, bandwidth, or the characteristics of circuit elements connected to the amplifier.

Although the ideal operational amplifier is a fictitious concept that can never be realized, there are several good reasons for using it to explore the properties of circuits employing it. First, the analysis of ideal circuits simplifies the issues involved in understanding unfamiliar op-amp circuit designs or creating new ones. Second, the circuit employing an ideal device is a good starting point for distinguishing between errors inherent in the circuit concept and those contributed by the amplifier itself. (For example, a feedback circuit which is inherently unstable, even if an ideal op amp is used, can be discarded or modified.) Finally, for many purposes, there do exist op amps that approach the ideal sufficiently closely with respect to the relevant desired key performance characteristics.

Precision amplification (follower with gain)

As the previous discussion implied, and Figure 30-5 shows, the op amp can be used as a high-input-impedance follower-with-gain. It could be used to buffer a high-

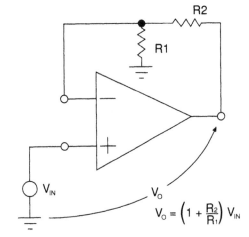

Figure 30-5. High input impedance follower with gain.

$$V_O = \left(1 + \frac{R_2}{R_1}\right) V_{IN}$$

impedance, low-level input source from a low-impedance output load and provide accurately determined gain for the input signal. (Here, and in the discussions that follow, we will assume that the amplifier's open-loop gain is so high that the input terminals must be at the same voltage). In Figure 30-5, the feedback consists of a voltage divider with ratio M; $M = R_1/(R_1 + R_2)$. Thus

$$V_{i^-} = MV_o$$

and the amplifier seeks to enforce this condition:

$$V_{i^-} = V_{in}$$

Therefore,

$$V_{out} = V_{in}/M$$
$$= (1 + R_2/R_1)V_{in}$$

The gain element may be a pair of fixed resistors, a variable potentiometer, a tapped divider consisting of many resistors, or even a stepped attenuator.

In practical circuits (i.e., those using nonideal amplifiers), offsets, noise, and many kinds of errors are amplified by the same factor, $1/M$. You will find that one of the steps in analyzing the performance of any op amp circuit is to boil down the feedback circuit—whatever its complexity—to an ultimate impedance ratio, which defines the circuit's "noise gain." An example of this was seen in the circuit of Figure 30-4, where $M = 0.5$; the error due to limited gain was doubled, i.e., multiplied by $1/M$.

Precision Voltage-to-Current Conversion

In the circuit of Figure 30-6, a precision current is generated to flow through a two-terminal load, indicated by the ammeter, A, and a series resistance, R_L. Since the voltage at V_i^- must follow the input voltage at V_i^+, the current that is caused to flow through resistor R, from V_i^- to common, must be equal to V_{in}/R. The current must come from the amplifier's output, via R_L and the ammeter, since there is ideally no current flowing through the amplifier's input terminals.

The output voltage, V_{out}, is whatever it has to be (within device limitations) to

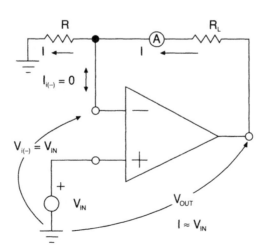

Figure 30-6.
Precision voltage-to-current conversion.

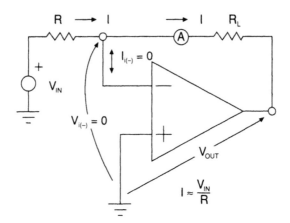

Figure 30-7.
Alternate method
of generating a
precision
current.

maintain V_i^- equal to V_{in}. The value of current is unaffected by the impedance of the load elements in the feedback circuit—only by V_{in} and R (and any leakage or bias currents in nonideal circuits). The amplifier adjusts the value of V_{out} to correct for any changes in the impedance of the load elements.

Voltage Null And Current Balance

Figure 30-7 shows another way of generating a precision current for the same two-terminal feedback pair. Here, V_i^+ is connected to common ("grounded"). V_{in} and R are connected in series with V_i^-, and the load (A and R_L) is connected as in Figure 30-6. In this configuration, V_i^- must be at zero (voltage null, or *virtual ground*); thus, the current flowing through R must again be equal to V_{in}/R, but in the opposite direction to that shown in Figure 30-6, for positive V_{in}.

As before, a current balance must be maintained: current through the feedback elements must be equal to the current generated by V_{in} and R. V_{out} again is adjusted by the amplifier to make it all happen, irrespective of the load impedance.

Besides differing in polarity, this circuit also differs significantly from Figure 30-6 in that V_{in} is not looking at high impedance—it is driving a load precisely equal to the resistance of R to ground, since the V_i^- terminal is at 0 V. It is important to note that although R appears to V_{in} to be grounded—because it is at ground potential—*it is not actually connected to ground*; this is called a *virtual ground*, a powerful concept in developing useful op-amp circuits.

Precision Voltage Inversion

In the previous example, we noted that V_{out} must be whatever value is necessary to maintain the current, I, established by V_{in}/R through the feedback path. If the feedback path consists of a fixed resistance, R_F, equal to R (Figure 30-8), the input current, I, flowing through feedback resistor, R_F, develops a voltage equal in magnitude to V_{in}, but of opposite polarity as seen from V_{out}. Since V_i^- must be at ground potential, $V_{out} = -V_{in}$. Thus, the circuit acts as a unity-gain sign inverter, and V_{in} and V_{out} are in push–pull.

Precision Voltage Amplification-Attenuation

Since the gain of the circuit of Figure 30-8 is equal to $-R_F/R$, it will function as a voltage amplifier if R_F is greater than R, and as an attenuator if R_F is less than R. (The noninverting circuit of Figure 30-5 can only provide active gain greater than

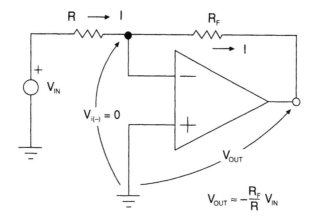

Figure 30-8.
Precision voltage
amplification-
attenuation.

$$V_{OUT} \approx -\frac{R_F}{R} V_{IN}$$

or equal to 1; but it could be used to buffer a passive attenuator for gains less than 1. The ratio, R_F/R, may be fixed, adjusted continuously with a potentiometer, or in steps with a stepped attenuator, from infinity (in principle) to zero. Gain accuracy depends on ratio accuracy. The input impedance for V_{in} is equal to R.

Precision Current-to-Voltage Conversion

In the circuit shown in Figure 30-9, the input element is an imperfect current source—for example, a photodiode—consisting of an ideal infinite-impedance source in parallel with a resistance, R_p. Since the amplifier maintains V_i^- precisely at 0 V, there is no voltage drop across R_p, hence no current flow through it. All the current generated by a positive I_{in} must flow through the feedback resistor, producing a negative output voltage equal to $-I_{in}R_F$. Thus, the input current is converted to a precisely determined output voltage that is independent of the current source's resistance. Another way of looking at it is to consider that an ideal load for a current source is zero impedance, provided in this case by the virtual ground at the op amp's inverting input.

Precision Independent Current Summing

In Figure 30-10, a second (positive) and a third (negative) current source, I_2 and $-I_3$, have been connected to the amplifier's inverting input terminal along with I_{in}. Since, all three current sources think they are connected to ground, the only path each (and

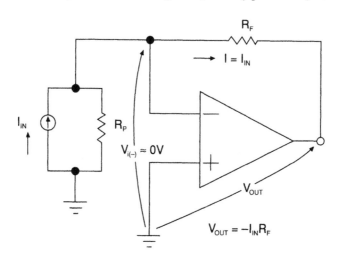

Figure 30-9.
Precision current-
to-voltage
conversion.

$$V_{OUT} = -I_{IN}R_F$$

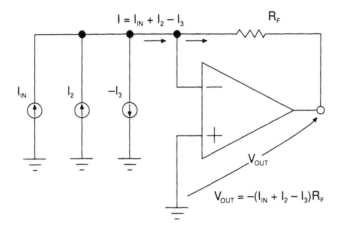

Figure 30-10.
Precision
independent
current summing.

$I = I_{IN} + I_2 - I_3$

R_F

$V_{OUT} = -(I_{IN} + I_2 - I_3)R_F$

therefore all) of the three currents can subsequently take is to be summed through the node at the inverting input—or "summing point"—and to flow through feedback resistor R_F. Again, the output voltage must be equal to the inverted voltage drop in R_F, or, in this case, $-(I_{in} + I_2 - I_3)R_F$.

Precision Voltage or Current Summing

In Figure 30-11, I_2 and I_3 have been replaced by voltage sources V_2 and $-V_3$, in series with resistors R2 and R3. Since R2 looks (to source V_2) as though it is grounded, the current through R2 is V_2/R_2; similarly, the current through R3 is $-V_3/R_3$. In the same way as for current sources, the voltage drop in R_F, and hence the output voltage, is equal to $-(I_{in} + V_2/R_2 - V_3/R_3)R_F$.

Thus all of the inputs are summed independently. Any of the input voltages, currents, or resistances can change without affecting the others. The gain (or attenuation) for V_2 and V_3 depends only on the ratios, R_F/R_2 and R_F/R_3. The respective input impedances are R_2 and R_3.

Precision Voltage to Current Conversion—Static or Dynamic, Linear or Nonlinear

So far, we have considered only resistive circuits. But op amps will work with many other kinds of elements. If, for example, in the inverting connection, the input element is a capacitor (Figure 30-12), the current flowing through the summing point is equal to $C \, dv/dt$, and the output voltage, the same magnitude as the drop

Figure 30-11.
Precision voltage
or current
summing.

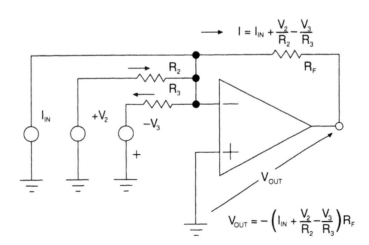

$I = I_{IN} + \dfrac{V_2}{R_2} - \dfrac{V_3}{R_3}$

R_2

R_3

R_F

$V_{OUT} = -\left(I_{IN} + \dfrac{V_2}{R_2} - \dfrac{V_3}{R_3}\right)R_F$

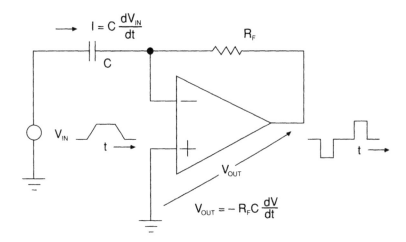

Figure 30-12.
Precision voltage-
to-current con-
version, static or
dynamic, linear or
nonlinear.

across R_F, is equal to $-R_F C\, dV/dt$. The circuit is, in effect, an ideal differentiator; the current through the feedback circuit is proportional to the derivative of the input voltage, and the output voltage will be proportional to (hence will measure) the rate of change of the input. In similar fashion, other, more complex circuits, with capacitors or inductors, may be connected to the input, to produce a current that is a dynamic function of the input voltage.

There is no restriction to linear circuit elements. For example, if we wanted to plot the reverse leakage current of a diode, it could be connected between a variable test voltage and the amplifier's inverting input, as shown in Figure 30-13. The diode leakage current, I_L, for each value of input voltage, V_{test}, would flow through the feedback resistance, producing a set of proportional output voltages, $-I_L R_F$, which measure the leakage current.

Precision Current-to-Voltage Conversion—Static or Dynamic, Linear or Nonlinear

Similarly, if R_F is replaced by a dynamic element, such as a capacitor, the output voltage will reflect its relationship between voltage and current. For example, if a capacitor, C, is used (Figure 30-14) with an input current determined by input voltage, V_{in}, and input resistance R, i.e., V_{in}/R, then the output voltage, proportional to the accumulated charge, is equal to $-(1/C)\int I\, dt$, or $-(1/RC)\int V_{in}\, dt$. Integrators are useful for measuring small currents, generating linear ramps, and performing integration operations in analog computing.

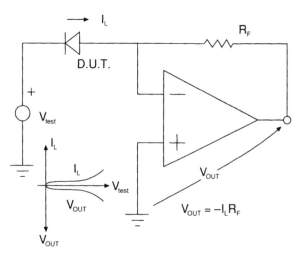

Figure 30-13.
Method of testing
the reverse
leakage
current of the
diode.

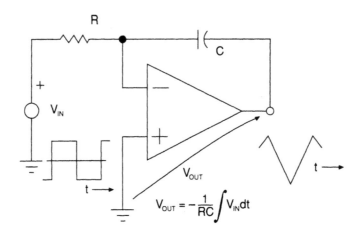

Figure 30-14.
Precision current
to voltage con-
version, static or
dynamic, linear
or nonlinear.

$$V_{OUT} = -\frac{1}{RC}\int V_{IN}dt$$

Again, there is no restriction to linear elements. For example, an op amp can be used to generate voltage proportional to the logarithm of an input by making use of the logarithmic relationship between the forward current and voltage of a diode-connected transistor. In Figure 30-15, a negative input current, I (perhaps supplied by a photomultiplier), flows through a feedback diode connected in the forward direction. The voltage developed across the diode, hence at the output, depends on the ideal voltage-current relationship of the diode, i.e., $(kT/q)\ln(I/I_o)$, where k is Boltzmann's constant, q the charge on an electron, T the absolute temperature, and I_o a reference current that depends on diode geometry and temperature.

Precision Transadmittance Functions and Precision Inverses

These last few examples have demonstrated two points that are worth some discussion. Figure 30-16 shows an op amp with two generalized nonlinear three-terminal networks connected to it, one in the input path, the other in the feedback path. Recognizing that input current must equal feedback current, *and assuming that the circuit is stable*, consider three cases:

1. If $g(\ )$ is represented by a resistor, R, the V_{out} is equal to $-R\,f(V_{in})$.
2. If $f(\ )$ is represented by a resistor, R, then V_{out} is equal to $g^{-1}(-V_{in}/R)$.
3. In general, the output is a direct function of $f(\ )$ and an inverse function of $g(\ )$, that is, $g^{-1}(-f(V_{in}))$. If $f(\ ) = g(\ )$, then $V_{out}=-V_{in}$.

Figure 30-15.
Generation of a
voltage
proportional to
the logarithm of
an input.

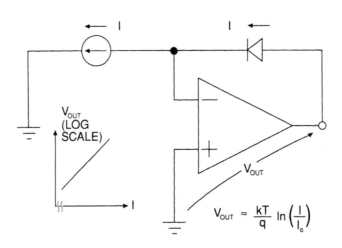

$$V_{OUT} \approx \frac{kT}{q}\ln\left(\frac{I}{I_o}\right)$$

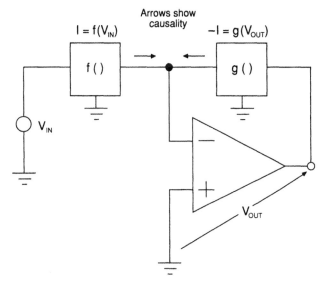

Figure 30-16.
Op amp with generalized nonlinear three-terminal networks.

One is sorely tempted to conclude that, in general, the inverse of a function can be obtained by lacing it in the feedback path, as several of the preceding examples have demonstrated. But to do so would be wrong without considering the stability question. Suppose, for example, that V_{in} is a step and $f(\ )$ is a passive circuit having a pure time delay of 10 min; if g is a similar circuit, then the output must anticipate its response by 10 min and produce a step immediately, clearly impossible without some other means of communication between input and output. Such a circuit would be unstable.

The common sense approach to anticipating "nonstarters" is to remember that an op-amp circuit is not simply an algebraic entity but involves time *causality*. The input signal to the op amp causes the output to change, which in turn causes the feedback network to deliver a correction signal to the input. If a stable voltage null and current balance cannot exist at the input, the circuit is not workable.

Precision Differential Operations

Figure 30-17 shows how an op amp circuit can subtract two input signals, with attenuation or gain. Input signal V_1 produces a voltage $V_1 R_{B1}/(R_{A1} + R_{B1})$ at the noninverting input, V_{in}^+. V_2 and the output combine linearly to produce a voltage

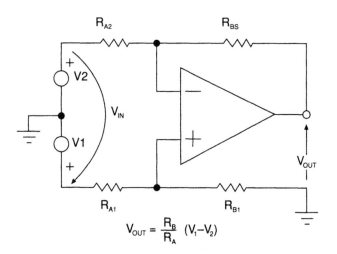

Figure 30-17.
Precision differential operations.

$$V_{OUT} = \frac{R_B}{R_A} (V_1 - V_2)$$

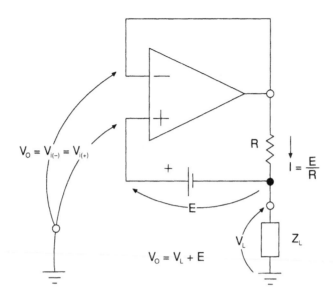

Figure 30-18.
Constant current source employing a floating voltage source.

$V_2 R_{B2}/(R_{A2} + R_{B2}) + V_{out} R_{AZ}/(R_{A2} + R_{B2})$ at the inverting output, V_{in}^-. Assume that the resistor ratios are matched, i.e., $R_{B1}/R_{A1} = R_{B2}/R_{A2} = R_B/R_A$. Since the output of the amplifier must be whatever value is necessary to make $V_{in}^- = V_{in}^+$, the output voltage must therefore be equal to $(R_B/R_A)(V_1 - V_2)$.

Positive and Negative Feedback

The application principles discussed above all involve *negative* feedback. However, limited amounts of positive feedback can also be used constructively. Figure 30-18 shows an example of a constant current source, employing a floating voltage source (e.g., a battery) with terminal voltage, E, a resistance, R, and an op amp. The op amp is connected as a unity-gain follower, and a current, I, is supplied through R to a variable load, Z_L. Since the voltages, V_i^+ and V_i^-, must be equal, the voltage across R must be equal to E; hence I, the current through R, must be equal to E/R, irrespective of the value—or nature—of Z_L (Z_L may even include voltage sources, inductors, capacitors, switches, and nonlinear elements, as long as the configuration is stable). If the voltage source, E, is a battery, it should have a lifetime approaching its shelf life, because no current is drawn from it.

Properties of Non-Ideal Op Amps

As noted earlier, the value of the operational amplifier is embodied in the two statements that embrace the fundamental source of its usefulness:

1. When negative feedback is applied to an ideal differential-input operational amplifier, the differential input approaches zero.
2. No current flows into either input terminal of the ideal differential-input amplifier.

Real amplifiers, components, and circuit environments introduce departures from the ideal; these are conveniently treated as errors. With the wide selection of amplifiers currently available, errors can be made very small for an application serving a specific purpose.

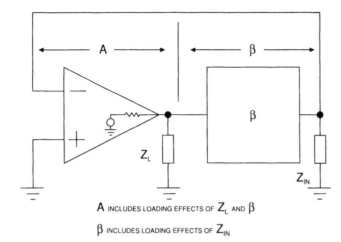

Figure 30-19.
Attenuation (β)
as a function of
frequency. A
includes loading
effects of Z_L and
β. β includes
loading effects
of Z_{IN}.

A INCLUDES LOADING EFFECTS OF Z_L AND β

β INCLUDES LOADING EFFECTS OF Z_{IN}

The departures from the ideal, or error sources, which become the device's *specifications*, can be grouped into several overlapping categories, which are related to the assumptions made for ideal amplifiers (in parentheses):

- Transfer function–related errors (infinite gain and bandwidth)
- Offsets and noise (no voltages or currents unrelated to signals)
- Large-signal dynamics (no limitations on signal range)
- Thermal sensitivities (independence of temperature)
- Physical limitations and performance boundaries (no limitations on signal swings and loading)
- Excitation (power supply) requirements and sensitivities (no limitations due to power supply)

Transfer Function Related

The transfer function of an operational amplifier circuit has the general form

$$V_o = \text{(ideal relationship to inputs)(gain error)}$$

Some typical ideal relationships have been discussed earlier. The gain error factor can often be best considered by dividing the circuit into two portions, the amplifier and the feedback circuit. Between the inputs of the amplifier and its output terminal is a gain, A, a function of frequency; between the output and the feedback terminal is an attenuation (usually), β, also a function of frequency (Figure 30-19). The error factor, referred to above, depends on the *loop gain*, the product of A and β. It is

$$\text{Error factor} = 1/(1 + (1/A\beta))$$

A is the open-loop gain of the amplifier, taking into account its external load and the loading effect of the feedback circuit; β is the "gain" of the feedback circuit, including all loading effects, such as amplifier input impedance, with all input sources set to zero. A simplified circuit demonstrating the error caused by a gain reduction of 0.5, due to loading (or a β of 0.5) is shown in Figure 30-4. The unity-gain inverter circuit (see Figure 30-8) also has a β of 0.5 when $R_F = R$.

Several "gain" ratios can be identified that are often used in discussions of feedback circuit performance. The ideal relationship, A, and β have already been discussed. The reciprocal of β, i.e., 1/β, is often called the *noise gain*; it is the (approximate) closed-loop gain factor by which errors referred to the input are amplified (as

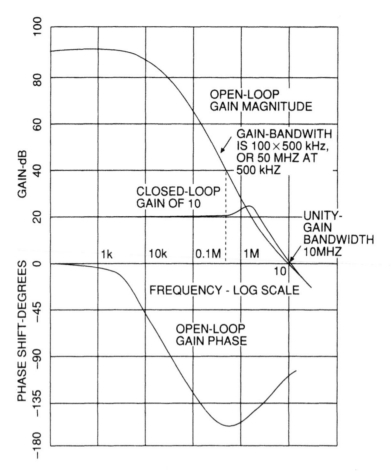

Figure 30-20.
Open-loop
frequency
response of a
typical op amp.

demonstrated in the discussion associated with Figure 30-4). The product, $A\beta$, is called the *loop gain*; it is a measure of the net gain available for limiting the amplification of errors; it also provides a measure of the loop's stability.

Gain

The DC gain, A_{vol}, is the large signal voltage gain, with the effects of rated output load (including the loading effect of the feedback network). The amplifier's open-loop gain is the ratio of specified full-scale output voltage range to the maximum input required at a specified value of load resistance. Gain may be expressed as a ratio, in volts/volt, or as a logarithmic quantity, in decibels; dB -20 $\log_{10}A$.

Gain as a Function of Frequency

Operational amplifiers have high, but limited, DC gain, generally ranging from about 10,000 to well over 1,000,000 at DC, and rolling off at higher frequencies. The error caused by the principal assumption above, that gain approaches infinity, appears mainly in the form of overall gain error, described by the error factor. As the error signified by the error factor increases, nonlinearity will also become increasingly perceptible. In practical amplifier applications, the open-loop gain of the amplifier must be great enough to ensure adequate loop gain, $A\beta$, and hence small gain error.

Even if β is independent of frequency, frequency dependence of the amplifier's open-loop gain causes the loop gain to depend on frequency, with a consequent

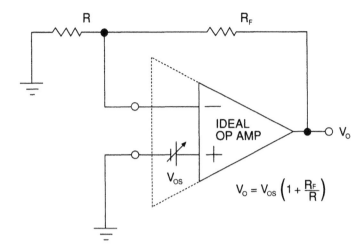

Figure 30-21. Follower-with-gain configuration.

$$V_O = V_{OS}\left(1 + \frac{R_F}{R}\right)$$

effect on gain error factor. Amplifier phase shifts with frequency can cause the gain error factor to take on values greater than unity, which could result in peaking of the output at certain frequencies, or even instability. As a rule, if $A\beta$ is equal to or more negative than -1.0, the circuit will tend to be unstable; this can happen if the sum of the phase shifts of A and β is $180°$ for frequencies at which the gain magnitude is greater than 1, or vice versa.

To avoid potential instability, many op amps are designed to have an approximately constant $90°$ phase shift at high frequencies for which the open-loop gain is greater than 1; thus, if the feedback circuitry—including the effects of stray capacitance—adds substantially less than $90°$ additional phase shift in that frequency range, the circuit promises stability. Figure 30-20 shows the open-loop frequency response of a typical op amp and its closed-loop response when connected for a gain of $+10$ V/V.

Two terms sometimes encountered are the *unity gain bandwidth* and the *gain bandwidth product* at a given frequency (if the gain were inversely proportional to bandwidth); they are illustrated in the figure.

Because the amplifier has limited ability to drive internal or external capacitive loads at high frequencies, there is a tendency towards voltage saturation at the highest frequencies. The gain plotted here as a function of frequency is called *small-signal gain*. It is the ratio of output to input for peak-to-peak voltage levels that are small enough to provide an approximately linear relationship (output proportional to input). In the error equation, both A and β are treated as complex functions of frequency.

Offsets and Noise

The ideal amplifier does not introduce any signals that were not present at the input, and the op amp circuit is considered to be sufficiently shielded and guarded to prevent it from picking up any noise from the environment.

Offset

Real op amps have errors that behave like voltage or current inputs. For example, in the follower-with-gain configuration (Figure 30-21), if a sufficiently sensitive high impedance meter is used, a DC output will be found to exist—even if there is no signal input—and it will be proportional to $1/\beta$ (i.e., $1 + R_f/R$), as if it were produced by a battery in series with the input. this voltage is called *offset*; although

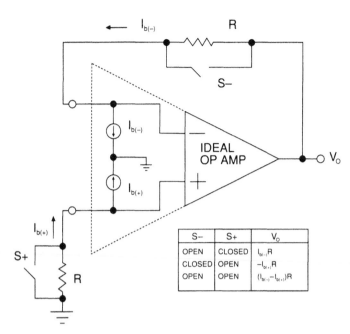

S–	S+	V_o
OPEN	CLOSED	$I_{b(-)}R$
CLOSED	OPEN	$-I_{b(+)}R$
OPEN	OPEN	$(I_{b(-)}-I_{b(+)})R$

measured at the output, it is always *referred to the input*, i.e., divided by the "noise gain," $1/\beta$. It generally drifts with time, temperature, and power-supply changes.

A key specification that depends on the design of the amplifier, offset can range from tens of microvolts to tens of millivolts. Many op amps are provided with a set of terminals that a pair of resistors or a potentiometer can be connected to for "zeroing" the offset. A few amplifier types reduce the offset automatically by means of schemes involving internal high-frequency switching and AC coupling or amplification. Because these offsets are very small, great care must be taken in measuring them, to ensure that they are not produced by external phenomena, such as thermocouple effects in the wiring or rectification of high-frequency noise in the vicinity.

Bias and Offset Currents

If an amplifier, whose offset has been zeroed out, is connected with a high-value feedback resistance, R, as shown in Figure 30-22, a DC voltage proportional to R will appear at the output. Because it is proportional to R, it seems to come from a current source, although none is connected to the amplifier. Indeed, the source of the current is the input stage of the amplifier; it may be a transistor base bias current, a FET leakage current, or a more complex phenomenon. In any event, it is called *bias current*. Depending on the device design, I_b can range from microamperes down to several femtoamperes (10^{-15}A). A bias current of 100 nA (0.1 µA) flowing through a 1 MΩ resistor produces an output offset voltage of 0.1 V. Amplifiers with junction transistor input stages may have bias current as low as 1 nA, while FET-input op amps can have bias currents below 1 pA at 25°C.

For most op amp types, if the amplifier is connected as a follower, with a low value feedback resistor, and R is reconnected from V_i^+ to ground, a voltage of about the same magnitude but opposite polarity will appear at the output, indicating that the bias currents tend to be similar in magnitude and flow in the same direction in relation to the amplifier.

If the impedances connected to the two input nodes are equal, the net effect of

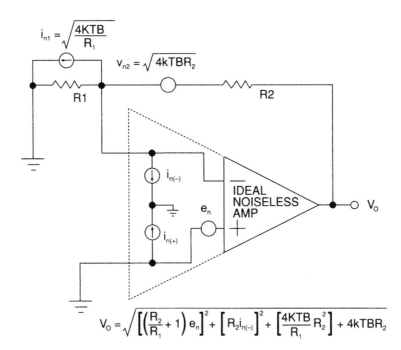

Figure 30-23.
Noise sources in
op amp circuit.

$$V_O = \sqrt{\left[\left(\frac{R_2}{R_1}+1\right)e_n\right]^2 + \left[R_2 i_{n(-)}\right]^2 + \left[\frac{4KTB}{R_1}R_2^2\right]} + 4kTBR_2$$

bias current at the amplifier output will depend on the difference between the two bias currents. The specified value of this difference between the bias currents is called *offset current*. Offset current is usually from one-half to one-tenth bias current. As might be expected, for amplifiers having low offset current and used in symmetrical circuits, there is a tendency for the effects of bias current to be cancelled.

Offset and bias currents are functions of time and temperature. They must be measured carefully in circuits and environments that ensure that stray DC leakages (for example, to the power supply across circuit boards) are negligible compared with the expected values. The maximum bias current specification should refer to the magnitude of the larger of the two currents; and its 25°C "room temperature" value should be specified at thermal equilibrium, i.e., warmed up. Since not all manufacturers specify bias current in this way, the reader should be cautious.

Noise

When all noise picked up from the environment is eliminated, there will be an irreducible minimum of noise due to the amplifier itself. There will be a voltage component of noise, analogous to offset voltage, and two current-noise sources, analogous to bias current, but independent and uncorrelated with one another. These noise signals, generally Gaussian, but sometimes including random shifts between fixed levels (popcorn noise), are translated into output voltages in the same way as offsets; voltage noise is amplified by about $1/\beta$, and the current noises are converted to voltage by the associated impedance levels. Being independent, their effects combine by the square root of the sum of the squares within the bandwidth of interest.

In addition, the contribution of Johnson noise in the resistors that make up the external op amp circuit must be considered for the bandwidth of interest. Noise from a resistor combines with amplifier noise, and the noise from resistors elsewhere in the circuit by root-sum-of-squares (Figure 30-23).

Amplifier noise is usually measured in terms of the noise energy in a 1 Hz band-

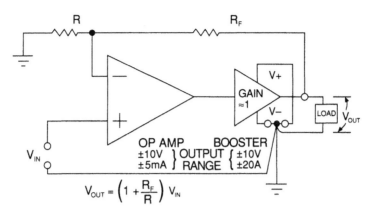

Figure 30-24.
Unity gain
follower
connected inside
the loop.

$$V_{OUT} = \left(1 + \frac{R_F}{R}\right) V_{IN}$$

width at a given frequency, or *spot* noise voltage-per-root-hertz. If the spot measurements indicate unchanging RMS noise over a range of frequencies, it is described as "white noise." If the spot measurements are in some sort of inverse relationship with frequency, it is called "pink noise" in general, and "$1/f$ noise" in the particular case where the measurements vary inversely with the square root of a ratio of two frequencies, indicating noise energy inverse with frequency.

Noise voltage and current specifications may be in RMS spot noise at specific frequencies, in RMS noise over a specified frequency range, or in peak-to-peak noise over a specified frequency range (usually at the lowest frequencies). Typical voltage noise specifications for a low noise amplifier are 0.35 µV peak-to-peak from 0.1 Hz to 10 Hz, about 10 nV/root Hz at frequencies from 10 Hz to 1 kHz.

Large-Signal Dynamics

As Figure 30-2 shows, the amplifier's range of linear performance with differential input voltage is limited. There are also limitations on available output voltage and current swings and loading, imposed by the circuit design, the available excitation from the power supply, and considerations of maximum power dissipation. Unless the amplifier's input circuit includes optical or magnetic isolation from the output and power supply, the input common-mode swing (both inputs together) is also severely limited. These values are usually specified under a given set of conditions; in addition, most manufacturers provide curves showing typical performance over a range of nonspecified conditions.

Even though most op amps have relatively small amounts of output power, they can control much larger amounts of power with high accuracy. For example, if a unity gain booster follower is connected inside the loop, as shown in Figure 30-24, its output becomes V_{out}, and the op amp itself must provide whatever drive the booster requires to enforce an accurate load voltage.

This suggests an important principle—the op amp is really the analog "brain" of a feedback control circuit; it can provide (analog) computations and the ability to force errors to be corrected to maintain the relationship programmed by the feedback circuit components. In principle, it doesn't really matter whether the "booster" block represents amperes or kiloamperes; it might even be the "muscle" of a control system that includes nonelectronic servo elements or the control rods for a nuclear power plant.

In addition to steady-state saturation effects, there are also nonlinear dynamic response problems resulting from fast signal swings that produce temporary overloads. The maximum rate at which charge on a capacitor can change (dV/dt) is

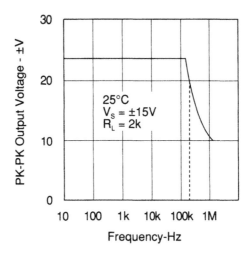

Figure 30-25.
Typical plot of
large-signal
frequency
response.
(Courtesy of
Analog Devices,
Inc.)

equal to I/C, where I is the maximum available current. Thus, for example, a 10 mA current from an output stage will be insufficient to charge a 0.0025 μF capacitive load faster than 4 V/μsec to follow a step change of input; similarly, a maximum internal current of 40 μA will drive a 10 pF capacitance at the same limited rate. This means that, if the amplifier is trying to follow a sine wave with a 10 V amplitude ($10 \sin 2\pi ft$), the maximum frequency that it can produce without additional distortion is determined by the maximum slope of the wave form:

$$d\text{V}/dt_{\text{max}} = 2\pi f \times 10$$
$$f_{\text{max}} = 4 \times 10^6/(20\pi)$$
$$= 64 \text{ kHz}$$

Since this point is difficult to establish, there are a few commonly used specifications to characterize the amplifier's output rate limitation. The maximum rate at which the amplifier can drive its output in either direction, in response to a step overdrive, is called the *slew rate* or *slewing rate*. The maximum peak-to-peak output available at any frequency, with a given load, irrespective of distortion, is called the *large-signal frequency response* of the amplifier. The *full power response* is the highest frequency at which rated peak-to-peak output is available. Typical values of slew rate range from thousands V/μsec for very fast amplifiers, to below 1 V/μsec for some widely used amplifier types.

Figure 30-25 is a typical data sheet plot of large-signal frequency response. Note the conditions of temperature, power supply, and load; the full power response for 20 V peak-to-peak output can be seen to be 200 kHz.

Settling Time

Often—for example, in digital-to-analog converter applications—it is important to know when the output of the op amp in a given circuit can be expected to have reached its final value. In response to a large step change of the input, the amplifier goes through several phases (Figure 30-26): first, there is a propagation delay, during which there is no response; then the output may slew at the maximum rate; it may overshoot the final value and bounce several times; finally, if stable, it will settle to within a permitted error band *and stay there* until the input changes.

The time from the application of a step change to the final entry of the error band, in a prescribed configuration, is the *settling time* to within $X\%$ of the final value.

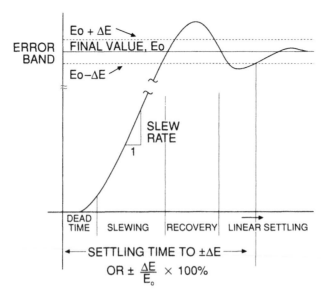

Figure 30-26.
Settling time of
an op amp.
(Courtesy of
Analog Devices,
Inc.)

Typical values for settling to 0.01% in a unity-gain inverting configuration, are 80 nsec—for a very fast amplifier—and 1 μsec for a moderately fast amplifier. Settling time is difficult to measure for fast amplifiers and very small error bands.

Temperature

Temperature affects all characteristics of an op amp. Three levels of ambient temperature effects are commonly specified—storage, operating, and rated operation with a given set of specifications. In addition, there may be maximum specifications for junction temperature, operating case temperature, and lead temperature during soldering.

The accepted design center ambient temperature is +25°C. For rated operation, the most popular temperature ranges are 0°C to +70°C ("commercial"), –40°C to +85°C ("extended industrial"), and –55°C to +125°C ("military").

While all characteristics are subject to temperature specification, the most critical ones are offset voltage and bias current. Offset drift with temperature can range from sub-microvolt-per-degree sensitivities to 100 or more μV/°C. As a rule, bias current, for bipolar transistor input types decreases at about 1% per Celsius degree; for FET-input types, the bias current tends to double for each 10°C increase.

Drifts are specified in two ways, either by providing limits to total offsets over the specified temperature range, or as a sensitivity (e.g., microvolts per degree Celsius), based on measurements at two or more temperatures. Temperature sensitivity is more informative, but it is often provided only as a "typical" specification, since it is more difficult to guarantee and verify.

Excitation and Common Mode

Op amps require power supplies for excitation. For many applications, two power supplies (typically at +15 V and –15 V) are necessary, because the inputs and/or output must be capable of swinging positive and negative with respect to common. The two power supplies may be paired identical supplies or they may be created from a single supply and regulator to provide a common connection to follow the voltage at a tapped divider midway between the two "rails." Although they are usually symmetrical, it is not an inherent necessity. For AC applications, or for applications in which no input or output level must go below the negative supply "rail," the amplifier may be operated with a single supply.

Quiescent Current

With some exceptions, op amps are usually designed to have efficient output stages. This means that the amplifier draws a small amount of quiescent current to operate the input stages and keep the output stage alive. To this is added the load current, i.e., V_o/R_L. Quiescent current is specified as a maximum quantity. When fully loaded, the amplifier may draw supply currents well beyond 100% of the quiescent current. To avoid excessive noise and pickup on the power supply rails, the power supplies used for op amps should have low dynamic output impedance.

Offset Sensitivity to Power Supply

In addition to low dynamic output impedance, power supplies should have low noise and ripple and be well regulated, because the internal references for the op amp circuitry are to some extent sensitive to supply voltage. The *maximum offset versus supply* specification is a measure of this sensitivity. Usually about 100 μV/V of change (ΔV_s), it can be as low as 10 μV/V for some types, and as high as 1 mV/V for others. This specification's reciprocal is sometimes called *power supply rejection ratio* (PSRR) and is specified in logarithmic form

$$PSR = 20 \log_{10}(PSRR)$$

with values of 80, 100, and 60 dB corresponding to the above ratios.

Common Mode Rejection

In the equations on page 355, there is an implicit assumption that the ideal amplifier's response to V_i^+ and V_i^- is perfectly balanced and that they are of equal weight; i.e., the input signal is recognized by the amplifier as +1 mV, whether V_i^+ and V_i^- are 10.000 V and 9.999 V, +0.0005 V and −0.0005 V, or −9.999 V and −10.000 V, with respect to common.

The difference signal, $V_i^+ - V_i^-$, is usually defined as the *normal mode signal*; and the average of the two inputs, $(V_i^+ + V_i^-)/2$, is defined as the *common mode voltage* (CMV). For op amps, where the two inputs are usually quite close together, we can think of the *CMV* as being essentially equal to V_i^+ or V_i^-.

If the common mode voltage at an amplifier's input changes by +10 V, it means that the positive supply has, in effect, been reduced by 10 V and the negative supply has been increased by 10 V. An ideal op amp, with a perfectly balanced input stage, would not respond to such changes. For real op amps, however, the inputs are not perfectly matched. The amplifier responds as though there is a variable offset voltage between the inputs, a function of *CMV;* if that function can be assumed to be linear, the amplifier responds as though there is a slight gain difference between the two inputs. For example, if the offset due to *CMV* is +0.1 mV at +10 V, 0 at 0 V, and −0.1 at −10 V, the effective maximum *common mode error* (CME) for a 20 V CMV swing is 0.2 V; and the apparent gain difference is $10^{-4}/10$, or 10^{-5} V/V. Thus, the amplifier responds to $1.000005 V_i^+ - 0.999995 V_i^-$.

For a normal mode difference of 1 mV at a common mode level of 10 V, the error of the amplifier in this example, 0.1 mV, although 10% of the tiny signal, is only 0.00001% of the common mode voltage. That is, there is a *common mode rejection ratio* (CMRR) of 100,000; the ratio is usually expressed in logarithmic form, *CMR* (dB) $= 20 \log_{10}(CMRR)$, or 100 dB, in this case.

Minimum values of *CMR* for nonisolated op amps vary from about 74 dB

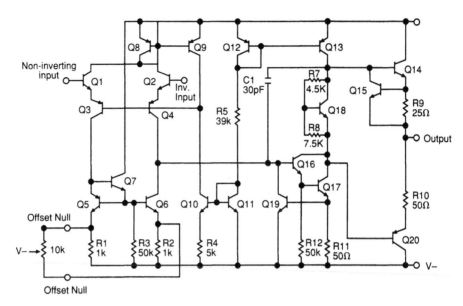

Figure 30-27.
Internal
schematic of the
741 op amp.
(Courtesy of
Analog Devices,
Inc.)

($CMRR$ = 5000) up to beyond 110 dB ($CMRR$ = 300,000) and can be as low as 50 dB ($CMRR$ = 300). For inverting-only amplifier types (including some chopper-stabilized types), common mode voltage is zero and CMR is irrelevant. Common mode rejection decreases with increasing frequency.

To investigate the effect of op amp common mode errors in op amp circuits, it is convenient to treat the maximum common mode error as an offset or noise, rather than to calculate the effects of the errors using the gain equation. Operational amplifier *circuits,* as well as op amps, have common mode error, but in general the common mode voltage is different for the circuit and for the op amp itself. This can be seen in the subtraction circuit of Figure 30-17, where the op amp's common mode voltage is as defined above, while the *circuit's* common mode voltage is equal to $(V_2+V_1)/2$.

Common Mode Voltage Specification

An op amp's minimum common mode voltage is the lowest level of common mode voltage swing for which the specified CMR can be obtained. That is, the specified common mode rejection may exist at higher voltages, but it is guaranteed at the specified minimum voltage swing.

Other Relevant Considerations

We have discussed above the most important departures from the ideal, which constitute the op amp's specifications. There are many other factors that are important in choosing an op amp for a particular op amp circuit.

Design and technology

Figure 30-27 shows the schematic diagram of the 741, a typical widely used low cost op amp, first produced by Fairchild, but now available from many sources. Although all 741s tended to be specified in the same way and to be used interchangeably, units available at present from various producers may have circuits that differ in many details from this schematic. Since this chapter largely seeks to inspire readers to design *circuits* using the devices, we will mention only those aspects of the design that affect op-amp *circuits* significantly.

The 741 is essentially a two-stage amplifier. The input stage comprises the transistors on the left-hand side of the diagram; the output and driver stage are at the right. The signal flow through the amplifier is as follows: the differential input signal, applied to input transistors, Q1 and Q2, unbalances the distribution of current, as set by Q8, between the collector circuits of pnp transistors, Q3 and Q4. The collector of Q4 controls the driver stage via Q16, and the driver's output, the collector of Q18, drives follower-connected output transistor, Q14, with complementary load transistor, Q20. The other transistors serve to provide reference, bias, and load voltages and currents, and to increase and decrease impedances as necessary.

Capacitor C1 provides frequency compensation. Dynamically, you can think of the first stage as a voltage to current converter; its output is the difference between the collector current of Q4 and the current through Q5, reflected by the current mirror, Q6. The difference current flows from the collectors of Q4–Q6 towards Q16, the input transistor of a rudimentary op amp that is connected as an integrator (see Figure 30-14). C1 is the integrating capacitor.

Following the flow of causality backward, and thinking in terms of ideal op amps, we can see that, in order for the output of the op amp to stay at a constant value, the charge on C1 must be constant, i.e., no current flows. In order for that to happen, the differential input voltage must be zero so that there is no unbalance current in the Q4–Q6 collector circuit. When a change in input calls for an output change, under feedback, in a typical op amp application, a difference voltage appears, and enough current flows to charge C1 by the required difference.

Open loop gain is limited by the transconductance of the input stage and by the transresistance of the driver stage. Small-signal bandwidth is limited by the transconductance of the input stage (especially the pnp's, Q3 and Q4) and C1. Large-signal bandwidth (slew rate) is limited by the current available to charge C1. Offset, drift, and common mode errors come from unbalanced V_{be}s, due to mismatched transistors and collector currents in the input stage; offset can be nulled externally, as shown. Bias and offset current errors are due to the base currents of Q1 and Q2 and their unbalance, which in turn depend on the magnitude of transistor current gains, their match, and the collector current balance.

Higher gains, wider bandwidths, more stable offsets, higher common mode rejection, etc., are obtained by substantial variations in circuit design, layout for better thermal balance, and improvements in processing. Lower bias currents can be obtained by using opposite-polarity current compensation or "superbeta" bipolar input transistors, but FETs are by far the most common approach. In monolithic op amps, the FETs are usually junction devices, resulting from an ion-implantation process. However, op amps with FET inputs are also manufactured using a CMOS process.

Although FET input amplifiers have orders of magnitude less bias current and much higher input impedance than do types with bipolar input transistors, FET devices tend to have higher offset and drift. This can be greatly improved by employing laser trimming of op amps on the wafer; the process is automated to keep costs low.

With C1 on the chip, this *internally compensated* op amp tends to be stable for resistive feedback ratios; however, internal compensation is not always needed—especially at high closed loop gains—and it tends to restrict bandwidth and slew rate. 101, 201, and 301 type op amps, while otherwise generally similar to the 741, permit an arbitrary compensation capacitor to be connected externally when necessary.

General purpose op amps, such as the 741, are quite low in cost and suitable for many "garden variety" applications. However, the variety of op amp designs is quite diverse, and devices are available to meet almost any specific need (and in many cases, combinations of needs), whether it be extremely wide bandwidth, extremely low drift, extremely low bias current, high output power, extremely low quiescent power, gated operation, etc. Dual and quadruple amplifiers on a single silicon chip are also available to bring the cost per channel down even further.

Chopper-stabilized op amps obtain low drift by using CMOS switches and capacitors to sample input offsets, then amplify them without drift, demodulate the amplified signal, and provide a feedback correction signal.

As noted earlier in this chapter, we have dealt with op amps as voltage to voltage devices. However, there do exist other forms of amplifiers designed for high gain and feedback operation, employing an op amp philosophy. Prominent among these are *operational transconductance amplifiers* (OTAs), or "Norton" amplifiers, which accept current inputs to control output voltage.

High-speed operational-amplifier circuits often use the *transimpedance amplifier,* a voltage-output op amp whose inverting input has a very low impedance and responds to *current.* It is designed so that a normal output voltage range can be achieved with an amplifier input current much less that the input or output signal currents, i.e., it has very high *transimpedance.* In contrast to the ordinary voltage-input op amp, with its high input impedance, leakage current approaching zero, and voltage kept near zero through feedback, a transimpedance amplifier in the same operational circuit configuration has low inverting input impedance (approaching zero), which keeps the voltage low, and the input *current* is kept near zero through feeback. Since it uses feedback and the voltage and current at its summing node are near zero, it can be expected to behave like an op amp—and it does in many circuits, but it is less sensitive to capacitance at the summing node and its closed-loop band-width decreases only slightly as the gain setting is increased.

There are many specialty integrated circuit devices being designed and manufac-tured that are complete op amp circuits on a chip, using analog medium-scale inte-gration (MSI) with one or more op amps as part of the circuit.

For performance levels that are difficult to obtain with monolithic technologies, hybrid op amps are manufactured, employing combinations of IC chips, discrete devices, and discrete components. Typical hybrid designs are used to provide the extremes of bandwidth, output power, or low noise.

In 40+ years, the op amp has emerged from a restricted, little-known role as the key component of analog computers to become one of the most popular of general purpose analog circuit building blocks. It is a well deserved role because of the easy predictability, precision, repeatability, and flexibility that can be had in circuits designed with it.

We have mentioned that, as analog MSI and LSI (large scale integration) emerge, the op-amp is becoming an on-chip component of monolithic devices. Looking ahead to the next few decades of integration, one wonders whether most op amp circuits will once again disappear into end-use devices, even as transistors have in op amps. Meanwhile, there are many fruitful and profitable uses to be made of them, even as we speculate on their future role.

Index

Acoustic memories, 23–24
Amplifiers:
  Battjes F_t doubler, 113
  bipolar long-tailed pair, 182–183
  common mode rejection, 52
  current control, 159
  current feedback, 266–268
  differential, 144–145
  distributed, 108–109
  emitter follower, 195–196
  error, 162–163
  fast vertical, 108–120
  precision using operational amplifiers, 366–367
  reflex, 71–72
  transistor, 109–111
  two-stage triode, 179
  unity gain buffer (UGB), 118
  zero gain, 139–144
Analog to digital converters, oversampling, 292
Analog tree, the mighty, 9

Barometers, relationship to analog circuit design,
  3–4
Breadboards:
  in conjunction with computer-aided design
    tools, 101–104
  intuition when using, 41
  limitations on accuracy of results, 94
  use in design evaluation, 41, 160–161

Capacitive load compensation, 204–211
Cascomp circuit, 117–118
Circuit analysis, 127
Circuit simulation:
  limitations of, 254–258
  Simulation Program with Integrated Circuit
    Emphasis (SPICE), 253–254
Circuit synthesis, 127–128
Comparators, 363
Computer-aided design (CAD) tools for analog
    design:
  advantages of, 197
  limitations of, 101–104, 124, 252–253
  use with other design tools, 93
Computer-assisted tomography (CAT) systems,
  26
Crystal oscillator circuits:
  bridge, 339
  Butler common base, 338
  characteristics of, 333–334
  circuit loading effects on crystal Q, 341–343
  Colpitts, 336, 338
  common base, 337
  criteria for selection, 340–341

design principles, 334–339
drift in, 333–334
electrical properties of oscillator crystals,
  345–346
emitter coupled, 337
harmonic, 339
Miller, 336
Pierce, 338, 340
series resonant, 337–338
testing, 344–345
Current feedback amplifiers:
  circuit model of, 261–262
  modeling with SPICE, 318–331
  noise in, 274–275
Current mirror circuit, 132–135, 136–137,
  139–144, 194–195
Current summing, 369–370
Current to voltage conversion, 369, 371–372

Differential analyzer, 6
Digiphase synthesizer, 277–282
Digital-to-analog converters (DACs):
  construction techniques, 39–41
  development of, 33–39
  8-bit weighted diode-diode, 37
  Hybrid Systems DAC 371-8, 38
  output capacitance compensation using,
    273–274
  testing of, 42
  time-modulated, 280
  12-bit weighted current division, 35–37
  12-bit weighted transistor-diode, 34
  12-bit weighted transistor-transistor, 33

Education of analog designers:
  at the Massachusetts Institute of Technology,
    80–87
  through work-related experiences, 71–72

Feedback:
  as basis for analog design, 6–7
  as taught in classroom settings, 83–84
  methods of designing feedback loops, 213
  negative feedback in operational amplifiers,
    361, 374
  positive feedback in operational amplifiers, 374
  stabilizing feedback loops, 157–159

George A. Philbrick Researches, Inc.:
  locations of, 69
  Malter, Bob, 69–70, 72–75
  P2 operational amplifier, 75–78
  P7 operational amplifier, 67–69, 75
Gilbert multipliers, 111–113

Index

Hewlett, William R., 43–48, 54–55
Hewlett-Packard:
  founding in a garage, 54–55
  HP183A oscilloscope, 111
  HP200 oscillator, 43
Hot dogs, 55

Impedance, 159–160
Integrated circuit design, analog:
  bipolar versus CMOS processes, 237
  capacitors in, 242
  four-layer structures in, 242–243
  history of, 235–238
  inductors in, 242
  layout considerations, 174–175
  MESFETs in, 246–247
  mixed-signal technologies in, 243–244
  operational amplifiers, 145–147
  resistors in, 241–242
  transconductance models in, 135–139
  transistors in, 238–241
  trimming of, 244–245
Integrator circuits:
  current and voltage autozeroing, 27–28
  current feedback, 273

Lightning empiricism, 295–297
Low pass filters, 155–158

Mathematical wave form generators, 28–29
Mixed signal technologies, 243–244
Multivibrators:
  astable, 60–61
  linear integrating, 291

Noise:
  in operational amplifiers, 379–380
  phase noise, 282–283
  shot, 289
  systematic, 287–288
  thermal, 288–289

Operational amplifiers:
  absence of slew rate limiting, 269–270
  built from discrete components, 67–73
  common mode rejection, 383
  comparators using, 363
  current feedback amplifiers using, 266–268
  current feedback approach used by Comlinear
    Corporation, 321–324
  designed as integrated circuits, 145–147
  excitation of, 382–383
  feedback, 374
  feedback loop compensation, 208–212
  gain of, 362–363
  gain and bandwidth trade-offs in, 263–264
  history of, 361–362
  large-signal dynamics of, 380–382
  model of, 145–147, 262–263, 362
  modeling using SPICE, 304–318
  no gain-bandwidth trade-off, 268–269
  noise in, 379–380
  offsets, 377–379
  overshoot, 206–208
  P2 type, 73–76
  P7 type, 67–68, 75
  precision amplification using, 366–367
  precision current summing, 369–370

precision current to voltage conversion using,
  369
precision differential operations, 373–374
precision voltage amplification using, 368
precision voltage inversion using, 368
properties of non-ideal, 374–386
quiescent current, 383
second-order effects, 270–272
741, 384–386
settling time, 381–382
slew rate and bandwidth interaction, 205–206
slew rate limiting, 264–266
stray input capacitance compensation, 273–274
temperature effects on, 382
transadmittance functions, 372–373
transfer function, 375–377
voltage nulling and current balance in, 368
voltage to frequency converters using, 351–354
Oscillators:
  astable multivibrators, 60–61, 179
  bridge, 339
  Butler common base, 338
  Butler emitter follower, 338
  Colpitts, 336
  common base, 337
  crystal controlled, 333–340
  emitter coupled, 337
  harmonic, 338–339
  Miller, 336
  phase noise in, 282–283
  Pierce, 338, 340
  pulse generators, 63–65
  resistance-capacity, 43–46
  testing crystal controlled, 344–345
  voltage-controlled, 282–284, 349
  Wein bridge, 46–54
Oscilloscopes:
  Fairchild (DuMont) 766, 109–110
  Hewlett-Packard 183A, 111
  Tektronix 535, 99–100
  Tektronix 545, 108
  Tektronix 7000 series, 111
  Tektronix 7104, 116–117
  Tektronix 7904, 114–115

Packard, David, 54–55
Phase-lock loop (PLL) design, 284–287
Philbrick, George:
  remarks on his P7 operational amplifier design,
    75
  role in design of the P2 operational amplifier,
    67–70
  role in design of the P2A operational amplifier,
    71–72
  thoughts on analog design, 5–14
P-N junctions, 129–130
Process of analog design:
  "being the machine" method, 23
  circuit block method, 200
  equivalent circuits methods, 152–153
  role of "lateral" thinking in, 184–185
  role of luck and experience in, 126
  "search and explain" method, 94–95
  simplification of complex designs, 120–122
  structure of analog design process, 89–92
  visualization method, 60
Programmable voltage source, 202–204
P2 operational amplifier:

assembly of, 74
circuitry of, 69–72
long-term stability of offset voltage, 73–74
Pulse generators, 63–65

RC networks, 155–157
Rectifiers, half-wave, 62

Settling time tester, 161–167
Shielding, 288
Simulation Program with Integrated Circuit
    Emphasis (SPICE):
description of, 253–254
elements of, 301
limitations of, 254–258
modeling current feedback operational
    amplifiers, 318–331
modeling operational amplifiers with, 304–318
types of analysis available with, 299–300
Start-up circuits, 170–173
Stock parts values, 153–154

Tektronix:
535 oscilloscope, 99–100
545 oscilloscope, 108
7000 series oscilloscopes, 111
7104 oscilloscope, 116–117
7904 oscilloscope, 114–115
Time synthesizers, 293–294
Transistors:

base current errors, 193–194
bipolar types in integrated circuits, 238–239
circuits using five or fewer, 170–174
junction, 130–132
junction FET types in integrated circuits,
    239–240
MOSFET types in integrated circuits, 240–241
transconductance model of, 135–136
Troubleshooting:
at the integrated circuit level, 97, 249–251
base current errors, 193–194
crystal oscillator circuits, 344–345
early voltage errors, 194–195
emitter follower problems, 195–196
limitations of measuring devices, 100–101
transient problems, 196–197

Unity gain buffer (UGB), 118

Voltage to current conversion, 367–368, 370–371
Voltage to frequency converters:
charge dispensing, 217
current balancing, 217–218
early designs for, 351–355
4701 design, 353–354, 357–360
linearity and resolution in, 218
loop charge pump, 218
Pease design for, 219–224
temperature compensation of, 224–225
using CMOS inverters in, 225–230

Also from the EDN Series for Design Engineers,
published by Butterworth–Heinemann

Troubleshooting Analog Circuits

Robert A. Pease

Bob Pease of National Semiconductor Corporation is one of the legends of analog design. Over the years, he's developed techniques and methods to expedite the often-difficult tasks of debugging and troubleshooting analog circuits. Now Bob has compiled his battle-tested methods in the pages of this book.

Based on his immensely popular series in *EDN Magazine*, the book contains a wealth of all-new material, as well as generous helpings of Bob's unique humor and philosophy regarding electronic troubleshooting. Whether you are an old-timer or a neophyte, you'll find something here to make your troubleshooting jobs easier.

Look for this book at your technical bookstore, or order directly from Butterworth–Heinemann.

1991, 208 pages, ISBN 0-7506-9184-0

Printed and bound by CPI Group (UK) Ltd, Croydon, CR0 4YY

03/10/2024

01040331-0017